# FASTtrack

# Pharmaceutics –
# Dosage Form
# and Design

# FASTtrack

# Pharmaceutics – Dosage Form and Design

**David S Jones**
Pro-Vice Chancellor (Education and Students)
and Chair of Biomaterial Science,
Queen's University Belfast, UK

**Pharmaceutical Press**

Published by Pharmaceutical Press
66-68 East Smithfield, London E1W 1AW, UK

© Pharmaceutical Press 2016

(**P.P**) is a trade mark of Pharmaceutical Press
Pharmaceutical Press is the publishing division of the Royal
Pharmaceutical Society of Great Britain

First edition published 2008

Reprinted 2010, 2011, 2014, 2015 (twice)

Second edition published 2016

Typeset by Swales & Willis Ltd, Exeter, Devon, UK
Printed in Great Britain by TJ International, Padstow, Cornwall

ISBN 978-0-85711-078-7 (print)
ISBN 978-0-85711-300-9 (ePDF)
ISBN 978-0-85711-301-6 (ePub)
ISBN 978-0-85711-302-3 (mobi)

A catalogue record for this book is available from the British
Library.

# Contents

# Introduction to the FASTtrack series

*FASTtrack* is a series of study guides created for undergraduate pharmacy students. The books are intended to be used in conjunction with textbooks and reference books as an aid to revision to help guide students through their exams. They provide essential information required in each particular subject area. The books will also be useful for pre-registration trainees preparing for the General Pharmaceutical Council's (GPhC) registration examination, and to practising pharmacists as a quick reference text.

The content of each title focuses on what pharmacy students really need to know in order to pass exams. Features include*:

- concise bulleted information
- key points
- tips for the student
- multiple choice questions (MCQs) and worked examples
- case studies
- simple diagrams.

The titles in the *FASTtrack* series reflect the full spectrum of modules for the undergraduate pharmacy degree.

Current titles include:
*Applied Pharmaceutical Practice*
*Complementary and Alternative Medicines*
*Pharmaceutical Compounding and Dispensing*
*Pharmaceutics – Dosage Form and Design*
*Pharmaceutics – Drug Delivery and Targeting*
*Pharmacology*
*Physical Pharmacy* (based on Florence & Attwood's *Physicochemical Principles of Pharmacy*)
*Managing Symptoms in the Pharmacy*
*Therapeutics*

Additional questions are available at www.ontrackpharmacy.com by selecting the FASTtrack option.

* Note: not all features are in every title in the series.

# Preface

It is often difficult to define success within the academic domain. We are accustomed as academics to publishing research papers and attracting grant funding, to name two examples, and these measures of success are tangible. Within the education domain, academics receive feedback from students that assist us on the journal to becoming world-class educators and, hopefully, inspiring the next generation of students to become academics, professionals and role models. Academically, writing an undergraduate textbook is a challenging task and one that does not always guarantee success. It requires the author to have not only a thorough knowledge of the field but to understand the needs of students, to understand how students learn and the student approach to solving real-life problems related to the topic of the textbook. It was within this backdrop that I authored the first edition of *Pharmaceutics – Dosage Form and Design*. I hoped that this textbook would encapsulate the academically demanding field of formulation science, a domain steeped in theory and the application of these theories to solve problems and hence improve patient outcomes. Thus, as an academic, I was concerned that the textbook would address these challenges, not only in my eyes but also in the eyes of the readers. This allows me to return to the topic of success. I have been delighted by the response of the readership to this textbook since it was first published and the positive comments have inspired me to write a second edition to the initial successful edition. It is these comments that allow me to acknowledge the success of the first edition of *Pharmaceutics – Dosage Form and Design*.

Not wishing to change a winning formula, the main attributes of the first edition have been maintained in this second version. The chapters, each of which is dedicated to a formulation type/platform, have been written to ensure the necessary knowledge (and indeed balance of knowledge) between the required theory and practice are clearly provided. Many of the chapters have been expanded in length, have an increase in the number of self-assessment multiple-choice questions and include a series of diagrams that illustrate formulation choices (and commonly used excipients). This latter feature I believe will be of great interest and will help the readers to understand clearly the nature of dosage form design.

The readers will observe that a new chapter focusing on Pharmaceutical Engineering has been added to this second edition. In particular, this chapter describes the theory and practice associated with four key pharmaceutical unit operations, namely mixing, filtration, milling and drying, complementing aspects of Chapters 9 and 10. The material in Chapter 11 is challenging but will allow the reader to understand the basic processes of the production of pharmaceutical dosage forms. This chapter contains an excellent array of illustrations and a bank of multiple-choice questions to facilitate the understanding of this important area. The inclusion of this chapter will allow the reader to understand the process of dosage design and its influence on production.

In preparing this edition I would sincerely wish to thank my colleagues who have selflessly provided feedback on the first edition, have identified areas for improvement and have communicated these to me. In this regard I would particularly wish to thank Dr. Ahmad Yassine (Lebanese University). I would additionally like to thank all students who have contacted me regarding the first edition. Their praise inspired me to author, and their comments regarding improvement have been incorporated into this second edition.

Finally, as with the first edition, I am sincerely grateful for the love, encouragement and support of my mother May, my late Father Fred, my wife Linda and our children, Dary and Holly, without whom this academic journey would not have been possible. This book is dedicated to them.

David S Jones
May 2016

# About the author

David Jones is Pro-Vice-Chancellor for Education and Students and Professor of Biomaterial Science at Queen's University Belfast. He gained a BSc (1st class honours) in Pharmacy (1985), a PhD in Pharmaceutics (1988) and, in 2006, a DSc., all from Queen's University Belfast. More recently (2014) he gained a BA (1st class honours) in Mathematics and Statistics from the Open University.

He registered with the Pharmaceutical Society of Northern Ireland in 1989. From 1989 to 1992 he was a lecturer in Pharmaceutics at the University of Otago (Dunedin, New Zealand) and from 1992 to 1994 he held the position of Head of Formulations at Norbrook Industries Limited (NI). In 1994 he was appointed to a lectureship at the School of Pharmacy, Queen's University Belfast and was promoted to a Senior Lectureship in 1997. Since 1999 he holds a personal Chair (in Biomaterial Sciences) at the School of Pharmacy, Queen's University Belfast.

His research concerns the characterisation, formulation and engineering of pharmaceutical materials/dosage forms and biomedical devices. He is the author of three textbooks, 10 patents and over 400 research papers/communications and has been awarded the Lilly prize for pharmaceutical research and the British Pharmaceutical Conference Science Award. Professor Jones is both a Chartered Engineer and a Chartered Chemist and is an elected Fellow of the Pharmaceutical Society of Northern Ireland, a Fellow of the Institute of Materials, Minerals and Mining, a Fellow of the Royal Statistical Society, a Fellow of the Royal Society of Chemistry and a Member of the Institution of Engineers in Ireland. He is the editor of the Journal of Pharmacy and Pharmacology and is a previous holder of a prestigious Royal Society Industrial Fellowship.

and low toxicity of this ingredient. Under normal circumstances tap (drinking) water should not be used due to the possibility of chemical incompatibilities within the formulation. The main features of Purified Water USP are as follows:

- It is prepared by distillation, ion exchange methods or by reverse osmosis.
- The solid residue (obtained after evaporation) is less than 1 mg per 100 ml of evaporated sample.
- It must not be used for the preparation of parenteral formulations.

In the case of parenteral formulations *Water for Injections BP* must be used, the specifications and use of which are described in Chapter 5.

## Co-solvents

As defined previously, co-solvents are employed to increase the solubility of the therapeutic agent within the formulation. The main co-solvents that are used in the formulation of oral solutions are detailed below.

### Glycerol

Glycerol (also termed glycerin) is an odourless, sweet liquid that is miscible with water and whose co-solvency properties are due to the presence of three hydroxyl groups (termed a triol) (Figure 1.1). It has similar co-solvency properties to ethanol.

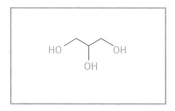

**Figure 1.1**   Structural formula of glycerol.

### Alcohol USP ($CH_3CH_2OH$)

Alcohol USP contains between 94.9 and 96.0% v/v ethyl alcohol (ethanol) and is commonly used as a co-solvent, both as a single co-solvent and with other co-solvents, e.g. glycerol. The known pharmacological and toxicological effects of this co-solvent have compromised the use of alcohol in pharmaceutical preparations. As a result there are both labelling requirements for preparations that contain alcohol and upper limits with respect to the concentration of alcohol that may be used in formulations.

### Propylene Glycol USP

Propylene Glycol USP is an odourless, colourless, viscous liquid diol that contains two hydroxyl groups (Figure 1.2). It is used in pharmaceutical preparations as a co-solvent, generally as a replacement for glycerin.

**Figure 1.2**   Structural formula of propylene glycol.

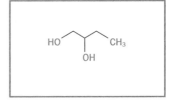

### Poly(ethylene glycol) (PEG)

PEG (Figure 1.3) is a polymer composed of repeating units of the monomer ethylene oxide (in parenthesis). The physical state of the polymer is dependent on the number of repeat units ($n$) and hence on the molecular weight. Lower-molecular-weight grades (PEG 200, PEG 400) are preferred as co-solvents in pharmaceutical solutions.

**Figure 1.3**   Structural formula of poly(ethylene glycol).

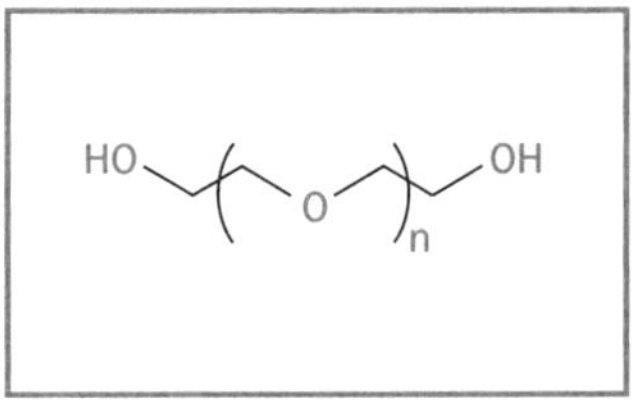

### Miscellaneous agents used to enhance the solubility of therapeutic agents

In addition to the use of co-solvents, other pharmaceutical strategies are available to the pharmaceutical scientist to increase the solubility of therapeutic agents in the chosen vehicle. These include the use of surface-active agents and complexation, as detailed below.

#### Surface-active agents

Surface-active agents are chemicals that possess both hydrophilic (water-liking) and hydrophobic (water-disliking) regions. At dilute concentrations surface-active agents will orient at the interface between two phases (e.g. water/oil, water/air), with the hydrophilic and hydrophobic regions of the molecule being positioned to the hydrophilic and hydrophobic phases, respectively. As the concentration is increased, the interface will become saturated with surface-active agent and the molecules that are present in the bulk aqueous phase will orient themselves in an attempt to shield the hydrophobic regions of the surface-active agent. This orientation is referred to as a *micelle* and the concentration of surface-active agent at this occurs is termed the *critical micelle concentration* (CMC).

For further details regarding the physicochemical properties of surfactants, the reader should consult the companion text by David Attwood and Alexander T. Florence (*FASTtrack: Physical Pharmacy*, 2nd edn (London: Pharmaceutical Press; 2012). The use of surface-active agents for the solubilisation of poorly soluble drugs occurs exclusively in the presence of micelles and hence at concentrations of surface-active agents in excess of the CMC. In this the core of the micelle represents a hydrophobic region into which the poorly water-soluble drugs may partition. The location in the micelle is related to the chemical structure of the drug. For example, if the therapeutic agent is poorly soluble the molecule will locate exclusively within the micelle, whereas if the drug is water-insoluble but contains polar groups, the molecule will orient within the micelle, with the polar groups at the surface of the micelle and the hydrophobic region of the molecule located within the hydrophobic core of the micelle. In so doing the drug is solubilised within the colloidal micelles; due to their small size, the resulting solution appears homogeneous to the naked eye.

### Complexation

Complexation refers to the interaction of a poorly soluble therapeutic agent with an organic molecule, e.g. surface-active agents, hydrophilic polymers, to generate a soluble intermolecular complex. One particular concern regarding the use of solution of drug complexes is the ability of the complex to dissociate following administration. This is particularly important in situations where the complexing agent is a hydrophilic polymer, as the high molecular weight of the drug–polymer complex would prevent drug absorption across biological membranes.

**Tip**

As the reader will have observed, there are several methods that may be used for the solubilisation of therapeutic agents. The choice of method should involve consideration of the stability of the formed solution, the pharmaceutical acceptability of the solubilisation strategy and cost.

## Common excipients in pharmaceutical solutions

There are several excipients that are commonly employed in the formulation of pharmaceutical solutions. These include: (1) buffers; (2) sweetening agents; and (3) viscosity-enhancing agents.

## Buffers

Buffers are employed within pharmaceutical solutions to control the pH of the formulated product and, in so doing, optimise the physicochemical performance of the product. Typically pH control is performed:

- to maintain the solubility of the therapeutic agent in the formulated product. The solubility of the vast number of currently available drugs is pH-dependent and, therefore, the solubility of the therapeutic agent in the formulation may be compromised by small changes in pH
- to enhance the stability of products in which the chemical stability of the active agent is pH-dependent.

The concentration (and hence buffer capacity) of buffer salts employed in the formulation of oral solutions should be selected to offer sufficient control of the pH of the formulation but yet should be overcome by biological fluids following administration.

Examples of buffer salts used in pharmaceutical solutions include:

- acetates (acetic acid and sodium acetate): circa 1–2%
- citrates (citric acid and sodium citrate): circa 1–5%
- phosphates (sodium phosphate and disodium phosphate): circa 0.8–2%.

It must be remembered that the buffer system used in solution formulations should not adversely affect the solubility of the therapeutic agent, e.g. the solubility of drugs may be affected in the presence of phosphate salts.

## Sweetening agents

Sweetening agents are employed in liquid formulations designed for oral administration specifically to increase the palatability of the therapeutic agent. The main sweetening agents employed in oral preparations are sucrose, liquid glucose, glycerol, sorbitol, saccharin sodium and aspartame. The use of artificial sweetening agents in formulations is increasing and, in many formulations, saccharin sodium is used either as the sole sweetening agent or in combination with sugars or sorbitol to reduce the sugar concentration in the formulation. The use of sugars in oral formulations for children and patients with diabetes mellitus is to be avoided.

## Viscosity-enhancing agents

The administration of oral solutions to patients is usually performed using a syringe, a small-metered cup or a traditional 5-ml spoon. The viscosity of the formulation must be sufficiently controlled in order to ensure the accurate measurement of the volume to be dispensed. Furthermore, increasing the viscosity of some formulations may increase the palatability. Accordingly there is a viscosity range that the formulation should exhibit to facilitate this operation. Certain liquid formulations do not require the specific addition of viscosity-enhancing agents, e.g. syrups, due to their inherent viscosity.

The viscosity of pharmaceutical solutions may be easily increased (and controlled) by the addition of non-ionic or ionic hydrophilic polymers. Examples of both of these categories are shown below:

- *non-ionic (neutral) polymers*
  - cellulose derivatives, e.g.:
    - methylcellulose
    - hydroxyethylcellulose
    - hydroxypropylcellulose
  - polyvinylpyrrolidone
- *ionic polymers*
  - sodium carboxymethylcellulose (anionic)
  - sodium alginate (anionic).

Some general comments regarding the suitability and use of these polymers to modify the viscosity of pharmaceutical solutions for oral use are as follows:

- The concentration of polymer required to increase the viscosity of the formulation to the required level is dependent on:
  - the type of polymer
  - the molecular weight (sometimes referred to as the grade) of the polymer
  - the pH of the solution (for ionic polymers).
- Typically the concentration of polymer required to modify the viscosity of solutions may range from 0.5% w/w for polymers that exhibit strong inter-chain interactions in solution to *circa* 10% w/v for linear polymers that exhibit weaker inter-chain interactions (e.g. polyvinylpyrrolidone).
- The physical state of ionic polymers in solution is strongly dependent on the pH of the formulation. At pH values above the $pK_a$ of the polymer the pendant groups attached to the polymer backbone ionise, leading to repulsion of the side groups and expansion of the polymeric chains. This chain expansion results in an increase in the viscosity of the polymer solution. As the polymer chain expansion is dependent on the degree of ionisation, changes in the pH of the formulation, e.g. during storage, will affect the resultant viscosity of the formulation.
- The ionic strength of the formulation may affect the viscosity of the pharmaceutical solutions containing hydrophilic

polymers due to a salting out process. Ionic polymers are more susceptible to this phenomenon than non-ionic polymers.

- The viscosity of ionic polymers is affected by the presence of multi-valence counterions. For example, in the presence of $Mg^{2+}$, $Ca^{2+}$, $Al^{3+}$, $Fe^{3+}$, the viscosity of solutions containing anionic polymers will be dramatically reduced due to salting out of the polymer. In this manner the charged groups of the polymer are shielded and the aqueous solubility of the polymer decreases.

- The viscosity of certain hydrophilic polymers may be affected by the presence of alcoholic co-solvents. In particular, at higher concentrations of alcohol the viscosity of polymer solutions may decrease due to the reduced solubility of the polymer in the hydroalcoholic solvent. Conversely, the viscosity of hydrophilic polymers in formulations containing polyols and water may be greater than in water alone.

Further details of the physicochemical properties of these polymers are provided in later chapters.

## Antioxidants

Antioxidants are included in pharmaceutical solutions to enhance the stability of therapeutic agents that are susceptible to chemical degradation by oxidation. Typically antioxidants are molecules that are redox systems which exhibit higher oxidative potential than the therapeutic agent or, alternatively, are compounds that inhibit free radical-induced drug decomposition. Antioxidants are oxidised (and hence degraded) within aqueous solutions in preference to the therapeutic agent, thereby protecting the drug from decomposition. Both water-soluble and water-insoluble antioxidants are commercially available, the choice of these being made according to the nature of the formulation. Examples of antioxidants that are commonly used for aqueous formulations include: sodium sulphite, sodium metabisulphite, sodium formaldehyde sulphoxylate and ascorbic acid. Examples of antioxidants that may be used in oil-based solutions include: butylated hydroxytoluene (BHT), butylated hydroxyanisole (BHA) and propyl gallate. Typically antioxidants are employed in low concentrations (<0.2% w/w) and it is usual for the concentration of antioxidant in the finished product to be markedly less than the initial concentration, due to oxidative degradation during manufacture of the dosage form. Antioxidants may also be employed in conjunction with chelating agents, e.g. ethylenediamine tetraacetic acid, citric acid, that act to form complexes with heavy-metal ions, ions that are normally involved in oxidative degradation of therapeutic agents.

## Preservatives

Preservatives are included in pharmaceutical solutions to control the microbial bioburden of the formulation. Ideally, preservatives should exhibit the following properties:

- possess a broad spectrum of antimicrobial activity encompassing Gram-positive and Gram-negative bacteria and fungi
- be chemically and physically stable over the shelf-life of the product
- have low toxicity.

A wide range of preservatives are available for use in pharmaceutical solutions for oral use, including the following (values in parentheses relate to the typical concentration range used in oral solutions):

- benzoic acid and salts (0.1–0.3%)
- sorbic acid and its salts (0.05–0.2%)
- alkyl esters of parahydroxybenzoic acid (0.001–0.2%). Usually a combination of two members of this series is employed in pharmaceutical solutions, typically methyl and propyl parahydroxybenzoates (in a ratio of 9:1). The combination of these two preservatives enhances the antimicrobial spectrum.

### Factors affecting preservative efficacy in oral solutions

The activity of a preservative is dependent on the correct form of the preservative being available in the formulation at the required concentration to inhibit microbial growth (termed the minimum inhibitory concentration: MIC). Unfortunately, in many solution formulations, the concentration of preservative within the formulation may be affected by the presence of other excipients and by formulation pH. Factors that directly affect the efficacy of preservatives in oral solutions include: (1) the pH of the formulation; (2) the presence of micelles; and (3) the presence of hydrophilic polymers.

#### The pH of the formulation

In some aqueous formulations the use of acidic preservatives, e.g. benzoic acid, sorbic acid, may be problematic.

The antimicrobial properties are due to the unionised form of the preservative, the degree of ionisation being a function of the pH of the formulation. The activity of the unionised form of the acid in this respect is due to the ability of this form to diffuse across the outer membrane of the microorganism and eventually into the cytoplasm. The neutral conditions within the cytoplasm enable the preservative to dissociate, leading to acidification of the cytoplasm and inhibition of growth.

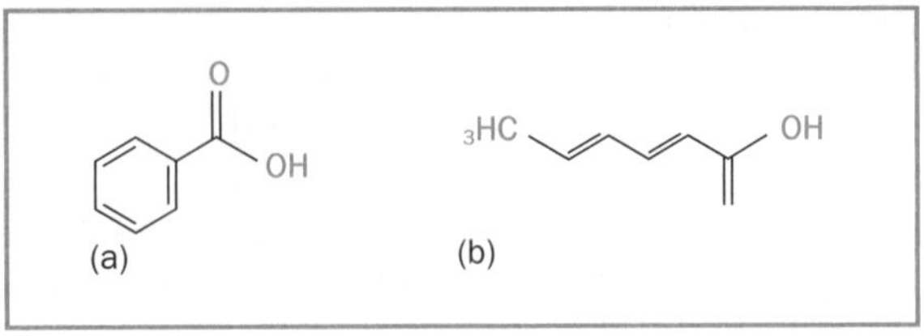

Figure 1.4   Structural formulae
of (a) benzoic acid and
(b) sorbic acid.

Figure 1.4   Structural formulae of (a) benzoic acid and (b) sorbic acid.

The fraction of acidic preservative at a particular pH may be calculated using a derived form of the Henderson–Hasselbalch equation, as follows:

$$\text{Fraction} = \left( \frac{1}{\left(1 + 10^{\text{pH}-\text{p}K_a}\right)} \right)$$

The use of this equation may be illustrated in the following example:

## Worked example

### Example 1.1

Assuming that the MIC for the unionised form of an acidic preservative (p$K_a$ 4.2) is 0.0185 mg/ml, calculate the required concentration to preserve an oral solution that has been buffered to pH 4.7.

The Henderson–Hasselbalch equation may be employed, as described above, to determine the fraction of unionised acid within the formulation.

$$\text{Fraction} = \left( \frac{1}{\left(1 + 10^{4.7-4.2}\right)} \right) = 0.24$$

The required concentration is then calculated by dividing the MIC for the unionised form of the preservative by the fraction of unionised preservative present, i.e. $\left( \dfrac{0.0185}{0.24} \right) = 0.07$ mg/ml.

In practice an overage is added and therefore the actual concentration of preservative required would be 0.1–0.15 mg/ml.

As the reader will observe, the p$K_a$ of the preservative is a vital determinant within the above calculations. Organic acids, e.g. benzoic acid, sorbic acid, have p$K_a$ values that are circa 4.2 and therefore, in solution formulations whose pH is neutral, a high concentration of preservative will be required to ensure that the required concentration of the unionised species is obtained. If the above calculation is repeated for an oral solution at pH 7.2, the following result is obtained:

$$\text{Fraction} = \left(\frac{1}{\left(1+10^{7.2-4.2}\right)}\right) = 0.00001$$

Therefore, the required preservative concentration is

$$\left(\frac{0.0185}{0.00001}\right) = 1850 \text{ mg/ml}.$$

Importantly, the preservative efficacies of parabens (alkyl esters of parahydroxybenozoic acid) and the phenolics are generally not affected by formulation pH (within a pH range between 4.0 and 8.0) due to the high $pK_a$ of the organic hydroxyl group. The structure of these preservatives is shown in Figure 1.5.

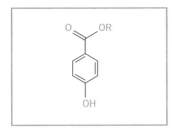

**Figure 1.5** Structural formula of the parahydroxybenzoate esters (parabens, where R refers to an alkyl group).

## The presence of micelles

The role of micelles for the solubilisation of lipophilic therapeutic agents was described above. If the preservative exhibits lipophilic properties (e.g. the unionised form of acidic preservatives, phenolics, parabens), then partition of these species into the micelle may occur, thereby decreasing the available (effective) concentration of preservative in solution. An equilibrium is established, as depicted in Figure 1.6.

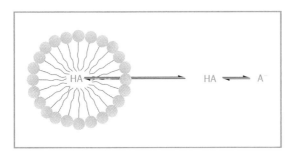

**Figure 1.6** A diagrammatic illustration of the equilibrium of an acidic preservative in the presence of micelles. HA and A⁻ refer to the unionised and ionised states of the preservative.

To correct this problem, the preservative concentration must be increased to ensure that the free concentration within the formulation is greater than or equal to the MIC of the preservative.

### The presence of hydrophilic polymers

It has been shown that the free concentration of preservative in oral solution formulations is reduced in the presence of hydrophilic polymers, e.g. polyvinylpyrrolidone, methylcellulose. This is due to the ability of the preservative to interact chemically with the dissolved polymer. As described above, this problem is addressed by increasing the concentration of preservative in the formulation. In certain circumstances the preservative may be incompatible with hydrophilic polymers in the formulation due to an electrostatic interaction. Therefore, cationic hydrophilic polymers should not be used in conjunction with acidic preservatives in oral solution formulations.

## Flavours and colourants

Unfortunately the vast majority of drugs in solution are unpalatable and, therefore, the addition of flavours is often required to mask the taste of the drug substance. Taste-masking using flavours is a difficult task; however, there are some empirical approaches that may be taken to produce a palatable formulation.

The four basic taste sensations are salty, sweet, bitter and sour. It has been proposed that certain flavours should be used to mask these specific taste sensations. In particular:

(a)  Flavours that may be used to mask a salty taste include:
- butterscotch
- apricot
- peach
- vanilla
- wintergreen mint.

(b)  Flavours that may be used to mask a bitter taste include:
- cherry
- mint
- anise.

(c)  Flavours that may be used to mask a sweet taste include:
- vanilla
- fruit and berry.

(d)  Flavours that may be used to mask a sour taste include:
- citrus flavours
- raspberry.

Usually a combination of flavours is used to achieve the optimal taste-masking property.

Certain excipients may be added to oral solution formulations, referred to as *flavour adjuncts* (e.g. menthol, chloroform) that add flavour to the formulation but, in addition, act to desensitise the taste receptors. In so doing these agents augment the taste-masking properties of conventional flavours.

Colours are pharmaceutical ingredients that impart the preferred colour to the formulation. When used in combination with flavours, the selected colour should "match" the flavour of the formulation, e.g. green with mint-flavoured solutions, red for strawberry-flavoured formulations. Although the inclusion of colours is not a prerequisite for all pharmaceutical solutions, certain categories of solution (e.g. mouthwashes/gargles) are normally coloured.

## Types of pharmaceutical solutions designed for administration to the gastrointestinal tract

There are three principal types of solution formulations that are administered orally: *oral solutions, oral syrups* and *oral elixirs.* In addition, other solution formulations are employed for a local effect, e.g. mouthwashes/gargles and enemas. Details of these are provided in the following sections.

### Oral solutions

Oral solutions are administered to the gastrointestinal tract to provide systemic absorption of the therapeutic agent. Due to the resilience of the gastrointestinal environment, oral solutions may be formulated over a broad pH range. However, unless there are issues regarding the solubility or stability of the therapeutic agent, the usual pH of oral solutions is circa 7.0. Typically the following classes of excipients are used in the formulation of oral solutions:

- buffers (e.g. citrate, phosphate)
- preservatives (e.g. parabens, benzoic acid, sorbic acid)
- antioxidants (water-soluble antioxidants are used, e.g. sodium metabisulphite 0.01–1.0% w/w)
- flavours and colours (the colour should be selected to complement the flavour of the formulation)
- viscosity-modifying agents (to affect the pourability of the formulation. For this purpose hydrophilic polymers are used, e.g. sodium alginate, hydroxyethylcellulose).

**Tips**

- The formulation of solutions for oral administration often requires the inclusion of several excipients.
- It is important that each excipient included in the formulation is necessary and justified.
- All excipients must be physically and chemically compatible (with each other and with the therapeutic agent).

**Figure 1.7** Diagrammatic representation of formulation considerations for aqueous pharmaceutical solutions.

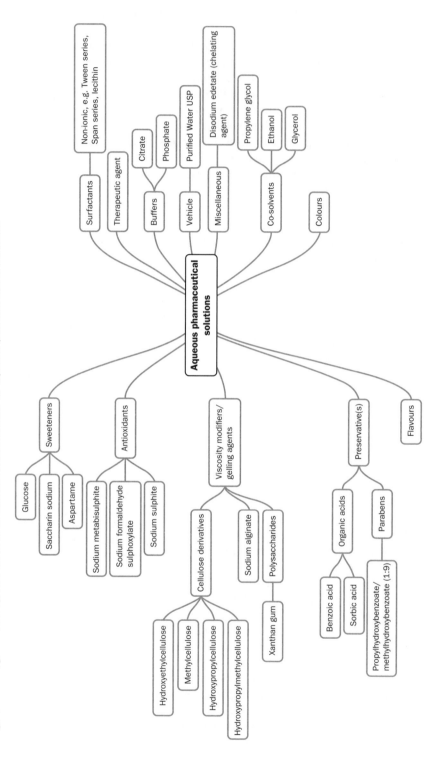

The reader should note that, to be classified as a solution, all components of the formulation (including the therapeutic agent) should be soluble, with no evidence of precipitation.

A diagrammatic representation of formulation considerations for oral solutions is shown in Figure 1.7.

## Oral syrups

Syrups are highly concentrated, aqueous solutions of sugar or a sugar substitute that traditionally contain a flavouring agent, e.g. cherry syrup, cocoa syrup, orange syrup, raspberry syrup. An unflavoured syrup is available that is composed of an aqueous solution containing 85% sucrose. Therapeutic agents may either be directly incorporated into these systems or may be added as the syrup is being prepared. If the former method is employed, it is important to ensure that the therapeutic agent is soluble within the syrup base.

It should also be remembered that the choice of syrup vehicle must be performed with due consideration to the physicochemical properties of the therapeutic agent. For example, cherry syrup and orange syrup are acidic and therefore the solubility of acidic or some zwitterionic therapeutic agents may be lowered and may result in precipitation of the drug substance. Under these circumstances, the physical stability of the preparation will have been compromised and the shelf-life of the product will have been exceeded. The use of acidic syrups may additionally result in reduced chemical stability for acid-labile therapeutic agents.

The major components of syrups are as follows:

- *Purified water.*
- *Sugar (sucrose) or sugar substitutes (artificial sweeteners).*
  Traditionally syrups are composed of sucrose (usually between 60% and 80%) and purified water. Due to the inherent sweetness and moderately high viscosity of these systems, the addition of other sweetening agents and viscosity-modifying agents is not required. In addition, the high concentration of sucrose and associated unavailability of water (termed low water activity) ensures that the addition of preservatives is not required. As the concentration of sucrose is reduced from the upper limit (e.g. through dilution), the addition of preservatives may be required.

  In some formulations, other non-sucrose bases may replace traditional syrup. One of the most popular is Sorbitol Solution USP, which contains 64% w/w sorbitol (a polyhydric alcohol, Figure 1.8) although other alternatives are available that are based on mixtures of sorbitol and glycerin. These non-sucrose bases may be mixed with traditional syrups, if required, in the formulation of oral syrups that possess a low concentration of sucrose in comparison to traditional syrups.

**Figure 1.8** Chemical structure of sorbitol.

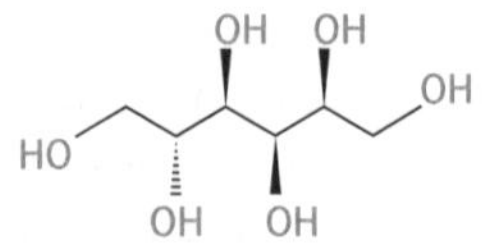

More recently, many products have been formulated as medicated sugar-free syrups due to the glycogenetic and cariogenic properties of sucrose. For the afore-mentioned reasons, all medicinal products designed for administration to children and to diabetic patients must be sugar-free. Syrup substitutes must therefore provide an equivalent sweetness, viscosity and preservation to the original syrups. To achieve these properties artificial sweeteners (typically saccharin sodium, aspartame), non-glycogenetic viscosity modifiers (e.g. methylcellulose, hydroxyethylcellulose) and preservatives (e.g. sodium benzoate, benzoic acid and parahydroxybenzoate esters) are included.

■ *Preservatives.* As highlighted above, preservatives are not required in traditional syrups containing high concentrations of sucrose. Conversely, in sugar-free syrups, syrups in which sucrose has been substituted at least in part by polyhydric alcohol and in traditional syrups that contain lower concentrations of sucrose, the addition of preservatives is required. Typical examples of commonly used preservatives include:

  • Mixtures of parahydroxybenzoate esters (usually methylhydroxybenzoate and propylhydroxybenzoate in a ratio of 9:1). The typical concentration range is 0.1–0.2% w/v. It is important to note that the preservative efficacy of these preservatives may be decreased in the presence of hydrophilic polymers (generally employed to enhance viscosity), due to an interaction of the preservative with the polymer. This effect is negated by increasing the overall preservative concentration.
  • Other preservatives that are employed include benzoic acid (0.1–0.2%) or sodium benzoate (0.1–0.2%).

■ *Flavours.* These are employed whenever the unpalatable taste of a therapeutic agent is apparent, even in the presence of the sweetening agents. The flavours may be of natural origin (e.g. peppermint, lemon, herbs and spices) and are available as oils, extracts, spirits or aqueous solutions. Alternatively, a wide range of synthetic flavours are available that offer advantages over their natural counterparts in terms of purity, availability, stability and solubility.

Certain flavours are also associated with a (mild) therapeutic activity. For example, many antacids contain mint due to the carminative properties of this ingredient. Alternatively other flavours offer a taste-masking effect by eliciting a mild local anaesthetic effect on the taste receptors. Examples of flavours in this category include peppermint oil, chloroform and menthol.

The concentration of flavour in oral syrups is that which is required to provide the required degree of taste-masking effectively.

■ *Colours.* These are generally natural or synthetic water-soluble, photo-stable ingredients that are selected according to the flavour of the preparation. For example, mint-flavoured formulations are commonly a green colour, whereas in banana-flavoured solutions a yellow colour is commonly employed. Such ingredients must not chemically or physically interact with the other components of the formulation.

A diagrammatic representation of formulation considerations for oral syrups is provided in Figure 1.9.

## Oral elixirs

An elixir is a clear, hydroalcoholic solution that is formulated for oral use. The concentration of alcohol required in the elixir is unique to each formulation and is sufficient to ensure that all of the other components within the formulation remain in solution. For this purpose other polyol co-solvents may be incorporated into the formulation. The presence of alcohol in elixirs presents a possible problem in paediatric formulations and, indeed, for those adults who wish to avoid alcohol. The typical components of an elixir are as follows:

■ *Purified water.*
■ *Alcohol.* This is employed as a co-solvent to ensure solubility of all ingredients. As highlighted above, the concentration of alcohol varies depending on the formulation. Generally the concentration of alcohol is greater than 10% v/v; however, in some preparations, the concentration of alcohol may be greater than 40% v/v.
■ *Polyol co-solvents.* Polyol co-solvents, e.g. propylene glycol, glycerol, may be employed in pharmaceutical elixirs to enhance the solubility of the therapeutic agent and associated excipients. The inclusion of these ingredients enables the concentration of alcohol to be reduced. As before, the concentration of co-solvents employed is dependent on the concentration of alcohol present, the type of co-solvent used

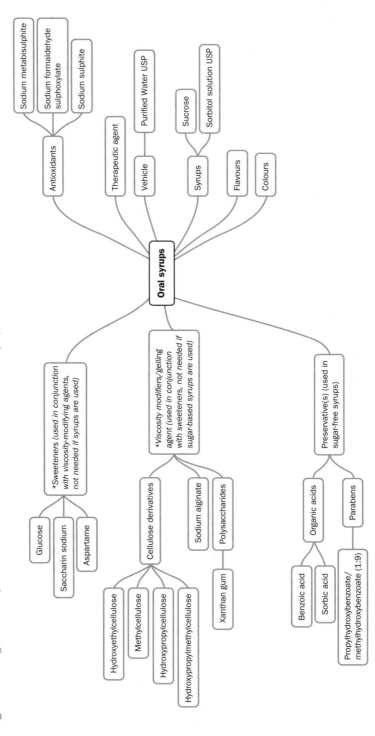

**Figure 1.9** Diagrammatic representation of formulation considerations for oral syrups.

and the solubility of the other ingredients in the alcohol/ co-solvent blend. The reader is directed to the pharmacopoeial monographs to observe the concentration of co-solvents in specific examples of the pharmaceutical elixirs. Two examples in the USP that illustrate the range of concentrations of co-solvents are Phenobarbital Elixir and Theophylline Elixir (Tables 1.1 and 1.2).

**Table 1.1**   Composition of Phenobarbital Elixir USP

| Ingredient | Pharmaceutical purpose | Concentration |
| --- | --- | --- |
| Phenobarbital | Therapeutic agent | 0.4% w/v |
| Orange oil | Flavour | 0.025% v/v |
| Propylene glycol | Co-solvent | 10% v/v |
| Alcohol | Co-solvent | 20% v/v |
| Sorbitol solution | Sweetener | 60% v/v |
| Colour | Colour | As required |
| Purified water | Vehicle | ad 100% |

**Table 1.2**   Composition of Theophylline Elixir USP

| Ingredient | Pharmaceutical purpose | Concentration |
| --- | --- | --- |
| Theophylline | Therapeutic agent | 0.53% w/v |
| Citric acid | pH regulation | 1.0% w/v |
| Liquid glucose | Sweetening agent | 4.4% w/v |
| Syrup | Sweetening agent | 13.2% v/v |
| Saccharin sodium | Sweetening agent | 0.5% w/v |
| Sorbitol solution | Sweetening agent | 32.4% v/v |
| Glycerin | Co-solvent | 5.0% v/v |
| Alcohol | Co-solvent | 20% v/v |
| Lemon oil | Flavour | 0.01% w/v |
| FDC yellow no. 5 | Colour | 0.01% w/v |
| Purified water | Vehicle | ad 100% |

- *Sweetening agents.* The concentration of sucrose in elixirs is less than that in syrups and accordingly elixirs require the addition of sweetening agents. The types of sweetening agents used are similar to those used in syrups, namely syrup, sorbitol solution and artificial sweeteners such as saccharin sodium (Figure 1.10).
- It should be noted, however, that the high concentration of alcohol prohibits the incorporation of high concentrations of sucrose due to the limited solubility of this sweetening

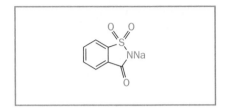

**Figure 1.10**   Chemical structure of saccharin sodium.

agent in the elixir vehicle. To obviate this problem, saccharin sodium, an agent which is used in small quantities and which exhibits the required solubility profile in the elixir, is employed.

■   *Flavours and colours.* All pharmaceutical elixirs contain flavours and colours to increase the palatability and enhance the aesthetic qualities of the formulation. The presence of alcohol in the formulation allows the pharmaceutical scientist to use flavours and colours that may perhaps exhibit inappropriate solubility in aqueous solution. For example, it may be observed that in the two formulations cited above, essential oils were used as the flavouring agents. As before, the selected colour should optimally match the chosen flavour.

■   *Ancillary comments*
   • Preservatives are not required in pharmaceutical elixirs that contain greater than circa 12% v/v alcohol, due to the antimicrobial properties of this co-solvent.
   • Due to the volatile nature of some of the components of elixirs, elixirs should be packaged in tight containers and not stored at high temperatures.
   • The addition of viscosity-enhancing agents, e.g. hydrophilic polymers, may be required to optimise the rheological properties of elixirs.

## Tips

■   The choice of liquid type for oral administration (solution, elixir or linctus) is often dependent on the physicochemical properties of the therapeutic agent. For example, if the drug has a bitter taste, linctuses are often used.
■   It should be noted that linctuses are now commonly formulated as sugar-free preparations.
■   The use of elixirs is not common.

A diagrammatic representation of formulation considerations for oral elixirs is provided in Figure 1.11.

## Miscellaneous solutions for administration to the gastrointestinal tract

In addition to conventional solutions, syrups and elixirs, there are other solution-based dosage forms that are administered to the gastrointestinal tract, notably *linctuses*, *mouthwashes/gargles* and *enemas*. These three subcategories are briefly described below.

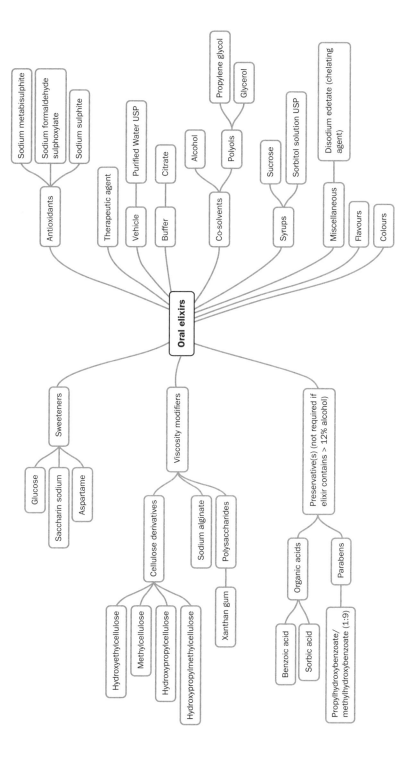

**Figure 1.11** Diagrammatic representation of formulation considerations for oral elixirs.

## Linctuses

Linctuses are viscous preparations that contain the therapeutic agent dissolved in a vehicle composed of a high percentage of sucrose and, if required, other sweetening agents. These formulations are administered orally and are primarily employed for the treatment of cough, due to their soothing actions on the inflamed mucous membranes. Linctuses may also be formulated as sugar-free alternatives in which sucrose is replaced by sorbitol and the required concentration of sweetening agent.

## Mouthwashes and gargles

Mouthwashes/gargles are designed for the treatment of infection and inflammation of the oral cavity. Formulations designed for this purpose employ water as the vehicle, although a co-solvent, e.g. alcohol, may be employed to solubilise the active agent. The use of alcohol as a co-solvent may act to enhance the antimicrobial properties of the therapeutic agent. Other formulation components are frequently required to enhance the palatability and acceptability of the preparation. These include preservatives, colours, flavouring agents and non-cariogenic sweetening agents.

## Enemas

Enemas are pharmaceutical solutions that are administered rectally and are employed to ensure clearance of the bowel, usually by softening the faeces or by increasing the amount of water in the large bowel (osmotic laxatives). Enemas may be aqueous or oil-based solutions and, in some formulations, the vehicle is the agent that promotes bowel evacuation, e.g. arachis oil retention enema. Aqueous formulations usually contain salts (e.g. phosphates) to alter the osmolality within the rectum, thereby increasing the movement of fluid to the rectal contents. Viscosity-enhancing agents, e.g. glycerol, may be included to aid retention of the formulation within the rectum and to reduce the incidence of seepage.

## Quality control of pharmaceutical solutions

Following manufacture (and packaging) and over the period of the shelf-life, the physicochemical properties of the pharmaceutical product must lie within specified boundaries. Departure from these properties will invalidate the pharmaceutical claims of the product. Therefore an essential aspect of product manufacture is the assurance and control of quality. Whilst it is outside of the scope of this textbook to discuss pharmaceutical stability, the reader should be aware that the quality of pharmaceutical products is assured through a thorough analysis of the physicochemical properties, both following manufacture and over the designated period of shelf-life (expiry date).

The types of assessments (analysis) that are applied to pharmaceutical products are dependent on the nature (type) of the product; however, as the reader will observe, there are several physicochemical parameters that are examined and quantified in all pharmaceutical products. These will be identified for each product in the subsequent chapters.

All pharmaceutical products have specifications regarding the properties of the product following manufacture (finished product specification) and over the period of storage. These specifications are defined as a range to allow both some flexibility in manufacture and to ensure that the product is safe and efficacious in clinical use. Typically, following manufacture and over the designed period of the shelf-life, the following tests are applied to pharmaceutical solutions:

1. **Concentration of therapeutic agent**: Following manufacture the concentration of therapeutic agent must lie within 95–105% of the nominal concentration. This range offers sufficient flexibility for the product to be successfully manufactured. Products whose drug concentration lies outside this range cannot be released for sale. Over the shelf-life of the product the concentration of drug must not fall below 90% of the nominal amount. If this occurs then the product has expired.

2. **Uniformity of content**: Whilst not strictly specified in the pharmacopoeias, following manufacture the uniformity of content of solutions should be analysed. A similar strategy as that used for oral suspensions may be employed. Thus the individual mass of the therapeutic agent in 10 units is determined. The content of each individual unit must lie within the range of 85–115% of the average content to successfully pass the test. If the mass of therapeutic agent in more than one unit lies outside this range or if the mass of therapeutic agent in one unit lies outside of an extended range of 75–125% of the average content then the batch has failed the uniformity of content test.

   However, if the mass of therapeutic agent of one unit lies outside the 85–115% range but within 75–125% of the average content, the drug contents of twenty further units are determined. The batch will pass uniformity of content then if not more than one of the individual contents of the 30 units is outside the range 85–115% and none is outside 75–125% of the average content range.

3. **Concentration of preservative**: The concentration of therapeutic agent must lie within 95–105% of the nominal concentration following manufacture and also upon storage.

4. **Preservative efficacy testing**: The efficacy of the preservative(s) within the formulation must be assessed using the appropriate pharmacopoeial method. These tests evaluate the resistance of the product to microbial challenge and are performed on products both after manufacture and during storage.

5. **Appearance**: Pharmaceutical solutions must be clear and devoid of particles. The presence of particles would invalidate the claim of the product to be a solution. The specification may include a statement of the colour of the product, changes to the colour of the product often resulting from drug degradation on storage. This test is performed on products both after manufacture and during storage.

   Furthermore, whilst not a pharmacopoeial requirement, it is common to examine the appearance, following exposure to a series of cycles of freezing and thawing. This storage regimen is useful to examine if there is precipitation of any formulation component. Exposure to extremes of temperature may occur as the product is transported from one country to another and consequently it is important to understand the effects of this process on the physicochemical properties of the product.

6. **pH**: The pH of the formulation is measured on products both after manufacture and during storage and is compared to the specified range, changes to the colour of the product being evidence of possible drug degradation.

7. **Viscosity**: The viscosity of the product is measured and compared to the specified range. Again this is performed on products both after manufacture and during storage.

8. **Uniformity of mass**: Following manufacture, the contents of twenty individual units are decanted, weighed separately and the average mass determined. To successfully pass the test the average masses of no more than two units must deviate by more than 10% of the average mass and no units should deviate by more than 20%.

   Alternatively, specifications may simply require that the container is completely emptied and the volume/mass measured. The test has been passed if the mass or volume is not less than the nominal volume/mass on the label.

9. **Uniformity of dose of oral drops**: This test examines the reproducibility and accuracy of dosing of oral drop solutions. In this the prescribed quantity for one dose is dispensed and weighed and this process is repeated for another nine doses, producing a total of ten masses. The average mass is

then calculated. To pass the test no single dose (mass) should deviate by greater than 10% of the average mass and none should be greater than 20% of the average mass.

The reader should note that the list of methods is indicative of the quality control methods that may be employed. The reader should consult the appropriate pharmacopoeias for a more detailed description of the above methods and others that have not been explicitly covered in this chapter.

## Multiple choice questions

1. **Regarding weakly acidic drug molecules, which of the following statements are true?**
   a. The solubility of weak acids increases as the pH is decreased.
   b. The solubility of weak acids increases as the pH is increased.
   c. The solubility of weak acids in pharmaceutical formulations may be affected by the presence of counterions.
   d. All weakly acidic therapeutic agents exhibit an isoelectric point.

2. **Regarding weakly basic drug molecules, which of the following statements are true?**
   a. The solubility of weak bases increases as the pH is decreased.
   b. The solubility of weak bases increases as the pH is increased.
   c. The solubility of weak bases in pharmaceutical formulations may be affected by the presence of counterions.
   d. All weakly basic therapeutic agents exhibit an isoelectric point.

3. **Propranolol (Figure 1.12) is a beta blocker that is used for the treatment of hypertension. Which of the following statements are true regarding this drug?**

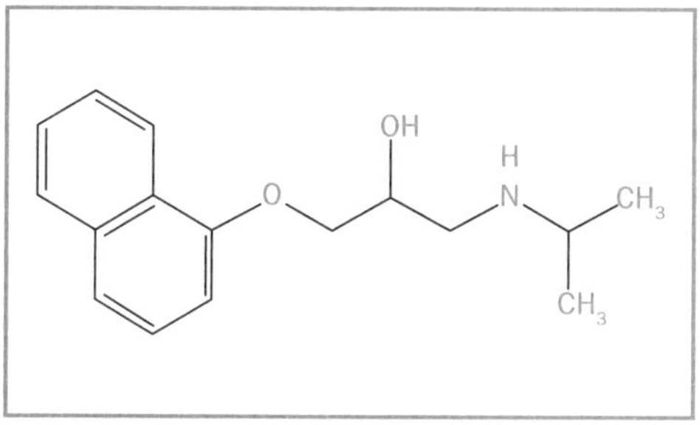

**Figure 1.12** Structure of propanolol.

a. The solubility of propranolol increases as the pH is decreased from pH 9 to pH 4.

b. The solubility of propranolol increases as the pH is increased from 4 to 9.

c. Propranolol is more soluble as a salt form.

d. Propranolol exhibits an isoelectric point.

4. **Regarding buffers for pharmaceutical solutions for oral administration, which of the following statements are true?**

a. Citrate buffer is commonly used as a buffer for pharmaceutical solutions.

b. Buffers are required solely to control the stability of therapeutic agents.

c. Buffer salts may affect the solubility of therapeutic agents.

d. The buffer capacity of a buffer system is increased as the concentration of buffer components is increased.

5. **Regarding the use of antioxidants in pharmaceutical solutions for oral administration, which of the following statements are true?**

a. Antioxidants are required in all solution formulations.

b. Antioxidants reduce the rate of oxidation of the therapeutic agent.

c. BHT and BHA are examples of antioxidants that are included in aqueous solutions.

d. The efficacy of antioxidants may be improved in the presence of ethylenediamine tetraacetic acid (EDTA).

6. **Regarding the use of co-solvents for the formulation of pharmaceutical solutions for oral administration, which of the following statements are true?**

a. Co-solvents are required in all pharmaceutical solution formulations.

b. Alcohols are commonly used as co-solvents in pharmaceutical solutions.

c. Glycerol may directly affect the pH of the formulation.

d. Co-solvents may affect the viscosity of the solution formulation.

7. **Regarding the use of preservatives in pharmaceutical solutions for oral administration, which of the following statements are true?**

a. Preservatives are required in all pharmaceutical solutions.

b. The presence of hydrophilic polymers in oral solutions may necessitate an increase the the required concentration of preservative.

c. Esters of parahydroxybenzoic acid are used as preservatives for pharmaceutical solutions for oral administration.

d. Preservatives render pharmaceutical solutions for oral administration sterile.

8.  **Regarding pharmaceutical elixirs, which of the following statements are true?**
    a.  Preservatives are required in all elixir formulations.
    b.  Elixirs generally require the addition of sweetening agents.
    c.  Elixirs generally contain < 10% Alcohol USP.
    d.  Colours are required for all elixir formulations.

9.  **Regarding pharmaceutical linctuses, which of the following statements are true?**
    a.  Preservatives are required in all linctus formulations.
    b.  Linctuses generally require the addition of synthetic sweetening agents.
    c.  Linctus formulations may contain high concentrations of sucrose.
    d.  Colours are required for all linctus formulations.

10. **Regarding oral syrups, which of the following statements are true?**
    a.  Preservatives are required in all oral syrups.
    b.  In certain syrups the concentration of sucrose may be ≤ 80% w/w.
    c.  Sugar-free syrups require the inclusion of a viscosity-modifying agent.
    d.  Colours are required for all oral syrups.

11. **You have been asked to formulate an oral solution containing two therapeutic agents, an acidic agent (p$K_a$ 4.5) and a basic agent (p$K_a$ 8.0), at pH 7.0. In a preliminary formulation study you have noticed that a small mass of precipitate forms upon storage. Which of the following statements are true?**
    a.  Due to precipitate formation the product has exceeded its shelf life.
    b.  Precipitate formation is most likely due to an interaction of one or both drugs with a formulation component.
    c.  Precipitation may be inhibited by increasing the pH of the formulation.
    d.  Precipitation may be inhibited by the use of a co-solvent.

12. **Which of the following statements are true regarding the preservation of oral solutions?**
    a.  The preservative efficacy of acidic preservatives, e.g. benzoic acid, sorbic acid, increases as the pH of the formulation increases from pH 5 to pH 8.
    b.  The preservative efficacy of acidic preservatives, e.g. benzoic acid, sorbic acid, decreases as the pH of the formulation increases from pH 5 to pH 8.
    c.  The preservative efficacy of esters of parahydroxybenzoic acid increases as the pH of the formulation increases from pH 5 to pH 8.

    **d.** The preservative efficacy of esters of parahydroxybenzoic acid decreases as the pH of the formulation increases from pH 5 to pH 8.

13. **Polyvinylpyrrolidone is a hydrophilic polymer that is used in pharmaceutical solutions for which of the following reason(s)?**
    **a.** To increase the viscosity of the solution and hence to enhance the pourability of the product.
    **b.** To enhance the efficacy of the preservatives.
    **c.** To enhance the solubility of poorly soluble drugs.
    **d.** To reduce the volatility of co-solvents.

14. **Ibuprofen (Figure 1.13) is a non-steroidal anti-inflammatory drug that is used to treat pain. Which of the following statements are true regarding this drug?**
    **a.** The solubility of ibuprofen increases as the pH is decreased from pH 9 to pH 4.

**Figure 1.13** Structure of ibuprofen.

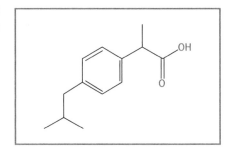

    **b.** The solubility of ibuprofen increases as the pH is increased from pH 4 to pH 9.
    **c.** When administered as an oral solution, ibuprofen may precipitate within the stomach.
    **d.** Propranolol exhibits an isoelectric point.

15. **Following manufacture, which of the following quality control tests should be applied to pharmaceutical solutions?**
    **a.** The concentration of therapeutic agent (and its comparison to the nominal drug concentration).
    **b.** The taste acceptability of the formulation.
    **c.** The pH of the formulation.
    **d.** The density of the formulation.

# chapter 2
# Pharmaceutical disperse systems 1: suspensions – general principles and the formulation of suspensions for oral administration

## Overview

**In this chapter the following points will be discussed:**

- the physical stability of pharmaceutical suspensions
- the advantages and disadvantages of pharmaceutical suspensions
- formulation considerations for pharmaceutical suspensions
- considerations for the manufacture of pharmaceutical suspensions.

## General description

Pharmaceutical suspensions are commonly referred to as dispersions in which the therapeutic agent is dispersed in the external phase (the vehicle). According to this definition the solubility of the therapeutic agent in the vehicle is low. The diameter of the disperse phase may range from circa 0.5 to 100 μm. Systems in which the particle size diameter falls below the above range are termed colloidal. In pharmaceutical dispersions (suspensions) the internal (drug) phase will separate upon storage; however, the main aim of the formulation scientist is to control the process of separation and, in so doing, optimise the stability of the formulation. A pharmaceutical suspension would be considered stable if, after agitation (shaking), the drug particles are homogeneously dispersed for a sufficient time to ensure that an accurate dose is removed for administration to the patient.

## KeyPoints

- Pharmaceutical suspensions are employed to deliver therapeutic agents of low (predominantly aqueous) solubility to the patient.
- Whereas pharmaceutical suspensions may be employed for the administration of drugs by many potential routes, this class of formulation is predominantly used for the delivery of drugs orally and parenterally (by injection) (see Chapter 5).
- Pharmaceutical suspensions are physically unstable; this instability is apparent by the presence of a solid cake and the resultant inability to redisperse the therapeutic agent. This leads to problems regarding the administration of the correct dosage of the therapeutic agent.
- In addition to enhancing the aesthetic properties, the excipients used in pharmaceutical suspensions are included to optimise the physical stability of the formulation.

The characteristics of an acceptable pharmaceutical suspension include the following:

- a low rate of sedimentation
- the disperse phase must be easily redispersed with gentle shaking
- the flow properties of the suspension should enable the formulation to be easily removed from the container (e.g. bottle)
- aesthetically pleasing.

## Advantages and disadvantages of pharmaceutical suspensions

### Advantages

- Pharmaceutical suspensions are a useful drug delivery system for therapeutic agents that have a low solubility. Although low-solubility therapeutic agents may be solubilised and therefore administered as a solution, the volume of the solvent required to perform this may be large. In addition, formulations in which the drug has been solubilised using a co-solvent may exhibit precipitation issues upon storage.
- Pharmaceutical suspensions may be formulated to mask the taste of therapeutic agents.
- Pharmaceutical suspensions may be employed to administer drugs to patients who have difficulty swallowing solid-dosage forms.
- Pharmaceutical suspensions may be formulated to provide controlled drug delivery, e.g. as intramuscular injections (see Chapter 5).

### Disadvantages

- Pharmaceutical suspensions are fundamentally unstable and therefore require formulation strategies to ensure that the physical stability of the formulation is retained over the period of the shelf-life.
- The formulation of aesthetic suspension formulations is difficult.
- Suspension formulations may be bulky and therefore difficult for a patient to carry.

## The physical stability of pharmaceutical suspensions

As detailed above, pharmaceutical suspensions are fundamentally unstable, leading to sedimentation, particle–particle interactions and, ultimately, caking (compaction). To gain an understanding of the physical stability of suspensions it is necessary to consider briefly two phenomena: the electrical properties of dispersed particles and the effect of distance of separation between particles on their subsequent interaction. It must be stressed that this is only a brief outline and the reader should consult the companion textbook in this series by David Attwood and Alexander T Florence (*FASTtrack: Physical Pharmacy*, 2nd edn: London: Pharmaceutical Press, 2012) for a more comprehensive description of this topic.

## Electrical properties of dispersed particles

Following dispersion within an aqueous medium, particles may acquire a charge due to either the ionisation of functional groups on the drug molecule and/or adsorption of ions to the surface of the particle. These are addressed independently below.

### Ionisation of functional groups

Insoluble drug particles may possess groups at the surface that will ionise as a function of pH, e.g. COOH, $NH_2$. In this situation the degree of ionisation is dependent on the $pK_a$ of the molecule and the pH of the surrounding solution.

### Adsorption of ions on to the surface of the particle

Following immersion in an aqueous solution containing electrolytes, ions may be adsorbed on to the surface of the particle. Furthermore, in the absence of added electrolytes, preferential adsorption of hydroxyl ions on to the surface of the particle will occur. Hydronium ions, by contrast, are more hydrated than hydroxyl ions and are therefore more likely to remain within the bulk medium. Following adsorption of ions on to the surface, a phenomenon referred to as the *electrical double layer* is established (Figure 2.1), the main features of which are as follows:

- Ions, e.g. cations, are adsorbed on to the surface of the particle, leaving the anions and remaining cations in solution. This generates a potential on the surface of the particle, termed the *Nearnst potential*. The ions responsible for this potential are termed *potential-determining ions*. Anions are then electrostatically attracted to the (positive) surface of the particle. The presence of these anions will repel the subsequent approach of further anions. This is referred to as the first section of the double layer and is therefore composed of adsorbed ions on the surface, counterions and bound hydrated solvent molecules.
- The boundary between the first and second layers of the electrical double layer is referred to as the *Stern plane*. The Stern plane is characterised by:

**Figure 2.1** Diagrammatic representation of the electrical double layer.

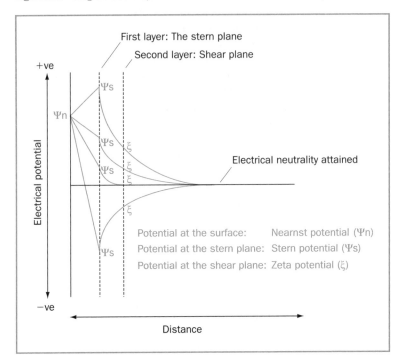

- the adsorbed ions on the surface of the particle
- the adsorbed counterions at the surface of the particle
- the Stern plane falls through the centre of the counterion layer
- normally the charge at the surface of the particle is greater than that at the Stern plane.

■ The second layer contains predominantly hydrated counterions that are loosely attracted to the surface of the particle, the features of which include:
- the boundary of this second layer will possess a potential, referred to as the *zeta potential*. The magnitude of this is generally less than that at the Stern plane.
- if the particle is rotated, this second layer forms the shear plane, i.e. the effective surface.

■ At a certain distance from the surface of the particle, electrical neutrality is restored.

A number of related alternative scenarios to that described above are possible:
■ *Scenario 1*. Electrical neutrality may be achieved at the boundary of the second plane of the electrical double layer,

i.e. the magnitude of the zeta potential is zero. In this the counterions are sufficiently present in this layer to neutralise the net positive charge on the surface of the particle.

■ *Scenario 2*. If the number of counterions in the electrical double layer exceeds the adsorbed ions, the zeta potential will exhibit an opposite charge to that of the Nearnst potential. For example, if the Nearnst potential is positive, the zeta potential may be negative.

■ *Scenario 3*. In certain circumstances molecules may interact with the charged particle surface via non-electrostatic mechanisms. For example, surface-active agents interact with surfaces via hydrophobic interactions. If the surface-active agents are charged, this will alter the Stern and zeta potentials. In this the magnitude of the Stern potential may be increased (i.e. exceeds the Nearnst potential) or may be totally reversed. There is a reduction in this charge at the shear plane (i.e. the zeta potential). Similarly, non-ionic surfactants may adsorb to the surface of the particle (again via hydrophobic interactions), thereby affecting both the Stern and zeta potentials.

The presence of electrolytes directly affects the above situations. As the concentration of electrolyte is increased there is a compression of the electrical double layer. The magnitude of the Stern potential is unaltered whereas the zeta potential decreases in magnitude. As the reader will discover, this approach may be used to stabilise pharmaceutical suspensions.

## The relationship between distance of separation and the interaction between particles

The interaction between suspended particles in a liquid medium is related to the distance of separation between the particles. In principle, three states of interaction are possible:

1. No interaction, in which the particles are maintained sufficiently distant from one another. In the *absence* of sedimentation this is the thermodynamically stable state.
2. Coagulation (agglomeration), in which the particles form an intimate contact with each other. This results in the production of a pharmaceutically unacceptable formulation due to the inability to redisperse the particles upon shaking.
3. Loose aggregation (termed floccules), in which there is a loose reversible interaction between the particles, enabling the particles to be redispersed upon shaking.

The interaction (attraction/repulsion) between particles that have been dispersed in a liquid medium has been quantitatively described by Derjaguin, Landau, Verwey and Overbeek. In the simplest form the 'DLVO' theory assumed that, when dispersed in a liquid medium, particles will experience (electrical) repulsive forces and attractive (London/van der Waals) forces. The overall energy of interaction between the particles ($V_t$) can therefore be described as an addition of the energies of attraction ($V_a$) and repulsion ($V_r$), i.e. $V_t = V_a + V_r$.

■   The *energy of attraction* ($V_a$) is due to London/van der Waals forces and it is inversely proportional to the distance between particles. The attractive forces between particles tend to operate at greater distances than the repulsive forces.

■   The *energy of repulsion* ($V_r$) is due to the overlap of or interaction between the electrical double layers of each particle and operates over a distance of approximately the thickness of the double layer.

Figure 2.2 displays a diagrammatic representation of the relationship between the overall energy of interaction between two particles and their distance of separation. In this, three main regions may be observed:

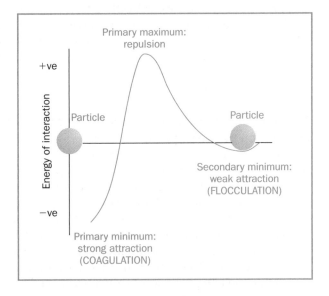

**Figure 2.2** Diagrammatic representation between the overall interactive energy between two particles and their distance of separation.

Primary maximum: repulsion

+ve

Energy of interaction

Particle

Particle

Secondary minimum: weak attraction (FLOCCULATION)

−ve

Primary minimum: strong attraction (COAGULATION)

1.  *The primary minimum.* This is a region of high attraction between particles. Particles that interact at distances corresponding to the primary minimum will irreversibly coagulate and the formulation so produced will be physically unstable.

2.  *The primary maximum.* This region is responsible for the repulsion between particles, the magnitude of which is

controlled by the zeta potential at the shear plane of the particles. This region prevents the particles from interacting at close distances (the primary minimum).

The magnitude of the primary maximum is affected by the presence and concentration of electrolytes. As detailed previously, increasing the concentration of electrolyte decreases the thickness of the double layer, thereby reducing the zeta potential. This leads to a reduction in the magnitude of the primary maximum and increases the magnitude of the secondary maximum. This effect is also observed following the addition of ionic surface-active agents.

3.  *The secondary minimum.* The secondary minimum is a region where attractive forces predominate; however, the magnitude of the attraction is less than that at the primary minimum. Particles located at the secondary minimum are termed floccules, this process being termed flocculation. This interaction increases the physical stability of the suspension by preventing the close approach to the primary minimum. Furthermore, the interaction between the particles may be temporarily broken by shaking, thereby enabling the removal of an accurate dose. The process by which particles are engineered to reside in the secondary minimum is referred to as *controlled flocculation.*

## Sedimentation, controlled flocculation and the physical stability of suspensions

Thermodynamically a disperse system may be considered to be stable whenever there is no interaction between particles. However, in terms of pharmaceutical suspensions, this state is physically unstable. Particles in a suspension will sediment under the influence of gravity and settle at the bottom of the container, the larger particles reaching the bottom initially and the smaller particles occupying the space between the larger particles. The particles at the bottom of the container are gradually compressed by the weight of those above and, in so doing, sufficient energy is available to overcome the primary maximum (repulsive forces) and the particles become sufficiently close to form an irreversible interaction at the primary minimum. This is referred to as *caking.*

In flocculated systems the rate of sedimentation of the flocs is high and the volume of the sediment produced is large due to the large void volume within the floccule structure. Generally the size of the drug particles used in the formulation of suspensions is sufficiently large to exhibit a useful secondary minimum. In addition the irregular (generally non-spherical) nature of suspended drug particles will enhance this property. Finally, the ability to undergo flocculation will increase as the concentration of suspended particles increases, due to the greater probability of particle–particle interactions. For particles in which the zeta potential (and hence the primary maximum) is high, manipulation of the magnitude of the secondary minimum is required (controlled flocculation). Generally this is performed by the addition of the required concentration of electrolyte or (ionic) surface-active agent. In performing controlled flocculation it must be remembered that if the reduction in the zeta potential (and hence the primary maximum) is too large, the resistance to particle–particle contact in the primary minimum is reduced. Therefore, irreversible coagulation of the particles may occur.

As caking in pharmaceutical suspensions is facilitated by sedimentation, it is of no surprise to note that controlling particle sedimentation may enhance the physical stability of pharmaceutical suspensions. The rate of sedimentation of particles (generally <2% w/w) may be defined by Stokes' equation. Although many pharmaceutical suspensions are composed of more than 2% w/v solids in which the sedimentation of drug particles is influenced by other particles, the equation may be used to provide an indication of factors affecting sedimentation. The equation is as follows:

$$\frac{dv}{dt} = \frac{d^2\left(\rho_s - \rho_l\right)g}{18\eta_l}$$

where $\frac{dv}{dt}$ refers to the rate of sedimentation; $d^2$ refers to the average particle diameter; $\rho_s$ and $\rho_l$ refer to the densities of the solid particles and vehicle, respectively; $\eta_l$ refers to the viscosity of the vehicle; and $g$ refers to gravity.

Therefore, as may be observed from the above equation, the rate of sedimentation may be decreased in practice by reducing the average particle diameter and increasing the viscosity of the vehicle. The former may be readily manipulated by milling the particles to the required size range (see Chapter 11) whereas the latter may be increased by the inclusion of hydrophilic polymers. These points are addressed at a later stage.

Assessment of the sedimentation of drug suspensions is usually performed by measuring the *sedimentation volume* and/or the *degree of flocculation*.

- *The sedimentation volume (F)*. This is the ratio of the volume of the sediment $(V_s)$ to the initial volume of the suspension $(V_i)$:

$$F = \frac{V_s}{V_i}$$

The sedimentation volume may range from less than unity to values that are greater than unity. The sedimentation volume of deflocculated suspensions is usually small, whereas the *F* value for flocculated systems is high (i.e. close to or greater than unity) due to the large volume occupied by the flocculated structure.

■ *Degree of flocculation (β).* The degree of flocculation is defined as the ratio of the ultimate sedimentation volume of the flocculated suspension to the ultimate sedimentation volume of the deflocculated suspension. This is usually the preferred measurement as it provides a point of reference, i.e. the suspension before and after flocculation.

## Formulation considerations for orally administered suspension formulations

The formulation of suspensions for oral administration requires consideration of both the physical properties of the therapeutic agent and the excipients required to ensure that the formulation is physically stable and suitable for administration to patients. General formulation considerations are as follows.

### Physical properties of the therapeutic agent

#### Particle size

As detailed previously in this chapter, the physical stability of suspensions may be enhanced by modifying (i.e. retarding) the rate of particle sedimentation. According to Stokes' law, the rate of sedimentation $\left(\dfrac{dv}{dt}\right)$

is directly proportional to the square of the average diameter of the particles (termed $d^2$). Therefore, as the average particle size of suspended particles is increased, there is a dramatic effect on the resultant rate of sedimentation, i.e. increasing the particle diameter twofold results in a fourfold increase in the rate of sedimentation. Therefore, the average particle diameter of therapeutic agent used in the formulation of suspensions has major implications in the physical stability of the formulation.

## Tips

■ Pharmaceutical suspensions are unstable systems that require the addition of excipients and knowledge of the particle size of the dispersed phase to ensure that a stable suspension may be formulated.

■ Generally the alternative choice to formulating suspensions is the production of solutions using solubilising agents (see Chapter 1). Therefore, the formulation scientist must carefully consider both options before making the final formulation choice.

■ Two key parameters that must be controlled are the electrolyte concentration (and type of ions) and the viscosity.

To optimise the stability of the formulation, the particle size should be minimised. This may be performed by either chemical (controlled precipitation) or physical methods (e.g. milling). It should be remembered that the energy required to reduce the average particle diameter to less than 10 µm is significant and therefore a compromise may have to be made concerning the actual particle size of therapeutic product used. Milling is further discussed in Chapter 11.

A phenomenon that may affect pharmaceutical suspensions and which influences the average particle size is *crystal growth* (sometimes referred to as *Ostwald ripening*). Small particles have a greater solubility (dissolution rate) than larger particles when dispersed in an aqueous vehicle. If there is a change (slight increase) in the storage temperature, this may enable the smaller particles to dissolve in the vehicle. Crystallisation of the dissolved drug may then occur on the surface of the larger particles, thereby increasing the average diameter of the suspended drug particles. This may therefore have implications regarding the physical stability of the suspension. One method that may be employed to reduce crystal growth is the inclusion of hydrophilic polymers within the formulation. These adsorb on to the suspended drug particles and offer a protective effect. In light of the potential problems associated with crystal growth, it is customary in formulation development to expose the suspension formulation to temperature cycling (e.g. repeated freeze–thaw cycles) and monitor changes in the average particle diameter and physical stability.

### Wetting properties of the therapeutic agent

Insoluble drug particles are hydrophobic and therefore may not be easily wetted, i.e. the vehicle will not readily form a layer around the suspended drug particle. To wet fully with an aqueous vehicle the contact angle ($\theta$), i.e. the angle at which the liquid/vapour interface meets the surface of the solid, must be low. The contact angle may be defined in terms of the interfacial tensions between the three phases, i.e. solid (drug)/vapour ($\gamma_{s/v}$), liquid (vehicle)/vapour ($\gamma_{l/v}$) and solid (drug)/liquid (vehicle) ($\gamma_{s/l}$) in the Young equation:

$$\gamma_{l/v} \cos\theta = \gamma_{s/v} - \gamma_{s/l}$$

Therefore, decreasing the interfacial tensions between the vehicle and the vapour and between the solid and the vehicle may reduce the contact angle. In practice this is achieved by the incorporation of surface-active agents into the formulation. These agents decrease the interfacial tension by adsorbing at the vehicle/vapour interface and at the solid/liquid interface.

It is important to ensure insoluble therapeutic agents are sufficiently wetted, as this will ensure that the particles are homogeneously distributed in the formulation and thereby enable the correct dosage of drug to be removed when required by the patient. Drug particles, if poorly wetted, will tend to aggregate spontaneously in an attempt to stabilise the system thermodynamically (i.e. lower the Gibb's free energy), thereby resulting in problems regarding the physical stability of the formulation.

## Excipients used in the formulation of suspensions for oral administration

As the reader will observe, there is a direct similarity between the types of excipients used for the formulation of suspensions and solutions for oral administration. The major difference between these two categories of formulations is the inclusion of excipients to physically stabilise suspensions. Many of the categories (and examples) of excipients used for suspensions are the same as for solutions.

### Vehicle
As in oral solutions (and related formulations), the most commonly used vehicle for the formulation of pharmaceutical suspensions for oral administration is Purified Water USP. In addition to purified water, the vehicle may contain buffers to control the pH of the formulation. Citric acid/sodium citrate is commonly used as a buffer system for oral suspension formulations.

### Excipients to enhance the physical stability of suspensions
As detailed previously, pharmaceutical suspensions may be stabilised by controlled flocculation and by the control of the rate of particle/floccule sedimentation. Further details of these approaches are shown below.

#### Addition of electrolytes
Electrolytes may be employed to control flocculation by reducing the zeta potential and hence the electrical repulsion that exists between particles (the primary maximum). In so doing the magnitude of the secondary minimum increases, thereby facilitating the interaction of particles at a defined distance. Buffers are electrolytes and may be used for this purpose; however, other salts can also be used.

To ascertain the correct concentration of electrolytes (ionic strength), a series of formulations containing different concentrations of electrolyte are prepared and the sedimentation volume or degree of flocculation determined. In flocculated systems the sedimentation volume and degree of flocculation are high. The addition of either insufficient or excess electrolyte will produce physically unstable suspensions that exhibit caking.

## Surface-active agents

Surface-active agents may influence the stability of pharmaceutical suspensions in several ways, as detailed in previous sections. These are briefly summarised below.

### Effect on wetting

Surface-active agents decrease the contact angle of insoluble particles, enabling greater wetting by the vehicle. This, in turn, assists product homogeneity and decreases aggregation.

### Effect on flocculation

Surface-active agents, both ionic and non-ionic, can interact with the suspended particles and, in so doing, can affect the magnitude of the zeta potential. This may lead to the lowering of the primary maximum and an increase in the size of the secondary maximum, thereby facilitating flocculation. The correct concentration of surfactant required to stabilise a suspension may be experimentally obtained in a fashion similar to that described above for electrolytes.

For oral suspensions non-ionic surfactants are preferred, e.g. polyoxyethylene fatty acid sorbitan esters, sorbitan esters or lecithin, further details of which are provided in Chapter 3. The greater toxicity of ionic surfactants precludes their use in oral suspension formulations. The concentrations of surfactants required to stabilise pharmaceutical suspensions are dependent on the physical properties of the dispersed particles (e.g. zeta potential); however, concentrations less than 0.5% w/v are generally employed in the formulation of oral suspensions.

## Hydrophilic polymers

Hydrophilic polymers are commonly used to enhance the physical stability and to affect the flow properties of oral suspensions. These two aspects are detailed below.

### Effects on the physical stability of suspensions

Hydrophilic polymers may adsorb on to the surface of suspended drug particles in pharmaceutical suspensions. Due to their large molecular weight, one section of the polymer chain will adsorb on to the particles, leaving the remainder of the chain to extend into the aqueous vehicle. As the concentration of polymer in the formulation increases, the thickness of the adsorbed layer of polymer increases. As two (polymer-coated) particles approach each other, there will be stearic repulsion due to an overlap of the adsorbed polymer chains. This will prevent the particles coming into close contact (at the primary minimum). It is important to note that the ability of hydrophilic polymers to stabilise suspensions stearically is dependent on several features, e.g. (1) the concentration of polymer; and (2) the type of polymer.

### The concentration of polymer

The concentration of polymer affects the density of the adsorbed polymer layer on the surface of the particles. The required concentration of polymer should be that which enhances repulsion but does not prevent the interaction of the particles in the secondary minimum (flocculation). Generally flocculation occurs at a distance which is approximately twice the thickness of the adsorbed polymer layer.

### The type of polymer

The type (and hence the chemistry) of polymer influences the stabilisation properties of hydrophilic polymers in two ways. Firstly, the chemical structure of the polymer will influence the nature of the adsorption on the surface of the drug particles which, in turn, influences the thickness and integrity of the adsorbed layer. Secondly, as the interaction between specific groups on adjacent polymer chains is responsible for the stearic stabilisation, the nature of the interacting groups on each chain is important. This ability to interact may effectively maintain the polymer-coated particles at a distance, resulting in the production of a structured floccule.

Finally if ionic (e.g. anionic) polymers are added to the formulation in the presence of divalent ions (e.g. $Mg^{2+}$, $Ca^{2+}$) or trivalent ion ($Al^{3+}$), the ions may act as a bridge by interacting both with the surface of the particle and with the charged moieties on the polymer. In this way the polymer is 'connected' to two particles and prevents the interaction of the particles at the primary minimum.

Secondarily, the addition of hydrophilic polymers to the aqueous vehicle will increase the viscosity of the formulation. According to Stokes' equation (and remembering the limitations of the equation!), increasing the viscosity of an aqueous vehicle will reduce the rate of sedimentation, thereby increasing the physical stability of the formulation.

### Effect on the rheological properties of oral suspension

As detailed in the previous paragraph, increasing the concentration of a hydrophilic polymer within an aqueous vehicle will alter the viscosity of the system. At very low polymer concentrations (frequently <0.01% for branched polymers), aqueous vehicles will behave as Newtonian systems, in which the shearing stress and rate of shear are proportional. However, at higher polymer concentrations, typically of those used in oral suspensions, the flow properties are *pseudoplastic* (shear thinning). This is a useful property for suspensions as the apparent viscosity (the reciprocal of the tangent of the flow curve) will be high under conditions of low shear stress (e.g. storage in the bottle) thereby lowering the rate of sedimentation of the particles, but under high shearing stress (e.g. those exerted whenever the contents of the bottle are shaken) the apparent viscosity will be low, thereby facilitating administration to the patient. In addition, pseudoplastic formulations may also exhibit *thixotropy*, a time-dependent recovery of the flow properties (illustrated in Figure 2.3 as the solid arrow). It is important to understand this property in particular, as this should ideally be minimised to enable the rapid recovery of the rheological properties of the formulation.

**Figure 2.3** Ideal flow curves illustrating Newtonian flow (dashed line) and pseudoplastic flow (with thixotropy, up curve direction shown).

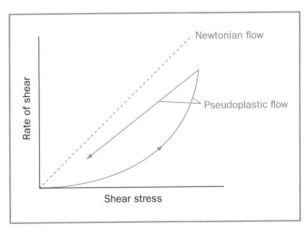

Flocculated oral suspensions may alternatively exhibit *plastic* flow in which a yield stress is required to initiate flow after which the flow properties are Newtonian. The yield stress corresponds to the stress required to overcome the interaction between flocculated particles. Plastic flow is illustrated in Figure 2.4.

Examples of polymers that are employed in oral suspensions to enhance the physical stability and to modify the flow (rheological) properties include:

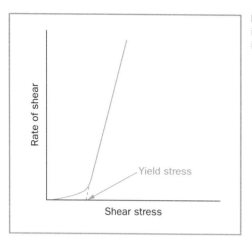

Rate of shear

Yield stress

Shear stress

**Figure 2.4**   Ideal flow curve illustrating plastic flow (and identifying the yield stress).

- Cellulose derivatives: these are branched polymers and therefore only low concentrations are required (<3% w/w). Specific examples that are commonly used include:
  - methylcellulose
  - hydroxyethylcellulose
  - hydroxypropylmethylcellulose
  - sodium carboxymethylcellulose (ionic).
- Polyvinylpyrrolidone: this is a linear polymer and will therefore require a larger concentration to enhance the rheological properties of the formulation.
- Sodium alginate (ionic).
- Acacia, tragacanth and xanthan gum: these polymers are natural polysaccharide-based polymers that may show batch-to-batch variability in their effects on formulation viscosity/rheology. Only low concentrations of these are required to enhance the viscosity of the formulation.

The concentration of each of these polymers will depend on the chemistry of the polymer, the molecular weight and, for ionic polymers, the pH of the formulation.

Rheological structuring of oral suspensions may also be achieved using the hydrated silicates, e.g. magnesium aluminium silicate. These materials swell in water, resulting in a vehicle that exhibits plastic flow. The concentration of magnesium aluminium silicate that is typically used is circa 5%. Caution should be issued when formulated in the presence of cationic compounds as this may induce flocculation of the dispersed, swollen silicates. Due to this susceptibility, hydrated silicates may be formulated along with hydrophilic polymers to produce formulations with the required consistency.

## Preservatives

Oral suspensions are *non-sterile*; however, there are restrictions on the number and type of microorganisms present in this type of dosage form. Whilst the presence of microorganisms within oral suspensions is allowed, it is essential that highly pathogenic microorganisms, e.g. *Escherichia coli*, are absent. Oral suspensions are multidose formulations and therefore inhibition of the growth/elimination of less pathogenic bacteria and fungi (the latter primarily causing spoilage) is required. Specifications are defined in the various pharmacopoeias regarding the number and type of microorganisms in oral products (solutions and suspensions). For example, the European Pharmacopoeia states that in oral products *E. coli* must be absent and, in addition, there should be not more than 1000 aerobic bacteria and not more than 100 fungi per gram or millilitre.

Examples of preservatives that are employed in oral suspensions include:

- Parabens (i.e. esters of parahydroxybenzoic acid), e.g. methyl and propyl parahydroxybenzoic acid are often used in combination in a ratio of 9:1 or 10:1. The concentration is usually 0.22% w/v (i.e. 0.2% w/v methylhydroxy benzoate and 0.02% w/v propylhydroxy benzoate).
- Organic acids, e.g. benzoic acid (circa 0.9% w/v).

When selecting the type and concentration of preservative for inclusion in an oral suspension formulation, the following points should be considered:

- To exert an antimicrobial effect, preservative must exist in solution within the formulation. Certain preservatives, e.g. the parahydroxybenzoate esters, will interact with hydrophilic polymers, e.g. polyvinylpyrrolidone, cellulose ethers and surface-active agents (e.g. polysorbate 80), thereby reducing the concentration of 'free' preservative in the formulation. As a result the antimicrobial efficacy of the preservative is decreased. To overcome this problem it is customary to increase the initial concentration of preservative to ensure that, after adsorption to the dissolved polymer, the required free concentration of preservative is available.
- The selected preservative must not adversely affect the chemical and physical stability of the suspension product.

## Sweetening agents/flavours

As is the case for oral solutions, sweetening agents (e.g. sucrose, liquid glucose, glycerol, sorbitol, saccharin sodium and aspartame) and flavours are used for taste-masking purposes. The type (particularly in the case of flavours) and the concentrations of these are selected, as before, to provide the necessary aesthetic properties. The details of these have been provided in Chapter 1.

## Antioxidants

Antioxidants are required in certain pharmaceutical suspensions for oral usage to enhance the chemical stability of the therapeutic agent, where this may be compromised by oxidation. As before (Chapter 1), the chosen antioxidants are compounds that are redox systems which exhibit higher oxidative potential than the therapeutic agent or, alternatively, are compounds that inhibit free radical-induced drug decomposition. Commonly used examples for incorporation into oral suspensions (usually at concentrations less than 0.1% w/w) include:

- sodium sulphite
- sodium metabisulphite
- sodium formaldehyde sulphoxylate
- ascorbic acid.

As for solutions, chelating agents, e.g. ethylenediamine tetraacetic acid, citric acid, that form complexes with heavy metal ions which are normally involved in oxidative degradation of therapeutic agents, may be included.

A diagrammatic representation of formulation considerations for aqueous pharmaceutical suspensions designed for oral administration is presented in Figure 2.5.

**Tips**

- The formulation of suspensions for oral administration will require the inclusion of several excipients.
- All excipients must be physically and chemically compatible (with each other and with the therapeutic agent).
- As the therapeutic agent is suspended in the formulation, the rate of degradation (e.g. by hydrolysis, oxidation) of therapeutic agents in pharmaceutical suspensions is lower than for comparator solution formulations. This is due to the low solubility of the drug substance in the aqueous vehicle.
- Possible interaction between components must be understood, e.g. preservatives and viscosity-enhancing agents.

## Manufacture of suspensions for oral administration

Suspensions for oral administration are usually manufactured by one of two methods: (1) direct incorporation; and (2) precipitation method.

### Direct incorporation

- In this method the soluble components are normally dissolved in the appropriate volume of diluent (vehicle).
- The solid therapeutic agent is then dispersed into the vehicle with the aid of mixing, prior to correction for volume.
- The mixing rate employed during the addition is an important determinant in the manufacture of the formulation. If the suspension is flocculated, high-speed

**Figure 2.5** Diagrammatic representation of formulation considerations for aqueous pharmaceutical suspensions.

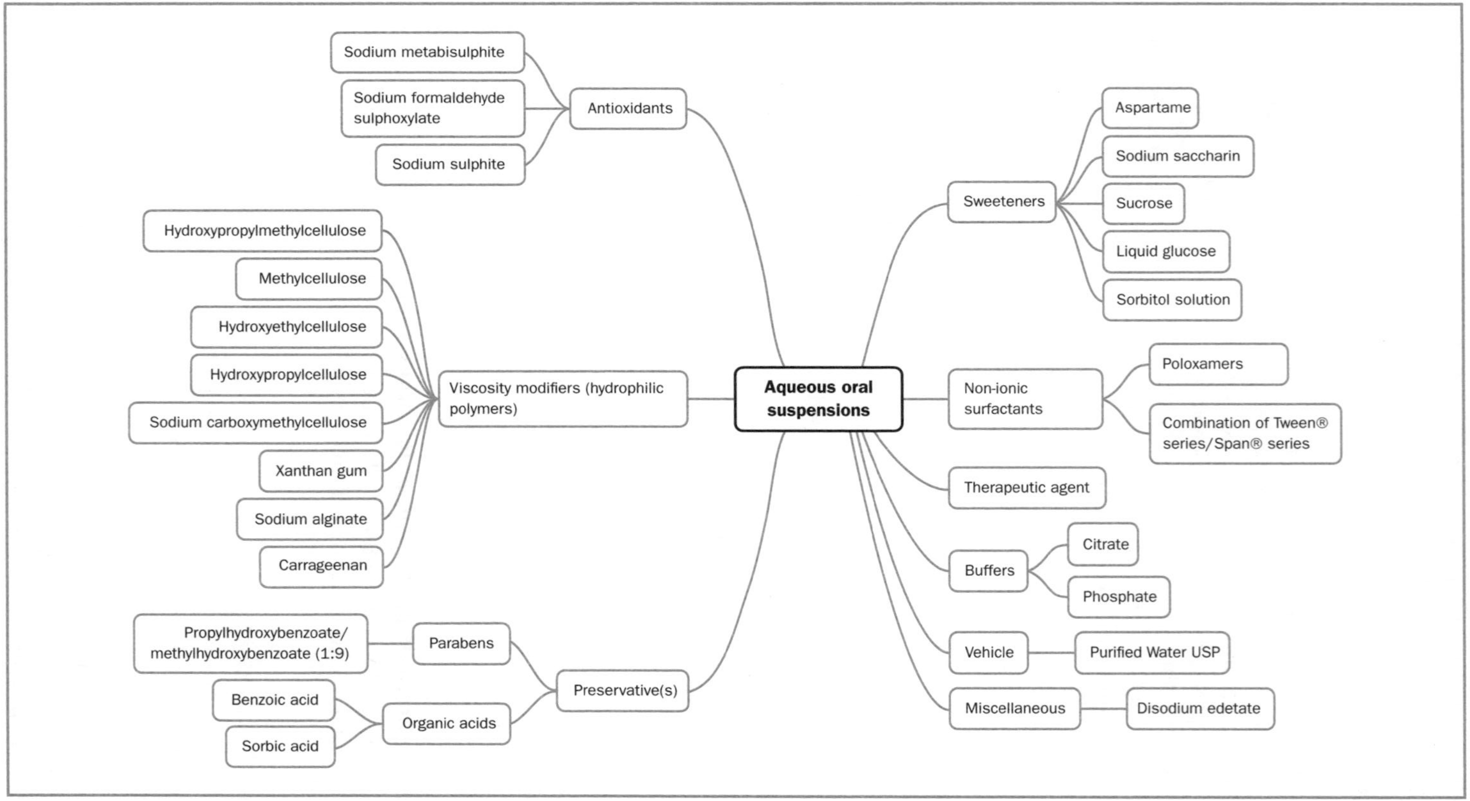

mixing may be employed as the flow properties of the system are pseudoplastic (shear thinning; Figure 2.3). However, if the formulation has been poorly designed and has poor flocculation properties, high-speed mixing will result in an increase in the viscosity of the product (termed *dilatant flow*). Ultimately this leads to issues regarding the quality of mixing as the increased viscosity may render the product difficult to mix homogeneously.

- The particle size of the suspended drug within the formulation may then be reduced using a ball mill. Alternatively, the particle size of the active ingredient may be optimised (by particle size reduction techniques, see Chapter 11) prior to incorporation into the vehicle.

### Precipitation method

- In this method the drug is dissolved in the vehicle (or a portion of the available volume), prior to precipitation following the addition of a counterion; the salt formed is insoluble.
- Such systems are frequently deflocculated and are therefore mixed at low shear rates.
- The excipients are then dissolved in the vehicle, or dissolved in a portion of the vehicle, which is then added to the suspension of drug.
- At this stage the formulation may be exposed to high shearing rates to ensure homogeneity.
- The volume of the formulation is then corrected by adding the required mass of diluent.
- One potential problem with this technique is the production of ionic by-products from the precipitation interaction. If the concentration of these is too high, then the precipitated therapeutic agent requires to be washed with an aqueous solvent.

## Quality control of pharmaceutical suspensions for oral administration

The rationale for the quality control of pharmaceutical products has been described in Chapter 1 and therefore will not be repeated. Typically, following manufacture and over the designed period of the shelf life, the following analyses are applied to pharmaceutical suspensions:

1. **Concentration of therapeutic agent**: Following manufacture the concentration of therapeutic agent must lie within 95–105% of the nominal concentration. This range offers sufficient flexibility for the product to be successfully

manufactured. Products whose drug concentration lies outside this range cannot be released for sale. Over the shelf-life of the product the concentration of drug must not fall below 90% of the nominal amount. If this occurs then the product has expired. One issue regarding suspensions is the redispersability of the solid materials following agitation. If the shaken product is not homogeneous upon shaking then the measured concentration of therapeutic agent may not be an accurate assessment of the concentration of drug in the dosage form. This has implications regarding the dose that a patient will receive within the clinical situation.

2. **Concentration of preservative**: The concentration of therapeutic agent must lie within 95–105% of the nominal concentration following manufacture and also upon storage.

3. **Preservative efficacy testing**: The efficacy of the preservative(s) within the formulation must be assessed using the appropriate pharmacopoeial method. These tests evaluate the resistance of the product to microbial challenge and are performed on products both after manufacture and during storage.

4. **Appearance**: Pharmaceutical suspensions must be homogeneous upon shaking the container, particularly after storage. The appearance is examined both following manufacture and during storage.

5. **pH**: The pH of the formulation is measured on products both after manufacture and during storage and is compared to the specified range, changes to the colour of the product being evidence of possible drug degradation.

6. **Viscosity**: The viscosity of the product is measured and compared to the specified range. In addition pharmaceutical companies may apply their own in-house assessment of redispersability of the product. This parameter is measure both after manufacture and during storage.

7. **Uniformity of content**: Akin to pharmaceutical solutions, following manufacture the uniformity of content of oral suspensions should be analysed. Typically, the individual mass of the therapeutic agent in 10 units is determined. The content of each individual unit must lie within the range of 85–115% of the average content to successfully pass the test. If the mass of therapeutic agent in more than one unit lies outside this range or if the mass of therapeutic agent in one unit lies outside an extended range of 75–125% of the average content then the batch has failed the uniformity of content test.

However, if the mass of therapeutic agent of one unit lies outside the 85–115% range but within 75–125% of the average content, the drug contents of twenty further units is determined. The batch will pass uniformity of content then if not more than one of the individual contents of the 30 units is outside the range of 85–115% and none is outside 75–125% of the average content range.

8. **Particle size of the dispersed drug**: The particle size distribution of the dispersed drug is characterised using an appropriate technique, e.g. Coulter counter, laser diffraction. Changes (increases) in the particle size distribution provide evidence of possible instability due to particle–particle interactions and/or crystal growth. Typically this parameter is examined both following manufacture and during storage and is compared to the product specification.

9. **Freeze–thaw storage**: Whilst not a pharmacopoeial requirement, it is common to examine the physical properties, in particular the particle size distribution and redispersability, following exposure to a series of cycles of freezing and thawing. This storage regimen is useful for two reasons. Firstly, exposure to extremes of temperature may occur as the product is transported from one country to another and, consequently, it is important to understand the effects of this process on the physicochemical properties of the product. Secondly, freeze–thaw cycling provides evidence of possible instability of the product upon storage.

10. **Uniformity of mass**: Following manufacture, the contents of twenty individual units are decanted, weighed separately and the average mass determined. To successfully pass the average masses of no more than two units must deviate by more than 10% of the average mass and no units should deviate by more than 20%.

    Alternatively, specifications may simply require that the container is completely emptied and the volume/mass measured. The test has been passed if the mass or volume is not less than the nominal volume/mass on the label.

The reader should note that the list of methods is indicative of the quality control methods that may be employed. The reader should consult the appropriate pharmacopoeias for a more detailed description of the above methods and others that have not been explicitly covered in this chapter.

## Multiple choice questions

1. **Regarding the stability of pharmaceutical suspensions designed for oral administration, which of the following statements are true?**
   a. Suspensions are inherently pharmaceutically unstable.
   b. The stability of pharmaceutical suspensions is affected by the particle size of the dispersed drug.
   c. The stability of pharmaceutical suspensions is affected by the concentration of buffer salts used.
   d. The particle size range of dispersed solids affects the stability of pharmaceutical suspensions.

2. **Regarding the rate of sedimentation of pharmaceutical suspensions designed for oral administration, which of the following statements are true?**
   a. The rate of sedimentation is increased as the diameter of the dispersed drug particles is increased.
   b. The rate of sedimentation is increased as the viscosity of the continuous phase is increased.
   c. The rate of sedimentation is affected by the concentration of buffer salts.
   d. The rate of sedimentation may be increased by centrifugation.

3. **Regarding the electrical double layer, which of the following statements are true?**
   a. The zeta potential is principally due to ionisation of the drug particle.
   b. The zeta potential for insoluble basic drugs is always positive.
   c. Manipulation of the zeta potential may be used to enhance the physical stability of suspensions.
   d. Increasing the concentration of added electrolyte enhances the thickness of the electrical double layer.

4. **Regarding the DLVO theory, which of the following statements are true?**
   a. The zeta potential acts as a repulsion barrier.
   b. Particles residing within the primary minimum produce pharmaceutically acceptable suspensions.
   c. Alteration of the magnitude of the secondary minimum may be performed by increasing the concentration of electrolyte.
   d. Increasing the concentration of hydrophilic polymer in a suspension increases the stability of the suspension by increasing the magnitude of the primary maximum.

5. **Regarding the role of surfactants in pharmaceutical suspensions for oral administration, which of the following statements are true?**
   a. Surfactants decrease the water contact angle of dispersed drug particles.
   b. Surfactants promote flocculation.
   c. Surfactants with low HLB are used to stabilise oral suspensions designed for oral administrations.
   d. Surfactants increase the viscosity of the continuous phase of pharmaceutical suspensions.

6. **Regarding the use of hydrophilic polymers for the stabilisation of pharmaceutical suspensions for oral administration, which of the following statements are true?**
   a. Hydrophilic polymers stabilise pharmaceutical suspensions by increasing the viscosity of the continuous phase and hence promoting the sedimentation of dispersed drug particles.
   b. Hydrophilic polymers may affect the zeta potential of the dispersed drug.
   c. The rheological properties of pharmaceutical suspensions containing hydrophilic polymers and 1–5% w/w dispersed drug may be described as dilatant.
   d. Pharmaceutical suspensions may exhibit thixotropy.

7. **Flocculated suspensions exhibit which of the following properties?**
   a. A low sedimentation volume.
   b. A high degree of flocculation.
   c. A high rate of sedimentation.
   d. Homogeneity of drug concentration per unit dose.

8. **Concerning the manufacture of pharmaceutical suspensions designed for oral administration, which of the following statements are true?**
   a. Suspensions for oral administration are prepared under aseptic conditions.
   b. Suspensions for oral administration are frequently sterilised following manufacture.
   c. High-speed mixing of concentrated suspensions may result in dilatant flow.
   d. The particle size distribution of the dispersed drug may be reduced post-manufacture using a ball mill.

9. **Concerning the use of pharmaceutical suspensions designed for oral administration, which of the following statements are true?**

a. Suspensions for oral administration are primarily used for administration to children or the elderly.
b. Many antacid formulations are suspensions.
c. Drugs with high aqueous solubility are frequently formulated as suspensions designed for oral administration.
d. Pharmaceutical suspensions designed for oral administration must be coloured.

10. **Concerning pharmaceutical suspensions for oral administration, which of the following statements are true?**
a. Suspensions for oral administration may require the addition of sweetening agents.
b. Suspensions for oral administration may require the addition of flavours.
c. The pH of the continuous phase may affect the stability of aqueous suspensions designed for oral administration.
d. The particle size of dispersed drug in pharmaceutical suspensions designed for oral administration must be <10 µm.

11. **Which of the following excipients may be used to stabilise the physical properties of pharmaceutical suspensions for oral administration?**
a. Methylcellulose
b. Sodium citrate
c. Benzoic acid
d. Patent Blue V.

12. **A poorly soluble drug has been formulated as a pharmaceutical suspension. You have decided to formulate the suspension by reducing the size of the suspended drug from circa 100 µm to circa 10 µm. This will result in which of the following?**
a. A reduction in the zeta potential of the suspended drug
b. Compression of the thickness of the electrical double layer
c. A reduced rate of sedimentation
d. A reduction in the rate of dissolution of the drug.

13. **A poorly soluble drug has been formulated as a pharmaceutical suspension. You have decided to formulate the suspension by elevating the concentration of hydroxyethylcellulose from 0.5%w/w to 2.5%w/w. This will result in which of the following?**
a. A reduction in the zeta potential of the suspended drug
b. Compression of the thickness of the electrical double layer
c. A reduced rate of sedimentation
d. A reduction in the rate of dissolution of the drug.

14. **A poorly soluble drug has been formulated as a pharmaceutical suspension. The concentration of the dispersed (drug) phase is 40% w/w. Which of the following statements are true regarding the properties of this suspension?**
   a. The suspension would be expected to exhibit pseudoplastic flow and therefore should be mixed during manufacture at a high shear rate.
   b. The suspension would be expected to exhibit pseudoplastic flow and therefore should be mixed during manufacture at a low shear rate.
   c. The suspension would be expected to exhibit dilatant flow and therefore should be mixed during manufacture at a high shear rate.
   d. The suspension would be expected to exhibit dilatant flow and therefore should be mixed during manufacture at a low shear rate.

15. **Typically quality control analysis of pharmaceutical suspensions designed for oral administration should include which of the following?**
   a. Concentration of drug (compared to the nominal concentration)
   b. Colour
   c. Sterility
   d. Viscosity.

# chapter 3
# Pharmaceutical disperse systems 2: emulsions and creams

## Overview

**In this chapter the following points will be discussed:**

- the physical properties of disperse systems in which an insoluble liquid is dispersed in a second liquid phase (generically termed emulsions)
- types of emulsions and creams
- factors affecting the stability and methods that may be used to stabilise pharmaceutical (liquid) disperse systems (emulsions and creams)
- formulation strategies for emulsions and creams
- the advantages and disadvantages and uses of emulsions and creams
- considerations for the manufacture of pharmaceutical emulsions and creams.

## General description

Pharmaceutical emulsions/creams are commonly used pharmaceutical products that are primarily prescribed for the treatment of external disorders. In addition to this use emulsions are clinically used for total parenteral nutrition (see Chapter 5), for the oral administration of therapeutic agents and for the rectal administration of antiepileptic agents. The terms emulsions and creams refer to disperse systems in which one insoluble phase is dispersed as droplets within a second liquid phase. The rheological properties (and hence the structure of the network within the formulation) of the two systems differ considerably. Creams are pseudoplastic systems with a greater consistency than, for example, oral or parenteral emulsions.

There are two principal types of emulsions/creams, termed oil in water (o/w)

## KeyPoints

- Emulsions and creams are disperse systems in which an insoluble liquid phase is dispersed within a second liquid phase. Creams are emulsions that offer greater consistency (viscosity) and are applied topically.
- Emulsions and creams are termed either oil in water (o/w), in which oil is the disperse phase and water the external phase, or water in oil (w/o), in which water is the disperse phase and oil is the external phase.
- The major use of emulsions is as cream formulations (for external application); however, emulsions may also be administered intravenously (see Chapter 5), rectally or orally.
- Emulsions/creams are physically unstable: the various excipients in the formulation are present primarily to stabilise the physical properties of the system.

and water in oil (w/o). In the former system the oil (or internal) phase is dispersed as droplets through the external aqueous phase. Conversely, in w/o emulsions, the internal phase is composed of water droplets and the external phase is non-aqueous. In addition to the emulsion types described above there are further more structurally complex types, termed *multiple emulsions*. These are termed water in oil in water (w/o/w) and oil in water in oil (o/w/o) emulsions. However, the pharmaceutical uses of these are extremely limited due to their possible reversion to the parent primary emulsion. For example, an o/w/o emulsion may revert to a w/o emulsion. As the reader will observe later in this chapter, the nature of the excipients and the volume ratio of the two phases used in the formulation of these systems determine both the type and consistency of the emulsion.

Emulsions and creams, akin to pharmaceutical suspensions, are fundamentally unstable systems, which, in the absence of *emulsifying agents*, will separate into the two separate phases. The emulsifying agents used are principally surface-active agents. O/w emulsions may be administered topically or orally whereas the use of w/o creams is principally (but not exclusively) limited to formulations designed for topical application.

The characteristics of an acceptable pharmaceutical suspension include the following:

- Physical stability (no phase separation).
- The flow properties of the emulsion/cream should enable the formulation to be easily removed from the container. Furthermore, if the formulation is designed for external application to, for example, the skin, it must be easily spread over the affected area.
- The formulation must be aesthetically and texturally pleasing. If the emulsion is designed for oral application, the flavour must be suitable, whereas if emulsions are to be externally applied, they must have the correct 'feel' (termed texture).

## Tips

- Emulsions are physically unstable systems and indeed are more unstable than suspensions
- The type of emulsion dictates the final use of the formulation
- Oil in water emulsions (but not water in oil emulsions) may be administered orally
- Both oil in water and water in oil creams are administered topically.

## Advantages and disadvantages of pharmaceutical emulsions

### Advantages

- Pharmaceutical emulsions may be used to deliver drugs that exhibit a low aqueous solubility. For example, in o/w emulsions the therapeutic agent is dissolved in the internal oil phase. Following oral administration the oil droplets (and hence the drug) may then be absorbed using the normal absorption mechanism for oils. Some drugs are more readily absorbed when administered as an emulsion than as other oral comparator formulations.

■ Pharmaceutical emulsions may be used to mask the taste of therapeutic agents, in which the drug is dissolved in the internal phase of an o/w emulsion. The external phase may then be formulated to contain the appropriate sweetening and flavouring agents.

■ Emulsions may be commonly used to administer oils that may have a therapeutic effect. For example, the cathartic effect of oils, e.g. liquid paraffin, is enhanced following administration to the patient as droplets within an o/w emulsion. The taste of the oil may be masked using sweetening and flavouring agents.

■ If the therapeutic agent is irritant when applied topically, the irritancy may be reduced by formulation of the drug within the internal phase of an o/w emulsion.

■ Pharmaceutical emulsions may be employed to administer drugs to patients who have difficulty swallowing solid-dosage forms.

■ Emulsions are employed for total parenteral nutrition.

### Disadvantages

■ Pharmaceutical emulsions are thermodynamically unstable and therefore must be formulated to stabilise the emulsion from separation of the two phases. This is by no means straightforward.

■ Pharmaceutical emulsions may be difficult to manufacture.

## Emulsion instability and theories of emulsification

### Emulsion instability and the role of surface-active agents

Emulsions are termed thermodynamically unstable systems. Following dispersion of an insoluble liquid, e.g. an oil into an aqueous phase, the oil phase will adopt a spherical (droplet) shape as this is the shape associated with the minimum surface area per unit volume. If the droplet contacts a second droplet, coalescence will occur to produce a single droplet of greater diameter and, in so doing, the surface area of the new droplet will be less than the surface areas of the two individual droplets prior to coalescence. This process will continue until there is complete phase separation, i.e. two liquid layers occur. An interfacial tension exists at the interface between the two phases due to the imbalance of forces at the interface. For example, at the interface between the two layers, there will be a net attractive force that is directed towards the bulk of each phase, due to the imbalance between the cohesive forces (oil–oil and water–water) within each phase and the oil–water attractive forces at the interface. The interfacial tension therefore acts both to stabilise the system into two phases and to resist the dispersion of one phase as droplets within the other phase.

Thermodynamically, this situation may be described in terms of the change in the interfacial Gibbs free energy ($\Delta G$), interfacial tension ($\gamma_{o/w}$) between the two phases and the change in surface area of the disperse phase when this is dispersed, albeit temporarily, as droplets within the external phase ($\Delta A$) as follows:

$$\Delta G = \gamma_{o/w} \Delta A$$

The dispersion of one phase within the other will cause a dramatic increase in the surface area of the interface between the two phases which, in turn, renders the system unstable (due to the increase in the interfacial Gibbs free energy). The system will therefore attempt to correct this instability; the subsequent coalescence of the droplets reduces the surface area of the interface, thereby reducing $\Delta G$. In this fashion the spontaneous coalescence of droplets of the internal phase may be explained. Accepting that a fundamental requirement for the formulation of pharmaceutical emulsions is the dispersal of one internal phase within a second external phase, this relationship provides an insight into one of the mechanisms of stabilisation of emulsions by emulsifying agents. As the reader will be aware, surface-active agents lower the interfacial tension and therefore, when present in emulsion systems, will partially negate the destabilising effects of the increase in surface area of the disperse phase. It is important to note that this is not the only mode of emulsification of these agents.

Classical studies on the stabilisation of emulsions have shown that the stability of the adsorbed layer was of primary importance. In these studies it was shown that whenever sodium cetyl sulphate (a hydrophilic surface-active agent) and cholesterol (a lipophilic surface-active agent) were employed as emulsifying agents, the two agents formed a stable film due to their interaction at the interface. The mechanical properties of this mixed surfactant film were sufficient to prevent disruption even when the shape of the droplets changed. Furthermore, the close-packed nature of the surface-active agents at the interface resulted in a greater lowering of the interfacial tension than could be achieved by either component when employed as a single emulsifying agent. The role of the interaction between surface-active agents (resulting in a mechanically robust interfacial film) was highlighted by replacing cholesterol with oleyl alcohol (a *cis* isomer), which resulted in a poor emulsion; however, the use of the *trans* isomer of oleyl alcohol, elaidyl alcohol, produced a stable emulsion. Further studies have shown that interfacial surfactant films form three-dimensional liquid crystalline layers of defined mechanical structure.

In addition to the mechanical properties of the adsorbed interfacial (liquid crystal) film, the adsorbed layer may carry a charge which, depending on the magnitude, may offer electrical repulsion between adjacent droplets. This is frequently observed whenever the droplets have been stabilised using ionic surface-active agents. Interestingly, flocculation of droplets of the disperse phase may lead to physical instability (see later) and, therefore, controlled flocculation (see Chapter 2) is not performed.

## Emulsion instability and the role of hydrophilic polymers

Hydrophilic polymers are frequently used as emulsion stabilisers in pharmaceutical emulsions. In contrast to surface-active agents, hydrophilic polymers do not exhibit marked effects on the interfacial tension. However, the stabilisation effect of these materials is due to their ability to adsorb at the interface between the disperse phase and the external phase to produce *multilayers* that are highly viscoelastic (gel-like) and can therefore withstand applied stresses without appreciable deformation. In so doing these polymers mechanically prevent coalescence. It should be noted that surface-active agents produce *monomolecular* not multimolecular films.

If the chosen hydrophilic polymer is ionic (e.g. gelatin, sodium alginate, sodium carboxymethylcellulose), then the multimolecular adsorbed film will be charged and therefore will exhibit a zeta potential. This may further protect the emulsion droplets from coalescence by offering an electrical repulsion, as described in the previous section (and in Chapter 2). Furthermore, it would be expected that stearic stabilisation of the droplets would occur due to the presence of the adsorbed polymeric layer. A similar phenomenon was described for suspensions in Chapter 2.

In addition, hydrophilic polymers will increase the viscosity of the external phase in an o/w emulsion and, in a similar fashion to suspensions, will affect the sedimentation rate of the droplets. This point is addressed in subsequent sections.

## Emulsion instability and adsorbed particles

Emulsions may also be stabilised by the addition of finely divided solid particles, if the particles are sufficiently wetted by both the oil and water phases (but preferentially wetted by one of the phases). The particles will accumulate at the interface between the phases and, if the particles show high interparticulate adhesion (thereby ensuring mechanical robustness to the adsorbed layer), the stability of the emulsion will be greatly enhanced. The type of emulsion produced by this method depends on the preference of the particles for each phase. For example, if the particles are wetted preferentially by the aqueous phase (i.e. the contact angle between the particle and water is less than 90°), an o/w emulsion will result. Conversely, if the finely divided solid is preferentially wetted by the oil phase, the resulting emulsion will be a w/o emulsion. Examples of finely divided solids that are employed in the formulation of o/w and w/o pharmaceutical emulsions are:

- o/w emulsions
  - aluminium hydroxide
  - magnesium hydroxide
  - bentonite
  - kaolin
- w/o emulsions
  - talc
  - carbon black.

## Type of emulsion

In the preparation of an emulsion, oil, water and the specified emulsifying agents are mixed together, resulting in the formation of droplets of each phase. At this stage theoretically either an o/w or a w/o emulsion may form. The resultant emulsion type is defined by the stability of the droplet phase; the phase of lower stability (i.e. the greater rate of coalescence) coalesces to form the external (or continuous) phase. There are several determinants of the type of emulsion produced, including: (1) phase volume of the internal phase; (2) the chemical properties of the film surrounding the internal phase; and (3) viscosity of the internal and external phases.

### Phase volume of the internal phase

Assuming that the internal phase is composed of spheres, it may be calculated that the maximum volume that may be occupied by the internal phase is 74%. This is termed the *critical value* and is dependent on the droplet size range and shape. Moreover, a large particle size range and irregular droplet shape may increase this value. In practice it is customary to use a phase volume ratio of 50% as this results in a stable emulsion (due to the loose packing of the internal phase). It should be remembered that the higher the phase volume of the internal phase, the greater the probability of droplet coalescence.

Interestingly, although the above description holds true for o/w emulsions, the critical value for w/o emulsions is markedly lower (circa 40%). This is due to the greater mechanical properties of hydrophilic polymers or polar surface-active agents (used to form o/w emulsions) than the hydrophobic groups that stabilise w/o emulsions. This point is extended in the next section.

### The chemical properties of the film surrounding the internal phase

As the reader will now appreciate, the adsorption of a mechanically robust film around the droplets of the internal phase is important to prevent droplet coalescence. The chemical composition of the surface-active agents (and hydrophilic polymers) at the droplet/external phase interface will dictate whether an o/w or w/o is formed. Typically oil droplets are stabilised by an adsorbed film composed of non-ionic, and especially ionic, surfactants or alternatively hydrated hydrophilic polymer chains. The surface-active agents and polymers that are responsible for this stabilisation are therefore predominantly (but not exclusively; the reader should recall that surface-active agents also possess hydrophobic groups) aqueous-soluble. Conversely, in w/o emulsions, the droplet is stabilised by the non-polar portion of the surface-active agent, which protrudes into the non-aqueous external phase. Furthermore, the length of this non-polar section plays an important role in the stabilisation of w/o emulsions, enhancing the mechanical integrity and reducing the tendency for the internal phase to coalesce.

The reader will therefore appreciate from this description that the solubility characteristics of the emulsifying agent define the type of emulsion that is formed. Therefore polymers and surface-active agents that are predominantly hydrophilic will form o/w emulsions, whereas predominantly hydrophobic surfactants will form w/o emulsions. Surface-active agents contain both hydrophilic and lipophilic groups and therefore it is the relative contributions of these that determine whether the agent is predominantly hydrophilic or lipophilic (hydrophobic). The contribution of these to the overall solubility is commonly referred to as the *hydrophile–lipophile balance* (HLB), a ratio scale that assigns a number to a surface-active agent, based on the contributions of the individual groupings on the molecule. This number can then be used when selecting surface-active agents for the formulation of either o/w or w/o emulsions.

The main features of the HLB scale are as follows:

- The HLB scale runs from circa 1 to 40; the water solubility of the surface-active agent increases as the HLB increases.

- Surface-active agents exhibiting an HLB between circa 3 and 6 are used to produce w/o emulsions and are therefore termed w/o emulsifying agents. These agents form poor dispersions in water but are soluble in the oil phase. Examples include:
  - sorbitan sesquioleate (e.g. Arlacel 83): HLB 3.7
  - sorbitan monooleate (e.g. Span 80): HLB 4.3
  - sorbitan monostearate (e.g. Span 60): HLB 4.7
  - glyceryl monostearate: HLB 3.8.
- Surface-active agents that exhibit an HLB between circa 6 and 9 form non-stable milky dispersions in water. Examples include:
  - sorbitan monopalmitate (e.g. Span 40): HLB 6.7
  - sorbitan monolaurate (e.g. Span 20): HLB 8.6.
- Surface-active agents exhibiting an HLB between circa 9 and 16 are used to produce o/w emulsions (termed o/w emulsifying agents). These agents form stable milky dispersions in water (HLB 9–10.5), translucent/clear dispersions in water (HLB 10.5–13) or clear solutions (HLB 13–16). Examples include:
  - polyoxyethylene sorbitan tristearate (e.g. Tween 65): HLB 10.5
  - polyoxyethylene sorbitan trioleate (e.g. Tween 85): HLB 11.0
  - polyoxyethylene sorbitan monostearate (e.g. Tween 60): HLB 14.9
  - polyoxyethylene sorbitan monooleate (e.g. Tween 80): HLB 15.0
  - polyoxyethylene sorbitan monopalmitate (e.g. Tween 40): HLB 15.6
  - polyoxyethylene sorbitan monolaurate (e.g. Tween 20): HLB 16.7.
- The HLB value of ionic surface-active agents is frequently greater than 16.

### Viscosity of the internal and external phases

The type of emulsion produced is affected by the viscosity of both the internal and external phases. If the viscosity is high the diffusion of the surface-active agent to the droplet surface will be reduced, as viscosity is inversely proportional to the diffusion coefficient of the surface-active agents. Furthermore, the increased viscosity will affect the process of coalescence of the droplets of the external phase. In general, if the viscosity of one phase is preferentially increased, there is a greater chance of that phase being the external phase of the emulsion.

## Tests to identify the type of emulsion

There are several tests that may be performed to identify the type of emulsion that has formed:

- *Electrical conductivity:* o/w emulsions conduct electric current whereas w/o emulsions do not.
- *Dilution with water:* o/w emulsions may be diluted with water (as this is the composition of the external phase) whereas w/o emulsions cannot be diluted with water.
- *Use of dyes:* oil-soluble dyes will stain the internal phase if the emulsion is an o/w emulsion whereas water-soluble dyes will dye the internal phase of a w/o emulsion.

## Emulsion instability

One of the goals of the pharmaceutical scientist is to formulate an emulsion that is physically stable, i.e. where the droplets of the internal phase remain discrete, retain their diameter and are homogeneously dispersed throughout the formulation. Fundamental to achieving this goal is the presence of the interfacial film (monomolecular or multilayered) at the interface between the droplet and the external phase. Emulsion instability may be either reversible or irreversible and is manifest in the following ways: (1) cracking (irreversible instability); (2) flocculation; (3) creaming; and (4) phase inversion.

### Cracking (irreversible instability)

Cracking refers to the complete coalescence of the internal phase, resulting in the separation of the emulsion into two layers, and occurs due to the destruction of the mono/multilayer film at the interface between the droplet and external phase. If an emulsion has cracked it cannot be recovered. This phenomenon may be due to:

- *Incorrect selection of emulsifying agents.* This results in the production of an interfacial film of insufficient mechanical properties. The importance of the role of complexation (interaction) between the surfactant molecules at the interface between the two phases has already been described earlier in this chapter.
- *Presence of incompatible excipients.* In the formulation of emulsions it is important that excipients do not interact with and destroy the interfacial film of surface-active agents. This will occur if, for example, a cationic surface-active agent (commonly used as a preservative in creams) is added to an emulsion in which the interfacial film of surface-active agents bears an anionic charge (e.g. due to sodium oleate, potassium

oleate or sodium lauryl sulphate). Similarly, if a therapeutic agent or a divalent ion bears an opposite charge to that exhibited by the interfacial film, disruption of the film will occur due to this ionic interaction.

- *Temperature.* Emulsions are generally unstable at high and low storage temperatures.
- *Microbial spoilage.* Microbial growth generally leads to destabilisation of the emulsion and is thought to be due to the microorganisms being able to metabolise the surface-active agents.

## Flocculation

The ability of emulsion droplets to flocculate has been introduced earlier in this chapter. In the flocculated state the secondary interactions (van der Waals forces) maintain the droplets at a defined distance of separation (within the secondary minimum). Application of a shearing stress to the formulation (e.g. shaking) will redisperse these droplets to form a homogeneous formulation. Although flocculation may stabilise the formulation, there is also the possibility that the close location of the droplets (at the secondary minimum) would enable droplet coalescence to occur if the mechanical properties of the interfacial film are compromised.

## Creaming

This phenomenon occurs primarily as a result of the density difference between the oil and water phases and involves either the sedimentation or elevation of the droplets of the internal phase, producing a layer of concentrated emulsion either at the top or bottom of the container. Creaming is predominantly an aesthetic problem as the resulting emulsion is rather unsightly; however, upon shaking the emulsion is rendered homogeneous. Patients often believe that an emulsion that shows evidence of creaming has exceeded its shelf-life.

It is therefore important to understand the physicochemical basis of creaming in emulsions and, in so doing, reduce the rate of or inhibit this phenomenon. The rate of creaming $\left(\dfrac{\mathrm{d}v}{\mathrm{d}t}\right)$ in an emulsion (in a similar fashion to suspensions) may be described by Stokes' equation:

$$\frac{\mathrm{d}v}{\mathrm{d}t} = \frac{2r^2(\rho_o - \rho_w)g}{9\eta}$$

where: $r$ refers to the average radius of the droplets of the internal phase; $(\rho_o - \rho_w)$ refers to the density difference between the oil phase and the water phase; $g$ refers to gravity (which is negative if upward creaming occurs); and $\eta$ refers to the viscosity of the emulsion.

As may be observed, creaming may be prevented if the density difference between the two phases is zero. In practice, however, this cannot be easily achieved. Therefore, the most straightforward methods by which the rate of creaming may be reduced are to:

- Reduce the average particle size of the disperse phase. This may be achieved by size reduction methods, e.g. the colloid mill (see Chapter 11).
- Increase the viscosity of the emulsion. This may be achieved by adding hydrophilic polymers to the external phase of o/w emulsions or by incorporating non-aqueous viscosity enhancers (e.g. aluminium stearate salts, Thixin) into w/o emulsions.

### Phase inversion

Phase inversion refers to the switching of an o/w emulsion to a w/o emulsion (or vice versa). This is a phenomenon that frequently occurs whenever the critical value of the phase volume ratio has been exceeded. In o/w emulsions the frequently cited phase volume ratio (o:w) is 74:26 and for w/o emulsions this value is 40:60.

## Formulation of pharmaceutical emulsions

In the formulation of pharmaceutical emulsions there are a number of questions that require to be initially addressed, including the type of emulsion required (o/w or w/o), the route of administration of the emulsion (e.g. oral or topical, the latter as a cream), the volume of the internal phase, the droplet size and the consistency required. These aspects are individually addressed below.

### Type of emulsion

As the reader is now aware, there are two types of primary emulsions; however the clinical uses of these types differ. Emulsions that are designed for oral or intravenous administration are o/w, whereas emulsions for topical administration (creams) may be either o/w or w/o. O/w creams are generally (but not exclusively) used for the topical administration of water-soluble drugs to the

**Tips**

- Emulsion instability may be classified as either reversible or irreversible.
- Cracking is irreversible.
- Phase inversion will result in the switching of the type of emulsion. Although the preparation may be stable, the nature (and hence the clinical performance) of the emulsion has changed and therefore this product will fail quality control processes.
- Creaming, whilst unsightly, is reversible upon shaking.

skin to achieve a local effect (e.g. for the treatment of infection or inflammation). They are typically easily applied to the surface, are non-greasy and may be washed from the skin. Conversely, w/o emulsions are greasy in texture and, following application, will hydrate the skin. Most moisturising formulations are w/o emulsions.

### Volume of the internal phase

The effect of the volume of the internal phase on emulsion stability has been addressed previously in this chapter. The ratio of the internal-to-external phase of o/w emulsions is typically 1:1; however, larger oil-to-water ratios are theoretically possible (up to circa 74%). Usually the concentration of the internal phase is restricted to circa 60% to ensure stability. The maximum concentration of internal phase of w/o emulsions is 30–40%. Higher concentrations will result in phase inversion.

### Droplet size

Previously it was shown that the rate of creaming of an emulsion may be reduced by reducing the average droplet size of the internal phase. In light of this it is customary when industrially processing emulsions and creams to reduce the droplet size (and reduce the polydispersity of the size distribution) by passage through a colloid mill (Chapter 11). The clinical importance of droplet size in parenteral emulsions, e.g. total parenteral nutrition, will be specifically addressed in Chapter 5.

### Viscosity of the internal and external phases

One of the major differences between traditional emulsions for oral or parenteral administration and creams is the increased viscosity of the latter. The superior viscosity of these formulations facilitates the location and spreading of the formulation on the skin. In addition, the viscosity of emulsion/cream formulations also affects the stability, controlling the rate of upward/downward sedimentation (as described by Stokes' law).

### Selection of type and concentration of emulsifying agents

All emulsion and cream formulations require the inclusion of emulsifying agents (principally surface-active agents) to ensure emulsion stability, the choice of which is determined by the type of emulsion required, clinical use and toxicity. For example, the use of anionic surfactants is restricted to external formulations. To determine the type of emulsifying agents used, reference is made to the HLB requirements of the internal phase of the formulation. If the HLB requirements are not known, it is common practice for the formulation scientist to prepare a series of emulsions using a mixture of surface-active agents that provide a range of HLB values using a weighted-mean approach. For example, an o/w emulsion may be prepared using a mixture of surface-active agents (1% w/w in total) that provides an overall HLB value of 10. A mixture of Span 60 (HLB 4.7) and Tween 80 (15.0) may be chosen for this purpose; the ratio of these two surfactants is calculated using the simple weighted-averages equation:

$$10 = 4.7x + (1 - x)15$$

where $x$ refers to the fraction of Span 60 and $(1 - x)$ is the fraction of Tween 80.

In this example the mixture of surface-active agents would be 0.485% w/w Span 60 and 0.515% w/w Tween 80. In practice a series of emulsions would be prepared, each differing in the HLB of the surfactant mixture but constant in terms of the overall concentration of surfactants. From this the most stable emulsion would be selected.

Alternatively, for certain non-aqueous components information is available regarding the required HLB to produce stable o/w or w/o emulsions. For example, to prepare a stable o/w emulsion using cottonseed oil as the internal phase requires a mixture of surface-active agents that produces an HLB value of 10.0, whereas the formation of a w/o emulsion in which cottonseed oil is the external phase requires a mixture of surface-active agents that produces an HLB value of 5. If the lipophilic component of the formulation is composed of more than one excipient, then the combined HLB value for this phase should be calculated and the ratio of surface-active agents in the mixture selected to provide this HLB requirement. For example, consider an o/w cream in which the internal phase (of total weight 35% w/w) has the composition shown in Table 3.1.

The HLB requirement for the formulation of a stable emulsion is 10.46. For this example we will choose a mixture of two non-ionic surface-active agents: polyoxyethylene sorbitan monostearate (Tween 60: HLB 14.9) and sorbitan monostearate (Span 60: HLB 4.7). Allowing $x$ to equal the fraction of sorbitan monostearate in the surfactant mixture and $(1 - x)$ to represent the fraction of the second surfactant, the following calculation may be performed:

$$10.46 = 4.7x + (1 - x)14.9$$

Therefore the ratio of Span 60 to Tween 60 is 0.44:0.56.

**Table 3.1** Calculation of required hydrophile–lipophile balance (HLB) values for an oil in water (o/w) cream

| Components (% w/w of final formulation) | HLB requirement to produce an o/w emulsion | Composition (fraction of the oil phase) | Calculated HLB requirement |
|---|---|---|---|
| Cottonseed oil (30%) | 10 | $\left(\dfrac{30}{35}\right) = 0.86$ | $(0.86 \times 10) = 8.6$ |
| Stearyl alcohol (3%) | 14 | $\left(\dfrac{3}{35}\right) = 0.09$ | $(0.09 \times 14) = 1.26$ |
| Beeswax (2%) | 12 | $\left(\dfrac{2}{35}\right) = 0.05$ | $(0.05 \times 12) = 0.6$ |
| | | | **Total: 10.46** |

Two further points should be noted regarding the use of surface-active agents for the stabilisation of pharmaceutical emulsions: (1) the choice of the mixture of surfactants and (2) the overall concentration of surfactant.

- In choosing the surfactant blend, the formulation scientist should examine the structure of the surfactants to ascertain whether the two components may interact at the interface between the internal and external phases. In addition, it is preferable that the surfactant blend used should not be composed of one surfactant with a low HLB and the second with a high HLB, as in this case emulsion stability may be problematic. However, the addition of a third surface-active agent of intermediate HLB may resolve this issue.
- The concentration of surfactant used should be the lowest concentration required to ensure stability.

## Types of surface-active agents used as emulsifying agents

There are four categories of surface-active agents used to stabilise emulsion/cream formulations: (1) anionic; (2) cationic; (3) non-ionic; and (4) amphoteric. Examples of these are provided in the section below. It should be noted that the cited examples are representative of the surface-active agents within each of the four classes. Moreover, the details provided for each example/category of surfactant are designed to provide an insight into the types and uses of the various agents. For a more comprehensive understanding of these agents, the reader should consult the companion text in the FASTtrack series by David Attwood and

Alexander T Florence (*FASTtrack: Physical Pharmacy*, 2nd edn: London: Pharmaceutical Press, 2012).

## Anionic surfactants

■ These dissociate to produce negatively charged ions with surface-active activity.
■ Whilst examples in this category are inexpensive, they are comparatively more toxic than for other categories of surface-active agent. This limits their use to external formulations, e.g. creams.

Examples of anionic surfactants include:

■ *Sodium/potassium salts of fatty acids, e.g. sodium oleate (Figure 3.1), sodium stearate and ammonium oleate*

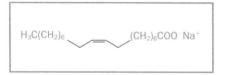

**Figure 3.1** Chemical structure of sodium oleate.

- These agents produce o/w emulsions (usually in combination with a second surface-active agent to ensure the formation of a mechanically robust film at the oil/water interface).
- Due to the effect of pH on the ionisation (and hence surfactant properties) of these molecules, the emulsifying properties are lost under acidic conditions. Similarly, the emulsifying properties are negated in the presence of di/trivalent cations.
- These surfactants may be formed in situ in the formulation by the co-addition of the fatty acid and a suitable counterion.

■ *Calcium salts of fatty acids*

- These are generally formed in situ by the interaction of a calcium salt, e.g. calcium hydroxide, with a fatty acid, e.g. calcium oleate. This approach is used in Zinc Cream BP and in certain lotions.
- Calcium salts of fatty acids form w/o emulsions, due to their limited dissociation (and hence solubility: Figure 3.2).

**Figure 3.2** Structural formula of calcium stearate.

- *Amine salts of fatty acids*
  - These are typically formed in situ in pharmaceutical emulsions, e.g. triethanolamine stearate.
  - These surface-active agents form o/w emulsions.
  - Akin to sodium/potassium salts of fatty acids, their emulgent properties are pH-dependent and may be negated in the presence of electrolytes.
- *Alkyl sulphates*
  - These are used to produce o/w emulsions (in conjunction with a second non-ionic surfactant of low HLB, i.e. <6). Fatty alcohols (e.g. cetyl, stearic alcohol) are frequently used for this purpose.
  - Examples of these include sodium lauryl sulphate (Figure 3.3) and triethanolamine lauryl sulphate.

**Figure 3.3**  Structural formula of sodium lauryl sulphate.

$$CH_3(CH_2)_{10}CH_2OSO_3{}^-Na^+$$

## Cationic surfactants

- These dissociate to produce positively charged ions with surface-active activity.
- They are primarily used pharmaceutically as preservatives of topical formulations; however, they may be used to form o/w emulsions (when combined with a second non-ionic surfactant of low HLB, i.e. <6).
- The main example used in topical formulations is cetrimide, a mixture of trimethylammonium bromide, with smaller amounts of dodecyltrimethylammonium bromide and hexadecyltrimethylammonium bromide (Figure 3.4).

**Figure 3.4**  Structural formula of hexadecyltrimethylammonium bromide (cetrimide).

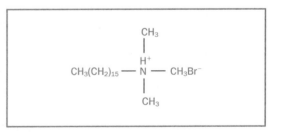

- Their emulgent properties are compromised in the presence of anionic agents (e.g. anionic surface-active agents, di/trivalent anions and polyelectrolytes, e.g. anionic polymers).

## Non-ionic surfactants

- These are by far the most popular category of surface-active agents used for the formulation of pharmaceutical emulsions.
- They may be used to formulate both o/w and w/o emulsions.
- Generally combinations of two non-ionic surfactants (one water-soluble and the other oil-soluble) are employed to ensure the formation of a stable interfacial film around the surface of the droplets of the disperse phase. In certain circumstances a single non-ionic surfactant may be used that is of intermediate HLB value.
- Non-ionic surface-active agents are more stable than ionic surfactants in the presence of electrolyte and/or changes in pH.
- Generally the hydrophobic portion of the molecule is composed of a fatty acid or fatty alcohol whereas the hydrophilic portion is composed of an alcohol or ethylene glycol moieties.

Examples of non-ionic surface-active agents include:

- *Sorbitan esters (e.g. Span series)*
  - This is a family of chemically related esters that are produced by esterifying a fatty acid to at least one of the hydroxyl groups of sorbitan.
  - Modification of the length of the fatty acid (denoted by the symbol R in Figure 3.5) will generate a range of surface-active agents with emulsifying properties (and low HLB values).
  - By themselves sorbitan esters will form w/o emulsions; however, when combined with the polysorbates (see below), both o/w and w/o emulsions may be formulated.

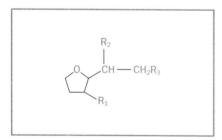

**Figure 3.5** Generic structure of the sorbitan fatty acid esters. In sorbitan monoesters $R_1$ and $R_2$ = OH, $R_3$ is the specific fatty acid derivative (e.g. lauric, stearate). $R_1$ = OH and $R_2$ and $R_3$ refer to the specific fatty acid derivative for sorbitan diesters. In sorbitan triesters $R_1$, $R_2$ and $R_3$ refer to the specific fatty acid derivative.

- *Polyoxyethylene fatty acid derivatives of the sorbitan esters (e.g. Tween series)*
  - This family of surface-active agents is prepared by forming polyoxyethylene esters of the sorbitan esters.
  - The emulsifying properties of the molecules in this series may be modified by altering the number of oxyethylene $(OCH_2CH_2)$ groups and the type of fatty acid (denoted as R in Figure 3.6).

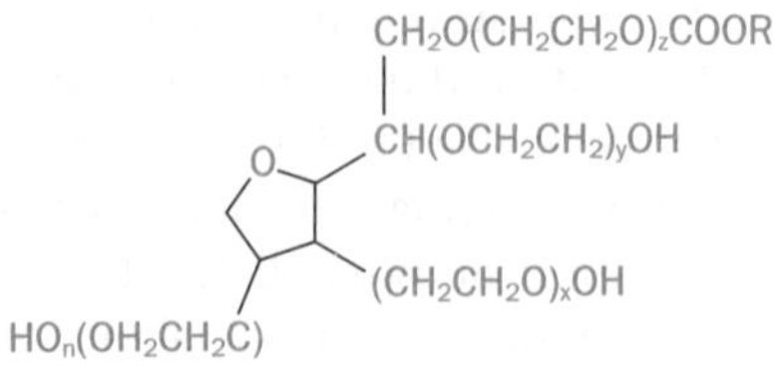

**Figure 3.6** Generic structure of the polyoxyethylene sorbitan fatty acid monoesters. Di- and triesters may be formed by esterification of the terminal alcohol groups.

$$CH_2O(CH_2CH_2O)_zCOOR$$
$$CH(OCH_2CH_2)_yOH$$
$$(CH_2CH_2O)_xOH$$
$$HO_n(OH_2CH_2C)$$

- These surfactants are used to form o/w or w/o emulsions in combination with a second surface-active agent, e.g. sorbitan esters, cetyl alcohol, glyceryl monostearate, to ensure emulsion stability.
- The emulsifying properties of this series are tolerant of changes in electrolyte concentration and pH.
- Generally they are non-toxic and are used in both parenteral and non-parenteral emulsions.
  - *Polyoxyethylene alkyl ethers (macrogols)*
    - These are ethers formed between polyethylene glycol and a range of fatty alcohols (lauryl, oleyl, myristyl, cetyl, stearyl). Two commercial series of these compounds are *Cremophor* and *Brij*.
    - The physicochemical properties of these *non-ionic* surface-active agents may be modified by altering the length of the polyoxyethylene group and the length of the aliphatic chain (denoted as $x$ and $y$ in Figure 3.7).

**Figure 3.7** Structural formula for polyoxyethylene alkyl ethers.

$$CH_3(CH_2)_x(OCH_2CH_2)_yOH$$

- The macrogols are used as emulsifying agents for both o/w and w/o emulsions. Combinations of the more lipophilic and hydrophilic examples of this series are used to produce stable emulsions. For example, cetomacrogol 1000 (Figure 3.8) is combined with cetostearyl alcohol to produce cream formulations.

**Figure 3.8** Structure of cetomacrogol 1000.

$$CH_3(CH_2)_{15-17}(OCH_2CH_2)_{20-24}OH$$

■ *Polyoxyethylene fatty acid esters*
  - These are a series of polyoxyethylene derivatives of fatty acids. The most commonly used derivatives are the stearate derivatives (the Myrj series). The surface-active properties of these compounds may be modified by varying the length of the oxyethylene substituent and, in addition, by mono- or diesterification of the acid, as shown in Figure 3.9.

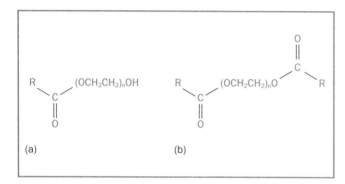

**Figure 3.9**  Generic structures of polyoxyethylene monoester and poly(oxyethylene) ester. The number of repeating oxyethylene groups is denoted by n whereas R refers to the chain length of the fatty acid.

  - Polyoxyethylene stearates (and related compounds) are non-ionic surfactants, offering a range of HLB values.
  - They are frequently combined with stearyl alcohol (or related fatty alcohols) in the formulation of o/w emulsions.
  - The emulsifying properties are tolerant of the presence of strong electrolytes.
■ *Fatty alcohols*
  - Examples include cetyl alcohol and stearyl alcohol (Figure 3.10).

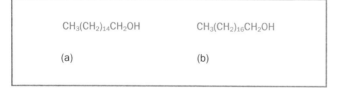

**Figure 3.10**  The structure of (a) cetyl alcohol and (b) stearyl alcohol.

  - In addition, cetostearyl alcohol (a mixture of cetyl (20–35%) and stearyl (50–70%) alcohols, although other alcohols, e.g. myrisitc alcohol, are present) is available (see later in this chapter).
  - Fatty alcohols are generally used in combination with more hydrophilic surfactants to produce stable o/w emulsions.

- When used alone, fatty alcohols act as w/o emulsifiers. Furthermore, the addition of these to a hydrophobic base will increase the water absorption properties of the formulation.
- In cream formulations excess fatty alcohols interact with the hydrophilic emulsifier to produce a viscoelastic external phase. In turn, this increases the viscosity of this phase (thereby decreasing upward/downward sedimentation), producing the consistency expected of cream formulations.
- Fatty alcohols may be used to enhance the viscosity of w/o creams.

■ *Amphoteric surfactants*
- These are compounds that possess both positively and negatively charged groups (cationic at low pH values and anionic at high pH values).
- The emulsifying properties are reduced as the pH approaches the isoelectric point of the surface-active agent.
- The most commonly used amphoteric surface-active agent is lecithin (Figure 3.11).
- Lecithin is used in emulsions (for intravenous and intramuscular administration) and creams, in which it acts as an o/w emulsifying agent.

**Figure 3.11** Structural formula for lecithin ($R_1$ and $R_2$ refer to either identical or different fatty acids).

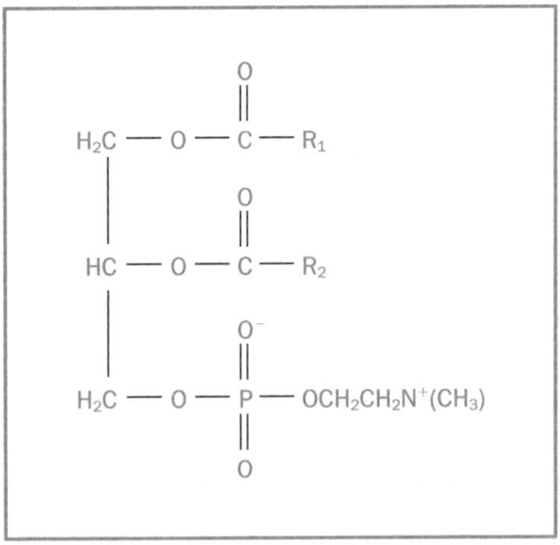

## Miscellaneous surfactants that are used in emulsion and cream formulations

The following emulsifying agents are derived from natural sources and are composed of mixtures of compounds. They are used as emulsifying agents either alone, or, preferably, along with a second emulsifying agent in the production of o/w and w/o creams.

## Lanolin (wool fat)

- Lanolin is a wax-like material that is derived from sheep's wool. It is a mixture of fatty alcohols, fatty acid esters of cholesterol and other sterols.
- On its own it may be used to produce w/o creams. Mixtures of this emulsifying agent with oils or soft paraffin produce emollient creams.
- Lanolin can absorb approximately twice its weight of water – this property is advantageous in the formulation of ointments.

## Lanolin alcohols (wool alcohols)

- This is a mixture of steroid alcohols and triterpene alcohols (including cholesterol).
- Lanolin alcohol is an excellent emulsifying agent for w/o emulsions.
- It has excellent water-absorptive properties.
- It is prone to oxidation (and therefore requires an antioxidant).

## Anionic emulsifying wax

- This is a mixture containing cetostearyl alcohol, water and sodium lauryl sulphate (or a related sulphated alcohol). The specific formula for anionic emulsifying wax in the British Pharmacoepeia (2004) is as given in Table 3.2:

**Table 3.2** Formula for Anionic Emulsifying Wax BP

| Component | Mass |
| --- | --- |
| Cetostearyl alcohol | 90 g |
| Sodium lauryl sulphate | 10 g |
| Purified water | 4 g (ml) |

- This mixture is used to produce o/w emulsions (using 2% w/w wax) or creams, at a higher wax concentration (10% w/w). However, the use of anionic emulsifying wax is predominantly for topical (cream) formulations.
- Due to the anionic nature of one of the ingredients, the emulgent properties of this wax are compromised by the presence of polyvalent metals and cations (e.g. quaternary ammonium compounds).

## Non-ionic emulsifying wax

- Non-ionic emulsifying wax (also called cetomacrogol emulsifying wax) is composed of cetostearyl alcohol and cetomacrogol 1000.

- This wax is used as an emulsifier for the preparation of emulsions (at concentrations up to 5% w/w). However, at higher concentrations this material additionally enhances the rheological structure of the preparation (15–25% w/w), thereby promoting stability. The primary use of non-ionic emulsifying wax is for the preparation of creams.
- The emulgent properties of non-ionic emulsifying wax are tolerant to the presence of electrolytes (and therapeutic agents).

### Beeswax (white and yellow)
- There are two forms of beeswax – white and yellow. White beeswax is a bleached form of yellow beeswax.
- Beeswax is a mixture of esters of monohydric alcohols ($C_{24}$–$C_{36}$, in even numbers only) and straight-chain acids (even numbers of carbon atoms up to $C_{36}$), long-chain hydroxyacids (e.g. $C_{18}$). The principal component is myricyl palmitate.
- Beeswax is used as a w/o emulsifying agent for creams and is also used to enhance the consistency of creams.

## Excipients used in pharmaceutical emulsions

One major category of excipients for pharmaceutical emulsions, namely surface-active agents, has been described above. However, as in other pharmaceutical formulations, other excipients are required to enhance the physical and chemical stability and to render the formulation aesthetically pleasing to the patient. These are described below.

### Vehicle
There are two liquid phases in pharmaceutical emulsions: an aqueous phase and an oil phase, each of which is formulated separately. The vehicle in the aqueous phase for pharmaceutical emulsions designed for oral or topical administration is usually *purified water*, the details of which have been provided in Chapter 1. When formulated for intravenous administration, *sterile water for injections* is used as the external aqueous phase, the details of which are provided in Chapter 5. If control of the pH of the aqueous

## Tips

- The type of surfactants used dictates the stability and type of the emulsion.
- In the choice of surfactant type for an emulsion, due attention must be paid to the toxicity of the surfactant. For example, non-ionic, but not anionic, surfactants may be administered orally. Certain surfactants are only used in the formulations for topical administration. Cationic surfactants are principally used as preservatives and not for the stabilisation of emulsions.
- Combinations of surfactants are used to stabilise emulsions. The elastic properties of the surfactant layer at the interface are important in the stabilisation of emulsions.

external phase is required, buffers, e.g. citrate, phosphate, may be included in the aqueous vehicle. In light of the ability of electrolytes to compromise the emulsifying properties of surface-active agents (and certain hydrophilic polymers), the concentration and type of buffer should be carefully chosen.

The oil phase of pharmaceutical emulsions is typically composed of vegetable oils, e.g. cottonseed oil, arachis oil, almond oil (mono-, di- and triglycerides of mixtures of unsaturated and saturated fatty acids). However, pharmaceutical emulsions for topical application (creams) may be formulated using a greater range of non-aqueous components. Alternative non-aqueous phases used in the formulation of creams and ointments include: (1) petrolatum and mineral oil; (2) isopropyl myristate.

## Petrolatum and mineral oil

Petrolatum and mineral oil are hydrophobic excipients that are derived from petrolatum. The former is a complex mixture of hydrocarbons (e.g. aliphatic, cyclic, saturated, unsaturated, branched hydrocarbons) that results in a wide range of chemical and physical specifications in the USP monograph. Mineral oil is a more purified fraction of petrolatum and is a mixture of aliphatic ($C_{14}$–$C_{18}$) and cyclic hydrocarbons. Both materials are employed as the internal phase in o/w emulsions and as the external phase in w/o emulsions (usually in combination with a fatty alcohol as the emulsifying agent).

## Isopropyl myristate (Figure 3.12)

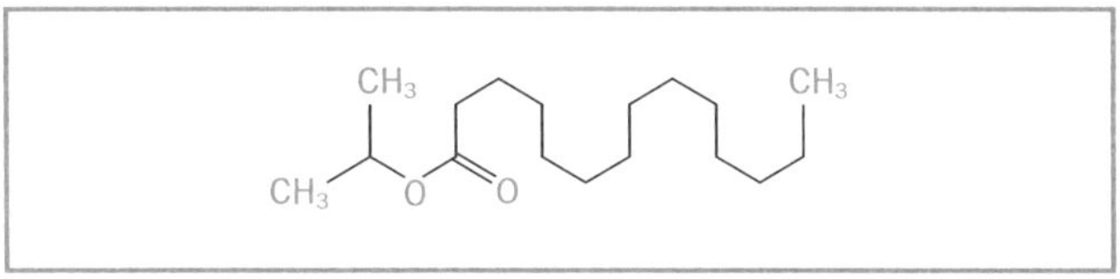

Figure 3.12   Structural formula of isopropyl myristate.

Isopropyl myristate is used as a non-aqueous component of cream formulations, either as the internal phase of o/w creams or as the external phase of w/o creams. Interestingly, isopropyl myristate has been additionally reported to enhance the permeation of drugs through the skin when applied topically.

## Antioxidants

As described in Chapter 1, antioxidants are included within pharmaceutical formulations to enhance the stability of drugs/components to oxidation. In emulsions and creams the two major components that may be liable to oxidise are the therapeutic agent and the oil selected for the oil phase, vegetable oils. Therefore the inclusion of lipophilic antioxidants within the oil phase may be required, e.g. butylated hydroxyanisole (circa 0.02–0.5% w/w), butylated hydroxytoluene (circa 0.02–0.5% w/w) and propyl gallate (<0.1% w/v) (Figure 3.13).

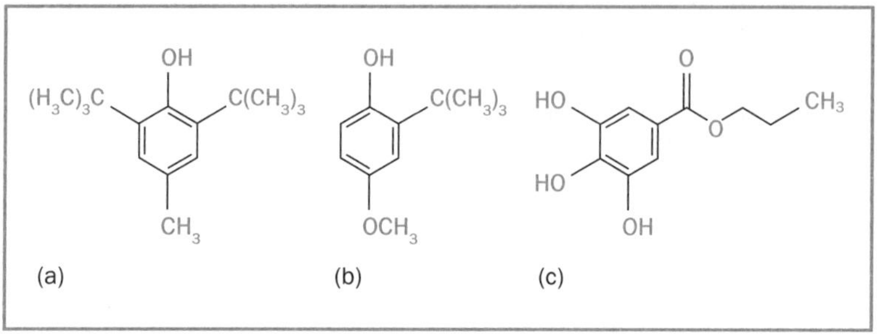

**Figure 3.13** Structural formula of (a) butylated hydroxytoluene, (b) butylated hydroxyanisole and (c) propyl gallate.

If the antioxidant is required in the aqueous phase of an emulsion or cream then a water-soluble example should be used, e.g. sodium metabisulphite (0.01–1.0% w/v) or sodium sulphite (0.1% w/v).

## Flavours and sweeteners
Flavours and sweeteners are commonly included in emulsions for oral administration to mask the unpalatable taste of the therapeutic agent or the internal oil phase. Suitable examples of these have been given in Chapter 1.

## Viscosity modifiers
The viscosity of emulsions and creams has been previously described to influence the physical stability of emulsions by decreasing the rate of creaming and therefore viscosity control within a formulation is an important attribute. The inclusion of hydrophilic polymers, e.g. methylcellulose, hydroxyethylcellulose, polyacrylic acid and sodium carboxymethylcellulose, to increase the viscosity of aqueous systems has been described in Chapters 1 and 2 and the same principles exist for emulsion formulations. It must be remembered that, as the viscosity of formulations increases, so does the difficulty in administration and, therefore, this must be borne in mind when the final viscosity of the o/w emulsion is selected. Furthermore, the ability of hydrophilic polymers to form a multimolecular layer around the surface of the dispersed-oil droplet is an important function of polymers within emulsion formulations.

## Preservatives for emulsions and creams

The concept of preservation of pharmaceutical systems, i.e. solutions and suspensions, has been discussed in previous chapters. In particular the effects of pH and the presence of hydrophilic polymers, dispersed particles and surfactant micelles on the available preservative concentration were highlighted. The preservation of o/w emulsions and creams becomes a challenging task to the pharmaceutical scientist due to the possible co-requirement for pH control of the external phase and the inclusion of hydrophilic polymers. However, the complexity of this issue is enhanced due to the presence of a dispersed-oil phase into which the antimicrobial active form of the preservative may partition and hence be unavailable to exert its antimicrobial effect. An equilibrium is therefore established, as depicted in Figure 3.14.

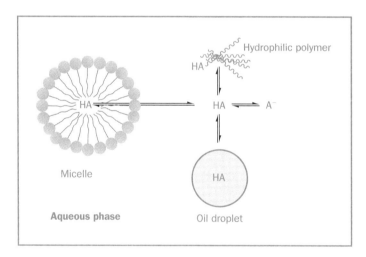

**Figure 3.14**
Diagrammatic representation of the equilibrium that exists between a disperse oil phase, the external aqueous phase, micelles of surface-active agent, a hydrophilic polymer (e.g. methylcellulose, polyvinylpyrrolidone) and the antimicrobially active form of an organic acid preservative (denoted as HA).

Figure 3.14 illustrates the partitioning of the unionised form of a weak acid preservative into micelles and oil droplets and the interaction with hydrophilic polymers that are present to stabilise the emulsions. The term HA may be replaced by a non-ionisable (or minimally ionisable) preservative, e.g. chlorocresol. As may be observed in Figure 3.14, the available (active) concentration is decreased by these various partitioning phenomena/interaction with hydrophilic polymers and therefore the concentration of preservative must be increased to ensure that the required concentration of free preservative is obtained. The concentration of preservative required to inhibit microbial growth in emulsions/creams may be estimated using the following formula:

$$C_w = C\left(\frac{\phi+1}{K_w^o\phi + R}\right)$$

where: $C_w$ refers to the concentration of 'free' preservative in the aqueous phase; $C$ refers to total concentration of preservative; $\varphi$ refers to the ratio of oil (internal phase) to water in the emulsion/ cream; $K^o_w$ refers to the partition coefficient of the preservative between the oil and water phases; and $R$ is the ratio of the total preservative to the free preservative.

With knowledge of the above parameters, the free concentration of preservative may be calculated, as illustrated in the example below.

## Worked example

### Example 3.1
Calculate the free concentration of chlorocresol in an emulsion in which the oil:water partition coefficient of the preservative is 1.5, the phase ratio in the emulsion is 1:1, the ratio of total to free preservative is 4, the pH is 7.2 and the initial concentration of preservative is 0.3% w/v.

According to the above equation the concentration of free preservative is:

$$C_w = C\left(\frac{\phi+1}{K_w^o\phi + R}\right) = 0.3\left(\frac{1+1}{1.5\times1+4}\right) = 0.1\%$$

Therefore, 33.3% of the preservative is available to the formulation to exert its antimicrobial effect within the formulation.

This situation becomes further complicated if the preservative ionises as a function of the pH of the formulation. To accommodate this, the degree of ionisation must be calculated (again using the Henderson–Hasselbalch equation). The $pK_a$ for chlorocresol is 9.2.

$$\text{Fraction} = \frac{1}{\left(1+10^{pH-pKa}\right)} = \frac{1}{\left(1+10^{-2}\right)} = 0.99$$

It can therefore be seen that at pH 7.2 chlorocresol is essentially unionised and no modification of the concentration is required. Conversely, if an organic acid is used ($pK_a$ 4.2), the fraction unionised will be 0.001 and therefore this must be considered in the calculation of the required concentration.

From the discussions to date, the role of the oil:water partition coefficient in the calculation of the free concentration of preservative is apparent. Therefore, to optimise preservative efficacy, the oil used as the internal phase should have low oil:water partition coefficient for the selected preservative. For example, the partition coefficients of methyl parahydroxybenzoic acid in mineral oil and vegetable oil are 0.02 and 7.5. Therefore an obvious method of minimising the concentration of preservative in the formulation (whilst retaining the required antimicrobial activity) would be to select mineral oil as the internal phase. Alternatively, if an internal phase is specified, the total concentration of preservative may be minimised by selecting an alternative preservative with a low oil:water partition coefficient.

A diagrammatic representation of formulation considerations for emulsions and creams is shown in Figure 3.15.

## Manufacture of emulsions

Generically the manufacture of emulsions involves the following steps:
1. Dissolution of the oil-soluble components in the oil vehicle and the (separate) dissolution of the water-soluble components in the aqueous phase
2. Mixing of the two phases under turbulent mixing conditions to ensure the dispersion of the two phases into droplets.

At the laboratory the manufacture of emulsions usually involves the use of a mechanical stirrer whereas the manufacture of creams involves mixing the two (heated) phases using a mortar and pestle. The emulsification of production-scale batches is normally performed using mechanical stirrers, homogenisers, ultrasonifiers or colloid mills. The use of colloid mills is usually reserved for formulations of higher viscosity, e.g. creams, due to the high running cost and slow production rate of this apparatus.

## Quality control of emulsions and creams

The rationale for the quality control of pharmaceutical products has been described in Chapter 1 and therefore will not be repeated. Typically, following manufacture and over the designed period of the shelf-life,

## Tips

- The formulation of emulsions requires the incorporation of several excipients to maintain stability and to ensure product performance.
- As in other dosage forms, all excipients must be physically and chemically compatible (with each other and with the therapeutic agent).
- The choice of preservative for an oil in water emulsion requires consideration of both the pH of the aqueous phase (if the preservative is acidic) and the solubility within the oil phase. Significant solubility of the preservative in the oil phase and/or excessive ionisation of the preservative lowers the preservative efficacy.

**Figure 3.15** Diagrammatic representation of formulation considerations for emulsions and creams.

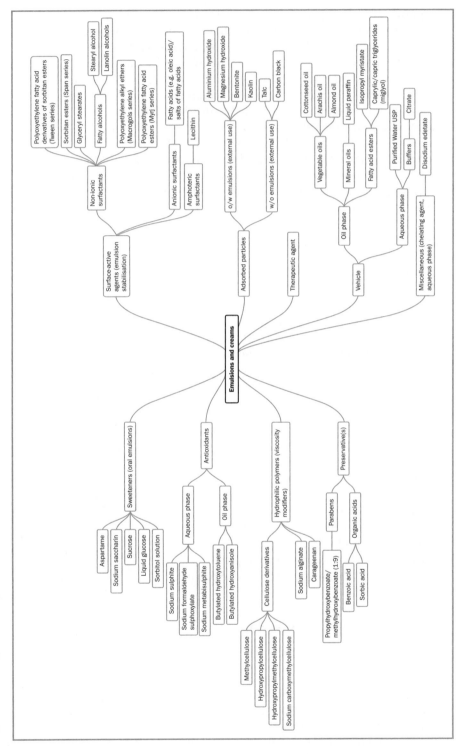

the physicochemical properties of emulsions and creams are determined using the same range of methods that would be applied to suspensions. Accordingly (whilst avoiding undue repetition), the methods used include:

1. **Concentration of therapeutic agent**: As with all pharmaceutical products, post-manufacture the concentration of therapeutic agent in emulsion and cream formulations must lie within 95–105% of the nominal concentration. Over the shelf-life of the product the concentration of drug must not fall below 90% of the nominal amount. If this occurs then the product has expired.

2. **Concentration of preservative**: The concentration of preservative must lie within 95–105% of the nominal concentration following manufacture and also upon storage.

3. **Preservative efficacy testing**: The efficacy of the preservative(s) within the emulsion/cream must be assessed using the appropriate pharmacopoeial method. These tests evaluate the resistance of the product to microbial challenge. It should be noted that creams are particularly susceptible to spoilage due to fungal growth. Preservative efficacy testing is performed both following manufacture and during storage.

4. **Appearance**: Emulsions and creams must be homogeneous and white in appearance unless a specific colour has been added to the product. In addition, the type of emulsion (o/w or w/o) must be confirmed to prove that phase inversion, either post-manufacture or during storage, has not occurred. The appearance is assessed on products following manufacture and during storage.

5. **pH**: The pH of the emulsions (both following manufacture and during storage) is examined if the emulsion is of the o/w type and is compared to product specifications. The measurement of pH of w/o emulsions and creams is not performed.

6. **Viscosity:** The viscosity of the product is measured, both following manufacture and during storage and compared to the specified range.

7. **Uniformity of content**: The uniformity of content of therapeutic agents in emulsions and creams is characterised as in previous chapters. For example, the individual mass of the therapeutic agent in 10 units is typically determined. The content of each individual unit must lie within the range 85–115% of the average content to successfully pass the test. If the mass of therapeutic agent in more than one unit lies outside this range, or if the mass of therapeutic agent in one unit lies outside an extended range of 75–125% of the average content, then the batch has failed the uniformity of content test. However, if the mass of therapeutic agent of one unit lies outside the 85–115% range but within 75–125% of the

average content, the drug contents of twenty further units is determined. The batch will pass uniformity of content then if not more than one of the individual contents of the 30 units is outside of the range 85–115% and none is outside 75–125% of the average content range.

8. **Uniformity of mass/volume**: The fill volume (emulsions)/weight (creams) of the product is measured and compared to the specified range.

9. **Droplet size of the dispersed phase**: The particle size distribution of the disperse phase of emulsions of low viscosity may be characterised using an appropriate technique, e.g. Coulter counter, laser diffraction. This property is characterised both following manufacture and during storage. The particle size of creams is rarely characterised

10. **Freeze–thaw storage:** Freeze–thaw storage analysis is applied to emulsions and creams for the same reasons as described for pharmaceutical suspensions. In particular this storage regimen will indicate the likelihood of emulsion instability by cracking. This 'in house' test enables the effect of fluctuations of temperature on the stability to be ascertained.

11. **Uniformity of mass**: If the formulation is a single dose preparation the following method is applied. Following manufacture, the contents of twenty individual units are decanted, weighed separately and the average mass determined. To successfully pass, the average masses of no more than two units must deviate by more than 10% of the average mass and no units should deviate by more than 20%. Alternatively (and for non-single dose preparations), specifications may simply require that the container is completely emptied and the volume/mass measured. The test has been passed if the mass or volume is not less than the nominal volume/mass on the label.

The reader should note that the list of methods is indicative of the quality control methods that may be employed. The reader should consult the appropriate pharmacopoeias for a more detailed description of the above methods and others that have not been explicitly covered in this chapter.

## Multiple choice questions

1. **Regarding the stability of pharmaceutical emulsions, which of the following statements are true?**
a. Emulsions are inherently pharmaceutically unstable.
b. The stability of pharmaceutical emulsions is affected by the size of the dispersed phase.

c. The stability of pharmaceutical suspensions is affected by the concentration of dispersed phase.

d. Phase volume of the internal phase directly affects the stability of pharmaceutical emulsions.

2. **Regarding the rate of creaming of pharmaceutical emulsions, which of the following statements are true?**

a. The rate of creaming is increased as the diameter of the internal phase is increased.

b. The rate of creaming is increased as the viscosity of the continuous phase is increased.

c. The rate of creaming is affected by the concentration and type of incorporated surfactants.

d. The rate of creaming is decreased by centrifugation.

3. **Regarding emulsions, which of the following statements are true?**

a. Multiple emulsions are more stable than primary emulsions.

b. Water in oil emulsions are commonly administered orally.

c. Oil in water emulsions are stable following dilution with water.

d. Dispersed globules of the internal phase do not possess a zeta potential.

4. **Regarding the role of surfactants in the formulation of emulsions, which of the following statements are true?**

a. Oil in water emulsions are promoted by the presence of surfactants with HLB values >8.

b. The elastic properties of the surfactant layer at the interface between the internal and external phases are important in the stabilisation of the emulsion.

c. Anionic surfactants may be used for the stabilisation of emulsions designed for oral administration.

d. Surfactants stabilise emulsions by increasing interfacial Gibbs free energy of the emulsion.

5. **Regarding the role of adsorbed particles in the stabilisation of pharmaceutical emulsions, which of the following statements are true?**

a. Adsorbed particles promote the formation of oil in water emulsions.

b. Examples of adsorbed particles that are employed pharmaceutically include kaolin.

c. The elastic properties of the adsorbed layer are primarily responsible for the stabilisation of emulsions.

d. The particle size of the adsorbed particles directly affects emulsion stability.

6. **Regarding the use of hydrophilic polymers for the stabilisation of pharmaceutical emulsions, which of the following statements are true?**
a. Hydrophilic polymers stabilise pharmaceutical emulsions by increasing the viscosity of the continuous phase.
b. Hydrophobic polymers may be employed to stabilise oil in water emulsions.
c. Hydrophilic polymers form a stable monomolecular layer at the interface between the internal and external phases.
d. Pharmaceutical emulsions may exhibit thixotropy.

7. **Regarding the preservation of pharmaceutical suspensions for oral administration, which of the following statements are true?**
a. Hydrophilic polymers that are present in pharmaceutical emulsions may reduce preservative efficacy.
b. Preservative efficacy may be reduced by partitioning of the preservative from the oil phase to the water phase of water in oil emulsions.
c. Parabens are commonly used as preservatives for emulsions.
d. The preservative efficacy of organic acids in oil in water emulsions may be affected by the type of oil used as the internal phase.

8. **Concerning the physical properties of emulsions, which of the following statements are true?**
a. Water in oil emulsions conduct electrical current.
b. Water in oil emulsions are more stable than oil in water emulsions.
c. High-speed mixing of water in oil emulsions results in dilatant flow.
d. The particle size distribution of the internal phase may be reduced post-manufacture using a colloid mill.

9. **Concerning the use of oil in water emulsions, which of the following statements are true?**
a. Oil in water emulsions may be used for the oral administration of therapeutic agents.
b. Oil in water emulsions may be formulated for topical administration.
c. Drugs with high aqueous solubility are frequently formulated as oil in water emulsions.
d. Oil in water emulsions designed for oral administration must be coloured.

10. **Concerning the formulation of pharmaceutical emulsions, which of the following statements are true?**
   a. Emulsions for oral administration may require the addition of sweetening agents.
   b. Emulsions for oral administration may require the addition of flavours.
   c. Emulsions for oral administration may require the addition of antioxidants.
   d. The optimum phase:volume ratio for emulsions is 50:50.

11. **You have been asked to prepare an emulsion for the oral delivery of a poorly soluble drug. Which of the following statements is/are valid?**
   a. The type of emulsion should be oil/water.
   b. The type of emulsion should be water/oil.
   c. The emulsion should be preserved using a quaternary ammonium compound.
   d. The product must be coloured.

12. **You have been asked to prepare a cream for the topical treatment of infection of the skin. Which of the following statements is/are valid?**
   a. The type of cream may be oil/water.
   b. The type of emulsion may be water/oil.
   c. As the cream contains an antimicrobial agent, a preservative is not necessary.
   d. The product must exhibit low viscosity.

13. **Which of the following statements is/are valid concerning the formulation of oil in water creams containing a therapeutic agent?**
   a. The cream may be diluted with purified water without a loss of physical stability.
   b. The therapeutic agent must be soluble in the external phase.
   c. The droplet size of the internal phase must be greater than 100 μm.
   d. The concentration of the internal phase must not exceed 40%.

14. **Which of the following phenomena are associated with instability in oil in water emulsions and will result in product failure?**
   a. Discolouration.
   b. Phase inversion.

   c. Creaming.
   d. Change in pH from 6.8 to 7.0.

15. **Which of the following statements are true regarding water in oil emulsions?**
   a. They are administered orally.
   b. They are always more stable than oil in water creams.
   c. Preservation of water in oil emulsions is not necessary.
   d. They must be formulated within a defined pH range.

# chapter 4
# Pharmaceutical disperse systems 3: ointments, pastes, lotions, gels and related formulations

## Overview

**In this chapter the following points will be discussed:**

- an overview/description of the physical properties and uses of ointments, pastes, lotions, liniments, collodions and gels
- formulation strategies for ointments, pastes, lotions, liniments, collodions and gels, including consideration of the excipients used
- the advantages and disadvantages and uses of ointments, pastes, lotions, liniments, collodions and gels
- considerations for the manufacture of ointments, pastes, lotions, liniments, collodions and gels.

## Introduction

This chapter deals with the formulation of several types of *disperse* systems, namely ointments, pastes, lotions, liniments, collodions and gels. In the vast majority of cases these formulations contain a therapeutic agent and are designed for the localised treatment of a designated area, e.g. haemorrhoids, infection, inflammation. Whilst the above formulation types may be classified as disperse systems, there are distinct differences in their uses and the strategies utilised in their successful formulation. Accordingly, in this chapter, each dosage form is discussed separately.

## KeyPoints

- Ointments, pastes, lotions, liniments, collodions and gels are further examples of disperse systems into which a therapeutic agent may be incorporated.
- Ointments, pastes, lotions, liniments, collodions and gels are topical formulations, being applied externally or into accessible body cavities (e.g. mouth, rectum, vagina).
- These formulations exhibit similar concerns as other disperse systems regarding physical stability.

## Advantages and disadvantages of pharmaceutical ointments, pastes, lotions, liniments, collodions and gels

### Advantages

■ Pharmaceutical ointments may be easily spread on skin, being retained at the site of application as an occlusive layer, thereby preventing moisture loss from the skin. This is particularly useful whenever restoration of the physical characteristics of the skin is required (e.g. due to inflammation).

■ Pharmaceutical ointments are associated with lubricating/emollient properties, properties that may be employed to reduce trauma of an affected site upon spreading.

■ In general, pharmaceutical ointments persist at the site of application, enabling the duration of drug release to be greater than for many other topical dosage forms. The increased viscosity of pharmaceutical pastes ensures that a thick film of the dosage form is applied to the site of action, which shows excellent persistence. This property is particularly useful if protection of an inflamed site is required, e.g. in eczema, psoriasis.

■ The hydrophobicity and retention of pharmaceutical ointments are useful attributes whenever applied to mucosa, e.g. inflamed haemorrhoids, eyelids, where fluid flow/ inflammation at these sites would normally serve to remove other formulations (e.g. oil in water creams) by dilution. It should be noted, however, that spreading of ointments on to moist surfaces may be difficult due to the hydrophobic properties of most ointments.

■ Due to the high solids content, pharmaceutical pastes are often porous, allowing moisture loss from the applied site. Furthermore, pastes may act to absorb moisture and chemicals within the exudates.

■ The opaque nature of pastes (due to the high solids content) enables this formulation to be used as a sunblock.

■ The chemical stability of therapeutic agents that are prone to hydrolysis will be dramatically enhanced by formulation within pharmaceutical ointments and pastes.

■ Pharmaceutical gels may be formulated to provide excellent spreading properties and will provide a cooling effect due to solvent evaporation. Similarly solvent evaporation from liniments will provide a cooling effect.

### Disadvantages

■ Pharmaceutical ointments are generally greasy and difficult to remove (and are therefore often cosmetically unacceptable).

Similarly, liniments and lotions may also be cosmetically unacceptable to the patient and difficult to use.

■ Pharmaceutical pastes are generally applied as a thick layer at the required site and are therefore considered to be cosmetically unacceptable.

■ Staining of clothes is often associated with the use of pharmaceutical pastes and ointments.

■ The viscosity of pharmaceutical ointments, and in particular pastes, may be problematic in ensuring spreading of the dosage form over the affected site. Conversely, the low viscosity of liniments and lotions may result in application difficulties.

■ Pharmaceutical ointments may not be applied to exuding sites (however, please note that this does not hold for pastes). Liniments may not be applied to broken skin.

■ Problems concerning drug release from pharmaceutical ointments may occur if the drug has limited solubility in the ointment base.

■ Pharmaceutical pastes are generally not applied to the hair due to difficulties associated with removal.

■ Therapeutic agents that are prone to hydrolysis should not be formulated into aqueous gels.

## Pharmaceutical ointments and pastes

### General description

Pharmaceutical ointments (termed *unguents*) are semisolid systems that are applied externally, primarily to the skin and also to mucous membranes, e.g. the rectum, the vagina/vulva, the eye. Typically, medicated ointments are used for the treatment of infection, inflammation and pruritus. However, non-medicated ointments are commonly used due to their emollient/lubricating properties. Pharmaceutical pastes are generally composed of ointment bases that contain a high concentration (frequently >50% w/w) of dispersed drug. The viscosity of pharmaceutical pastes is greater than that of pharmaceutical ointments.

### Introduction

The formulation of ointments and pastes involves the dispersal or dissolution of the selected therapeutic agent into an *ointment base* and, therefore, in addition to the physical properties of the dispersed/dissolved drug, the physicochemical properties of the ointment base are fundamental to the clinical and non-clinical performance of this type of dosage form. The choice of ointment

base is dependent on several factors, including: (1) the site of
application; (2) the required rate of drug release; (3) the chemical
stability of the drug; and (4) the effect of the therapeutic agent on
formulation viscosity.

### The site of application

In certain clinical conditions the site to which the ointment will
be applied may be dry, e.g. psoriasis, or moist. If the area is dry,
ointments are often used to occlude the site, thereby retaining
moisture. Indeed, this effect is considered to play an important
role in the treatment of certain clinical conditions. Conversely,
occlusive ointment bases are not applied to sites in which there is
fluid exudate.

### The required rate of drug release

Following application, the therapeutic agent must be released to
exert its pharmacological effect, either locally or, after absorption,
systemically. Drug release from the ointment base requires
solubility (albeit partial) of the therapeutic agent within the
formulation. This will allow diffusion of the therapeutic agent (a
molecular process) through the ointment base until it reaches the
biological substrate. Therefore the choice of the ointment base
is partially dictated by the physicochemical properties (and in
particular the solubility) of the therapeutic agent.

### The chemical stability of the drug

If a therapeutic agent is prone to hydrolysis, incorporation into
a water-based formulation, e.g. oil in water creams, may lead to
drug degradation and hence a shortened shelf-life. This problem
may be obviated by incorporating the drug into a hydrophobic
ointment base. For example, the shelf-life of hydrocortisone is
markedly greater in an ointment formulation than in an oil in
water cream formulation.

### The effect of the therapeutic agent on formulation viscosity

The effect of the physical incorporation of a therapeutic agent into
an ointment base on the rheological properties of the formulated
product will be dependent on the required drug concentration,
the physical properties of the therapeutic agent (e.g. particle
size, shape) and the chemical composition and viscosity of the
ointment base. Therefore, it is important that an ointment base is
selected that will produce a product that may be readily applied
to the required site. In light of the high drug content, this point is
particularly important in the formulation of pastes.

## Types of base for ointments and pastes

There are four types of base that are used to formulate pharmaceutical ointments and pastes: (1) hydrocarbon; (2) absorption; (3) water-miscible/removable; and (4) water-soluble.

### Hydrocarbon bases

Hydrocarbon bases are non-aqueous formulations, based on various paraffins, that have the following properties:

- emollient, thereby restricting water loss from the site of application due to the formation of an occlusive film
- excellent retention on the skin
- predominantly hydrophobic, and therefore difficult to remove from the skin by washing and difficult to apply to (spread over) wet surfaces (e.g. mucous membranes, wet skin)
- only a low concentration (<5%) of water may be incorporated into hydrocarbon bases (with careful mixing)
- chemically inert.

Hydrocarbon bases frequently contain the following components: (1) hard paraffin; (2) white/yellow soft paraffin; (3) liquid paraffin (mineral oil); and (4) microcrystalline wax.

#### Hard paraffin

This is a mixture of solid saturated hydrocarbons that are derived from petroleum or shale oil. Hard paraffin is a colourless or white wax-like material that is physically composed of a mixture of microcrystals. The melting temperature of hard paraffin is between 47°C and 65°C and, when solid, it is used to enhance the rheological properties of ointment bases.

#### White/yellow soft paraffin

This is a purified mixture of semisolid hydrocarbons (containing branched, linear and cyclic chains) that are derived from petroleum. White/yellow soft paraffin consists of microcrystals embedded in a gel composed of liquid and amorphous hydrocarbons that are themselves dispersed in a gel phase containing liquid and amorphous hydrocarbons. The melting range of the soft paraffins is between 38°C and 60°C. White soft paraffin and yellow soft paraffin (the former being a bleached form of yellow soft paraffin) may be used as an ointment base without the need for additional components, although it may be combined with liquid paraffin (see below).

#### Liquid paraffin (mineral oil)

This is a mixture of saturated aliphatic ($C_{14}$–$C_{18}$) and cyclic hydrocarbons that have been refined from petroleum. It is usually formulated with white/yellow soft paraffin to achieve the required viscosity for application to the required site.

Formulations containing liquid paraffin require the incorporation of an antioxidant due to the ability of this material to undergo oxidation.

### Microcrystalline wax

This is a solid mixture of saturated alkanes (both linear and branched) with a defined range of carbon chain lengths ($C_{41}$–$C_{57}$). This excipient is used to enhance the viscosity of ointments (and creams). One of the advantages of microcrystalline wax is the greater physical stability provided to formulations containing liquid paraffin (reduced bleeding of the liquid component).

## Absorption bases

Unlike hydrocarbon bases, absorption bases may be formulated to contain significant amounts of an aqueous phase. These may be either non-aqueous formulations to which an aqueous phase may be added to produce a water in oil emulsion (termed *non-emulsified bases*) or *water in oil emulsions* that can facilitate the incorporation of an aqueous phase (without phase inversion or cracking). Although absorption bases can accommodate a larger volume of aqueous phase than hydrophobic bases, they are still difficult to remove from the site of application by washing. This is due to the predominantly hydrophobic properties of this formulation class.

The key properties of both non-emulsified bases and water in oil emulsions that are relevant to the formulation of ointments and pastes are detailed below.

### Non-emulsified bases

These are hydrophobic formulations to which water may be added. Following application, a film is formed that offers occlusion (and hence emollient properties); however, the extent of occlusion is less than for hydrocarbon bases. The spreading properties of these formulations are more favourable than for hydrocarbon bases.

Typically non-emulsified bases are commonly composed of: (1) one or more paraffins (see previous section) and (2) a sterol-based emulsifying agent. Examples of the types of emulsifying agents used in absorption bases include: (1) lanolin (wool fat); (2) lanolin alcohols (wool alcohols); and (3) beeswax (white or yellow).

#### Lanolin (wool fat)

Lanolin is a wax-like material that is derived from sheep's wool. It is available in two forms, termed lanolin (wool fat) and hydrous lanolin (wool alcohols). Lanolin is typically mixed with vegetable oils or paraffins to produce an ointment base that can absorb approximately twice its own weight of water to produce water in oil emulsions. The usual concentrations of lanolin used in ointments (e.g. Simple Ointment BP) range from 5% to 10% w/w.

### Lanolin alcohols (wool alcohols)

Wool alcohol is a crude mixture of sterols and triterpene alcohols and contains at least 30% cholesterol and 10–13% isocholesterol. This is added to mixtures of paraffins (hard, so white/yellow, soft or liquid) to produce the required consistency. The inclusion of wool alcohols (5% w/w) results in a 300% increase in the concentration of water that may be incorporated into paraffin bases.

### Beeswax (white or yellow)

Beeswax is a wax that consists of esters of aliphatic alcohols $(C_{24}-C_{36}$ even numbers) and linear aliphatic fatty acids (up to $C_{36}$, even numbers) that is combined with paraffins to produce non-emulsified bases. White beeswax is the bleached form of yellow beeswax.

### *Water in oil emulsions*

Ointment bases in this category can accommodate a greater concentration of water but yet can still provide similar performance to that provided by non-emulsified bases with respect to, e.g. occlusion, spreading properties. A common excipient that is employed in the formulation of this type of ointment base is *hydrous lanolin*, which is a mixture of lanolin and circa 25–30% water. It is incorporated into paraffins and oils to produce a base that can incorporate the subsequent addition of an aqueous phase. The water content of bases that have been formulated using hydrous lanolin is significant, e.g. Oily Cream BP is a water in oil emulsion ointment base that is composed of wool alcohols (50% w/w) and water (50% w/w).

## Water-miscible/removable bases

These are water-miscible bases that are used to form oil in water emulsions for topical applications. The use of these bases offers a number of advantages, including:

- They are able to accommodate large volumes of water, e.g. aqueous solutions of drug, excess moisture at the site of application, e.g. exudate from abrasions and wounds.
- They are not occlusive.
- They may be easily washed from the skin and from clothing. Furthermore, they may be readily applied to (and removed from) hair.
- They are aesthetically pleasing.

The British Pharmacopoeia describes three water-miscible/removable bases:

1. emulsifying ointment
2. cetrimide emulsifying ointment
3. cetomacrogol emulsifying ointment.

Each of these contains:

- liquid paraffin 20% w/w
- white soft paraffin 50% w/w
- emulsifying wax (anionic, cationic or non-ionic) 30% w/w.

As may be observed, an important component of this ointment base is emulsifying wax, of which there are three types: (1) anionic; (2) non-ionic; and (3) cationic. The important properties of these waxes are as follows:

### Anionic emulsifying wax

- This is a waxy solid that, when incorporated into a paraffin base (producing emulsifying wax), may be used to produce an oil in water emulsion, e.g. Aqueous Cream BP (which contains 10% w/w anionic emulsifying wax).
- Anionic emulsifying wax is composed of:
  - cetostearyl alcohol 90 g
  - sodium lauryl sulphate 10 g
  - purified water 4 ml.

### Non-ionic emulsifying wax

- This is also referred to as *Cetomacrogol Emulsifying Wax BP* and is composed of:
  - cetostearyl alcohol 800 g
  - cetomacrogol 1000 (macrogol cetostearyl ether 22) 200 g.

### Cationic emulsifying wax

- This is also referred to as *Cetrimide Emulsifying Wax BP*.
- Cationic Emulsifying Wax BP is composed of:
  - cetostearyl alcohol 900 g
  - cetrimide 100 g.

## Water-soluble bases

The reader will have observed that the three previous ointment bases are predominantly hydrophobic, are hydrophobic with added surface-active agents or are water-miscible, containing both water-soluble and insoluble components. By contrast, water-soluble bases are composed entirely of water-soluble ingredients.

The advantages of the use of these bases include:

- They are non-greasy and may be easily removed by washing.
- They are miscible with exudates from inflamed sites.
- They are generally compatible with the vast majority of therapeutic agents.

Water-soluble bases are predominantly prepared using mixtures of different molecular weights of polyethylene glycol (Figure 4.1) to produce the required ointment consistency. Lower average molecular weights of this polymer (200, 400 and 600 g/mol) are liquids. As the average molecular weight increases ($\geq$1000 g/mol), the consistency of this polymer changes from a liquid to a waxy solid.

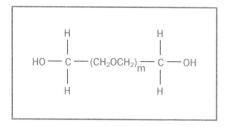

**Figure 4.1**   Structural formula of polyethylene glycol (m refers to the number of repeating ethylene oxide groups.

Blends of 60% w/w polyethylene glycol 400 (a liquid) and 40% w/w polyethylene glycol 4000 (a solid) have been used as a water-soluble ointment base. If required, the consistency may be increased by lowering the ratio of polyethylene glycol 400 to polyethylene glycol 4000 in the ointment base. Blending the two polyethylene glycol fractions is performed by heating the mixture followed by cooling of the homogeneous liquid at a controlled rate.

The main disadvantage associated with water-soluble bases is their inability to incorporate large volumes of aqueous solutions as these will soften and, if the concentration of water is large enough (>5% w/w), dissolve the ointment base. Therefore the use of these bases is usually reserved for the incorporation of solid therapeutic agents. However, these bases may incorporate up to 25% of an aqueous solution if a portion of the lower-molecular-weight polyethylene glycol is replaced with stearyl alcohol. This will enhance the mechanical properties of the ointment.

## Miscellaneous excipients used in the formulation of ointments and pastes

This chapter has described various strategies for the formulation of bases for ointment and paste formulations. In these the therapeutic

# Tips

- The choice of ointment base employed in the formulation of ointments is dependent on the proposed use of the dosage form and other formulation factors, e.g. stability of the therapeutic agent and the capacity of the formulation for water.
- The properties of ointment bases range from highly hydrophobic (e.g. hydrocarbon bases) to water-miscible systems.
- Hydrophobic ointment bases should not be applied to exuding areas due to poor water uptake capacity of these dosage forms. Conversely, ointments prepared using water-miscible bases may be applied to such sites.

agent may be directly incorporated as a solid component or, in the case of the absorption and water-miscible bases, the addition may be in the form of a solution. This solution may be aqueous, alcoholic (e.g. propylene glycol, glycerol) or hydroalcoholic and must not adversely affect the physical stability and/or appearance of the formulated product.

Other excipients may be included in ointments and pastes, including: (1) additional/alternative solvents; (2) preservatives; and (3) antioxidants.

## Additional/alternative solvents

These are hydrophobic liquid components that may be added to ointment bases (predominantly hydrophobic or absorption bases). Examples of these include: (1) liquid silicone; (2) vegetable oils; and (3) organic esters.

### Liquid silicone (polydimethylsiloxane)

This may be used in barrier ointments due to the water-repellent properties of this component.

### Vegetable oils

Vegetable oils may be used either to replace mineral oils or, alternatively, may be added to hydrophobic or absorption bases to increase the emollient properties of the formulated product. Examples of oils that are used for this purpose are coconut oil, cottonseed oil and sesame oil.

### Organic esters

These may be used partly to replace a mineral oil to enhance the spreadability and to enhance drug dissolution within the ointment base. One of the most commonly used examples is isopropyl myristate.

## Preservatives

Topically applied ointments and pastes are not sterile products; however, they are manufactured under clean conditions to minimise the microbial bioburden within the formulated product. Ointments/pastes that do not contain water do not usually require the addition of a preservative (due to the low water activity in the formulation). However, if the product contains water, then a preservative will be required. Preservatives that may be used in formulations designed for external use include:

- phenolics: phenol (0.2–0.5%), chlorocresol (0.075–0.12%)
- benzoic acid and salts (0.1–0.3%)
- methylparabens (methyl parahydroxybenzoic acid) (0.02–0.3%)
- propylparabens (propylparahydroxybenzoic acid) (0.02–0.3%) (and their mixtures)
- benzyl alcohol (≤ 3.0%) (Figure 4.2a)
- phenoxyethanol (0.5–1.0%) (Figure 4.2b).
- bronopol (0.01–0.1%, usually 0.02%) (Figure 4.3).

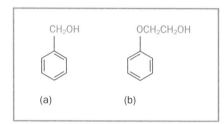

**Figure 4.2** Structural formulae of (a) benzyl alcohol and (b) phenoxyethanol.

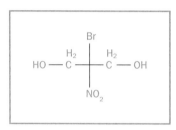

**Figure 4.3** Structural formula of bronopol.

In the preservation of ointments, the same physicochemical and microbiological principles exist and therefore partitioning of the preservative from the aqueous to the non-aqueous phase may occur. Under these circumstances it is important to ensure that the required concentration (≥ minimum inhibitory concentration) of the antimicrobial species is present within the *aqueous phase*.

### Antioxidants

The use of antioxidants has been described in previous chapters. In pharmaceutical ointments antioxidants are employed to prevent or reduce oxidation of either the non-aqueous components of the ointment base (e.g. mineral/vegetable oils)

## Tips

- Ointments, similar to emulsions and creams, contain a range of excipients that are required to stabilise the formulation.
- The formulator must ensure that no interactions occur between the excipients.
- The use of pastes is generally reserved for certain topical conditions, e.g. the treatment of warts.

and/or the therapeutic agent. The types of preservatives used for this purpose include:

- lipophilic antioxidants (to be dissolved within the non-aqueous vehicle), e.g. butylated hydroxyanisole (0.005–0.02%), butylated hydroxytoluene (0.007–0.1%), propyl gallate (≤1%)
- hydrophilic antioxidants (to be dissolved in the aqueous phase), e.g. sodium metabisulphate (0.01–0.1%), sodium sulphite (0.1%).

Diagrammatic representations of formulation considerations for ointments are shown in Figures 4.4–4.6.

### Manufacture of ointments and pastes

The manufacture of ointments and pastes is similar to that described for emulsions and creams. The most straightforward example involves the dispersal of the powdered therapeutic agent into the preheated hydrocarbon base using a mechanical mixer. Heat is required to lower the viscosity of the base, thereby facilitating the mixing of the solid drug.

If the therapeutic agent is incorporated into the ointment base as a separate liquid phase, the hydrophobic components and hydrophilic components are separately dissolved in the lipophilic and hydrophilic liquid phases, respectively (again with the aid of heating and mechanical mixing). In general (following dissolution of the various components), the two phases are maintained at circa 70°C and then mixed together (with stirring). The mixing of the two phases may be performed by:

- mixing the two phases simultaneously
- adding the aqueous phase to the non-aqueous phase.

Following complete mixing, the temperature of the formulation is gradually reduced to room temperature.

## Pharmaceutical lotions, liniments, collodions and paints

### General description

Pharmaceutical lotions, liniments, collodions and paints are external, liquid-based formulations that are applied externally for the treatment of local conditions, e.g. inflammation, acne, infection (bacterial, fungal, viral and parasitic). Although the clinical use of these classes of formulation is relatively minor, they still represent a formulation option to the pharmaceutical scientist.

**Figure 4.4** Diagrammatic representation of formulation considerations for ointments and pastes using hydrocarbon bases.

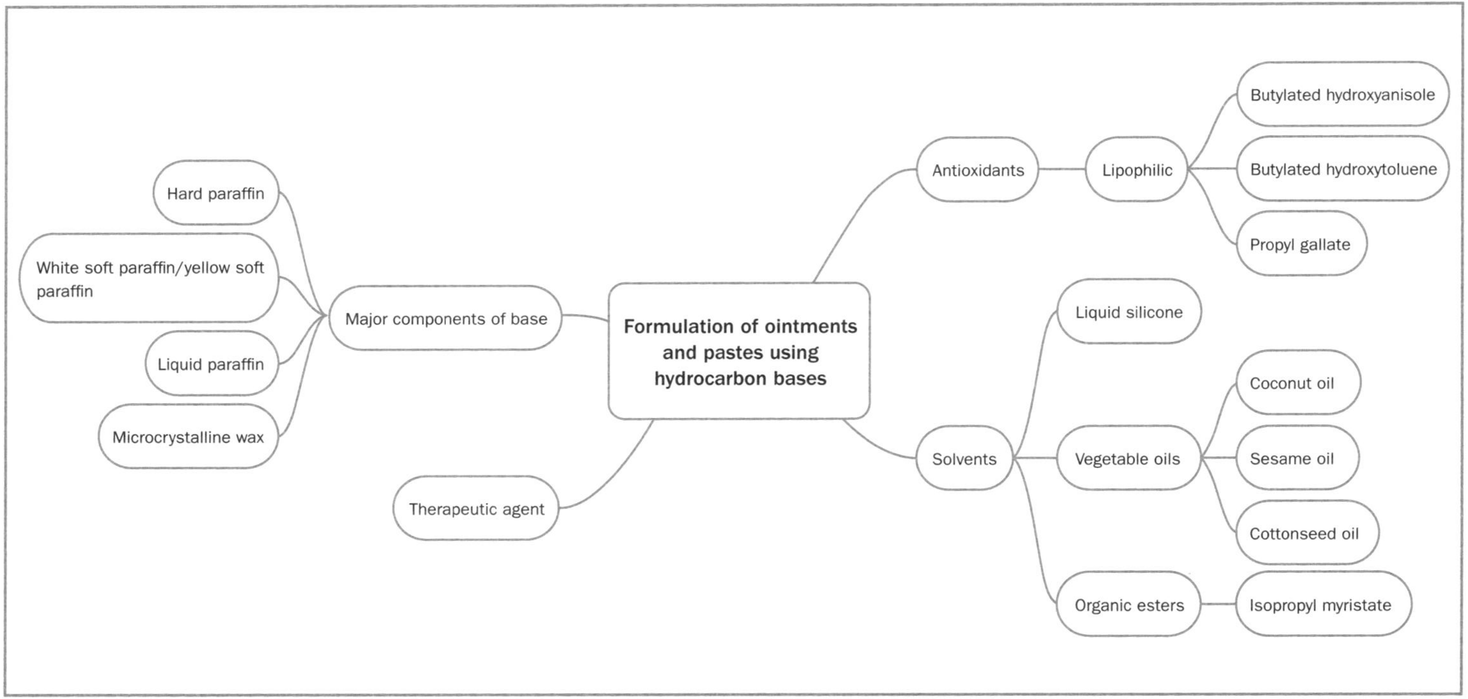

**Figure 4.5** Diagrammatic representation of formulation considerations for ointments using absorption bases.

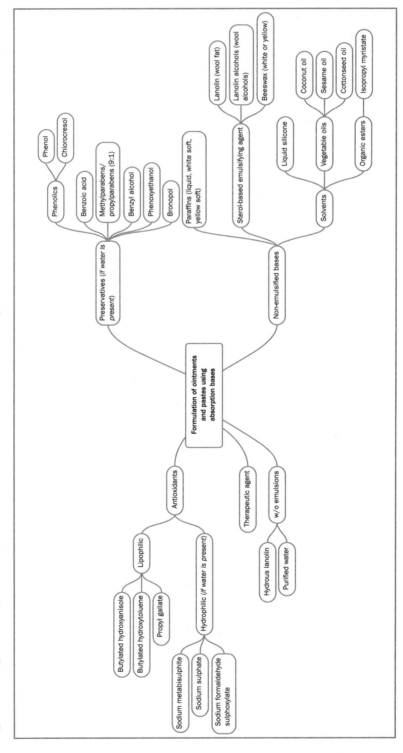

**Figure 4.6** Diagrammatic representation of formulation considerations for ointments using water miscible/removable bases.

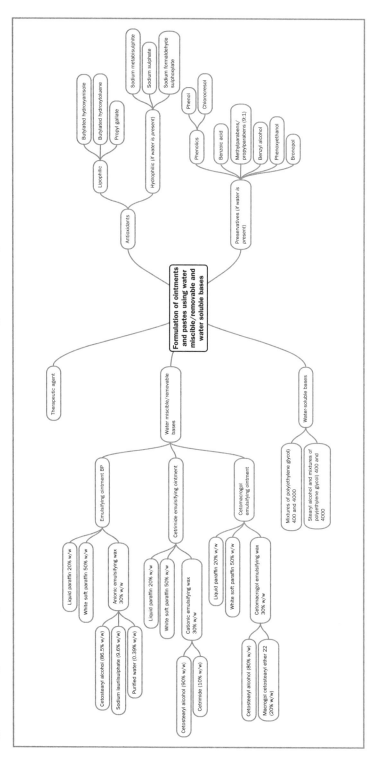

## Lotions

Lotions are formulated as either solutions or suspensions and, in addition to the therapeutic agent(s), may contain:

- *alcohol:* this acts as a coolant (due to evaporation following application) and as a co-solvent.
- *humectants:* these act to retain moisture on the skin after application. The most commonly used example is glycerol.
- *vehicle:* lotions are aqueous formulations and therefore will contain *purified water* (with or without the addition of buffer salts).
- *preservatives:* examples of these have been described previously in this chapter.
- *components to stabilise the suspended therapeutic agent:* if the lotion has been formulated as a suspension, agents are required to maintain the physical stability of the formulation (see Chapter 2).

## Liniments

Liniments are alcohol- or oil-based solutions that are applied externally to unbroken skin with gentle rubbing. There are two types of formulation bases that are used in the formulation of liniments: (1) alcohol-based liniments; and (2) oil-based liniments. Alcohol-based liniments act as counterirritants and rubefacients (causing reddening of the skin) and may act to increase the penetration of the drug through the skin. In addition, these formulations will provide a cooling effect due to evaporation of the alcohol base. Conversely, oil-based liniments are employed for conditions in which a massage effect is required. Typical oils used for this purpose are maize oil and cottonseed oil.

Liniments are normally employed for the treatment of inflammatory conditions, e.g. sciatica, fibrositis and neuralgia. Examples of oil-based liniments include Camphor Liniment BP and Methyl Salicylate Liniment BP. Soap Liniment BPC is an example of an alcohol-based liniment.

In general, no other excipients are used in the formulation of liniments.

## Collodions

Collodions are solutions of pyroxylin (a nitrated cellulose, predominantly cellulose tetranitrate, that is obtained following the treatment of defatted cellulose with nitric and sulphuric acids), castor oil and colophony dissolved in an organic solvent (composed of alcohol and ether). These are normally applied to dry skin using a brush applicator and, following the evaporation of the solvent, will form an occlusive film. Collodions may contain therapeutic agents, e.g. collodion, salicylic acid collodion.

### Collodion
This is a solution of pyroxylin in a solvent composed of ether (3 parts) and alcohol (1 part). This forms an inflexible, mechanically strong film on the skin and is normally used to seal abrasions. The film may be rendered more flexible by adding 2% camphor and 3% castor oil to the above formulation (termed flexible collodion). The oil acts as a plasticiser (thereby facilitating the use of the product over flexible areas) whereas the presence of camphor renders the films waterproof.

### Salicylic acid collodion
This is a solution of salicylic acid (10%) in flexible collodion that is used for the treatment of warts.

### Paints
These are aqueous, hydroalcoholic, alcoholic or organic solutions of a therapeutic agent that are applied topically. The use of paints nowadays is limited due to the emergence of more elegant dosage forms.

## Pharmaceutical gels

### General description
Pharmaceutical gels are semisolid systems in which there is interaction (either physical or covalent) between colloidal particles within a liquid vehicle. The vehicle is continuous and interacts with the colloidal particles within the three-dimensional network that is formed by the bonds formed between adjacent particles. The vehicle may be aqueous, hydroalcoholic, alcohol-based or non-aqueous. The colloidal particles may be dispersed solids, e.g. kaolin, bentonite or, alternatively, dispersed polymers. *Xerogels* are gels in which the vehicle has been removed, leaving a polymer network, e.g. polymer films.

There are two main categories of pharmaceutical gels, based on the nature of the three-dimensional network of particles: (1) dispersed solids and (2) hydrophilic polymers.

### Gels based on dispersed solids
As discussed in Chapter 2, under certain conditions dispersed solids will undergo flocculation. If flocculation extends throughout the system a continuous solid particle network is established, with the liquid vehicle dispersed in the void volume between the particles. The nature of the interaction between the particles in the network may be van der Waals interactions (at the secondary minimum), e.g. *Aluminium Hydroxide Gel USP*. However, for certain dispersed solids the nature of the interaction is electrostatic

bonding. Examples of the particles that exhibit this type of interaction include kaolin, bentonite and aluminium magnesium silicate. The particles exhibit a plate-like crystal structure in which there are electronegative regions along the flat face of the crystal (due to O⁻) and electropositive regions (due to the ionised aluminium and magnesium ions) at the edges of the plates. The interaction of these two regions facilitates the establishment of a structured 'house of cards'-type particle network. The bond strength between the particles is weak: interparticle bonds are broken by the application of relatively low shearing stresses (such as those that occur whenever the product is shaken), thereby liberating the individual particles. Following removal of the stress the bonds between the particles will reform and hence the rheological structure of these systems is recovered. This time-dependent recovery of the rheological structure (that was lost upon shaking) is termed *thixotropy*.

## Gels based on hydrophilic polymers

Pharmaceutical gels are most commonly (but not exclusively) manufactured by dispersing hydrophilic polymers within an appropriate aqueous vehicle. When dissolved within an aqueous phase, hydrophilic polymers behave as lyophilic colloids and their unique physical properties result from the self-association of the dissolved polymer and its interaction with the aqueous medium. There are two types of self-association (termed irreversible and reversible) that may be demonstrated by lyophilic colloids and this allows gels that are manufactured from lyophilic colloids to be classified as either type 1 or type 2 gels.

### Type 1 gels

In type 1 gels (often termed *hydrogels*) the interaction between the polymer chains is covalent and is mediated by molecules that croslink the adjacent chains (termed cross-linkers). A diagrammatic representation of the covalent interactions within a type 1 (chemical) is shown in Figure 4.7.

An example of the chemical structures of a cross-linked hydrogel and the monomer (hydroxyethylmethacrylate) and cross-linker (ethyleneglycol dimethacrylate) used in the synthesis of the hydrogel is provided in Figure 4.8. In this diagram the points of intersection of the polymer chains are covalent cross-links.

These gels exhibit unique physicochemical properties, including:

- The ability to absorb a considerable mass of aqueous fluid (often 100 times the original mass) whilst still retaining a three-dimensional structure. The effect of swelling on the dimension of a hydrogel composed of poly(hydroxyethylmethacrylate) that has been cross-linked using ethyleneglycol dimethacrylate is shown in Figure 4.9.

**Figure 4.7**   Diagrammatic representation of covalent interactions within a type 1 gel. The solid lines depict polymer chains and the intersections between these depict covalent bond formation.

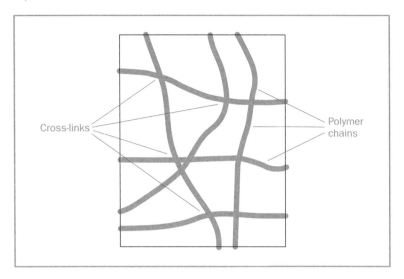

- Hydrogels exhibit robust mechanical properties, being resistant to fracture following exposure to stresses frequently up to 1 kPa. Moreover, hydrogels exhibit excellent flexibility.

Conversely, xerogels (hydrogels from which the aqueous phase has been removed by drying) are brittle (Figure 4.9b). In this case the absorbed solvent acts as a plasticiser. Unlike type 2 gels (see below), type 1 gels do not exhibit flow when exposed to an applied stress due to the inability of the stress to overcome (destroy) the covalent bonds. Under these conditions, the elastic properties of type 1 gels enable the applied energy to be stored and utilised (after the stress is removed) to return the polymer chains to their equilibrium position.

Due to this ability to absorb a large mass of fluid (whilst retaining their mechanical properties), hydrogels are clinically used as wound dressings, as lubricious coatings on urethral catheters and as soft contact lenses. In addition, hydrogels may be used for the controlled delivery of therapeutic agents at the site of implantation.

### Type 2 gels

In type 2 gels the interactions between the polymer chains are reversible and are facilitated by weaker bonds, e.g. hydrogen bonding, ionic association or van der Waals interactions. The application of stresses to type 2 gels will end in the temporary destruction of these bonds, thereby enabling the formulation

**Figure 4.8**  Chemical structures of (a) the monomer hydroxyethylmethacrylate and (b) cross-linker ethyleneglycol dimethacrylate used in the synthesis of (c) the hydrogel poly(hydroxyethylmethacrylate).

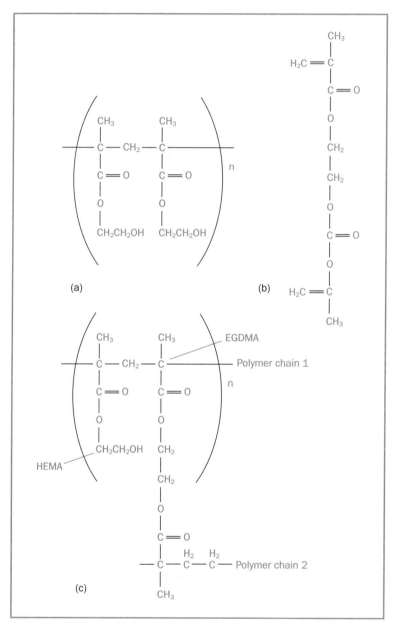

to flow. As a result, type 2 gels are rheologically referred to as *pseudoplastic (shear-thinning)* systems. Following the removal of the stress, the inter-macromolecular bonds are reformed and the viscosity of the formulation returns to its equilibrium value.

Figure 4.9   Two photographs showing the effect of swelling on the dimensions of a type 1 (hydrogel) gel prepared by crosslinking poly(hydroxyethylmethacrylate) with ethyleneglycol dimethacrylate: (a) swollen and (b) xerogel forms of the same section of hydrogel.

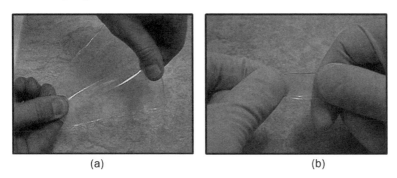

(a)                                                  (b)

A diagrammatic representation of the interactions that occur in type 2 (physical) gels is shown in Figure 4.10. As may be observed, the areas where adjacent polymer chains interact are referred to as junction zones and, in practice, a substantial fraction of the polymer is involved in polymer–polymer interactions at these zones.

The overwhelming majority of pharmaceutical gels are type 2 gels and typically the following polymers are employed in the formulation of these systems: (1) cellulose derivatives; (2) polysaccharides derived from natural sources; and (3) polyacrylic acid.

Figure 4.10   Diagrammatic representation of non-covalent interactions within a type 2 gel. The solid lines depict polymer chains and areas of alignment of the chain refer to the junction zones.

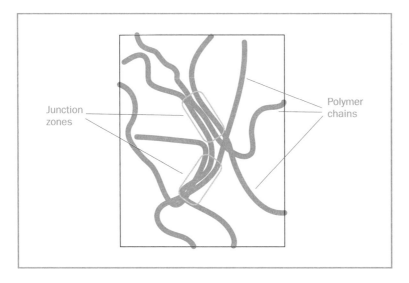

### Cellulose derivatives

The cellulose derivatives represent a family of chemically related polysaccharides that are structurally derived from cellulose (following the appropriate chemical substitution). The most commonly used examples from this series that are used to formulate pharmaceutical gels include:

- methylcellulose
- hydroxyethylcellulose
- hydroxypropylcellulose
- sodium carboxymethylcellulose.

The structural formulae of these polymers are presented in Figure 4.11.

### Polysaccharides derived from natural sources

Polysaccharides that have been derived from natural sources are commonly used as the basis for pharmaceutical gels. Examples of these include: (1) carrageenan; and (2) alginic acid/sodium alginate.

### Carrageenan

This is a family of polysaccharides that is derived from red seaweed. There are three chemically related carrageenans, termed lambda, iota and kappa, which differ according to the location of sulphate groups and the presence or absence of anhydrogalactose. Kappa carrageenan exhibits excellent gelling properties (due to the presence of a tertiary helical structure); iota carrageenan (but not lambda carrageenan) also displays gelling (albeit weaker) properties. The typical range of concentrations of kappa carrageenan used to form pharmaceutical gels is 0.3–1.0% w/w.

### Alginic acid/sodium alginate

Alginic acid is a polysaccharide that is derived from algae (Phaecophyceae family). Addition of calcium ions to a solution of alginic acid will result in an electrostatic interaction, producing a viscous gel at low concentrations of calcium and a cross-linked polymer at higher concentrations. Alginic acid is incompatible with basic drug molecules.

### Poly(acrylic acid)

Poly(acrylic acid) (Figure 4.12) is a synthetic polymer that is produced following the polymerisation of acrylic acid and cross-linking with either allyl sucrose or allyl ethers of pentaerythritol.

In water, polyacrylic acid exists as aggregated (coiled) colloidal particles of minimal viscosity (pH circa 3). However, if the pH of the system is neutralised by the addition of an appropriate base, e.g. triethanolamine, triethylamine or sodium hydroxide,

**Figure 4.11** Structural formulae of cellulose derivatives that are commonly used to formulate pharmaceutical gels: (a) methylcellulose; (b) hydroxyethylcellulose; (c) hydroxypropylcellulose; (d) sodium carboxymethylcellulose.

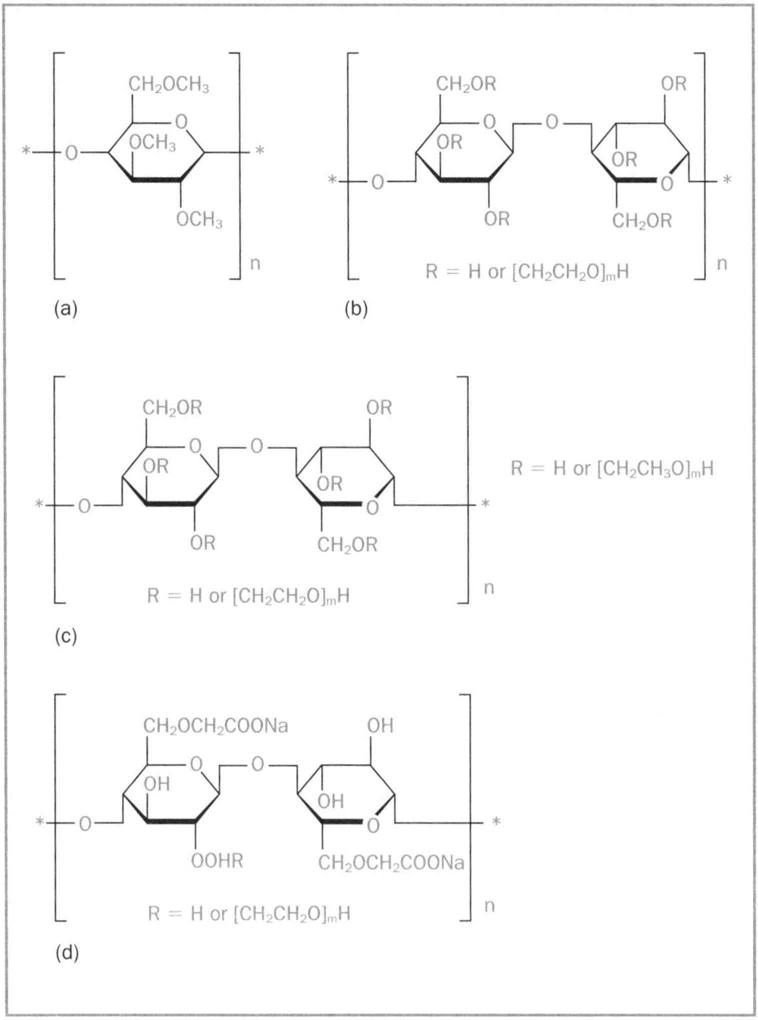

the pendant carboxyl groups will ionise, resulting in expansion of the polymer chains due to repulsion of the adjacent ionised groups. In so doing the viscosity of the formulation is dramatically increased. Typically pharmaceutical gels are produced using 0.5–2.0% w/w poly(acrylic acid) that has been neutralised with an appropriate base. Incompatibilities exist between poly(acrylic acid) and basic therapeutic agents. Furthermore, the viscosity of gels prepared using poly(acrylic acid) is adversely affected by medium/high concentrations of electrolytes.

**Figure 4.12**   Structural formula of poly(acrylic acid). R refers to allylsucrose or pentaerythritol. The subscripts refer to a number of repeating units of acrylic acid that reside between the cross-links.

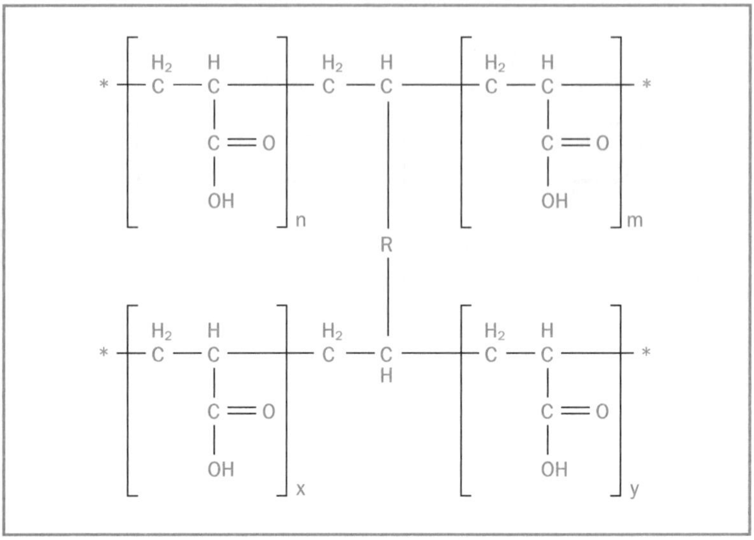

## Factors affecting gelation of type 2 gels

Gelation in type 2 gels occurs whenever a sufficient number of polymer–polymer interactions (junction zones) occur. However, both the mechanism of gelation and the number (frequency) of interactions are affected by physicochemical and environmental factors, as outlined below.

### Concentration of hydrophilic polymer

At low concentrations, solutions of hydrophilic polymers exhibit Newtonian flow due to the limited number of polymer–polymer interactions. As the concentration of polymer increases, the number of polymer–polymer interactions increases and eventually, at a defined polymer concentration, the flow properties of these systems become non-Newtonian (termed the gel point). Further increases in the concentration of polymer lead to an increase in the number of junction zones and hence the resistance to deformation from an applied stress (the viscosity) increases. Therefore, the physicochemical and rheological properties of a pharmaceutical gel may be readily manipulated by altering the concentration of hydrophilic polymer.

### Molecular weight of the polymer

As the molecular weight of the hydrophilic polymer increases (at a defined concentration of polymer), there are a greater number of available sites on the polymer chains that may engage in polymer–polymer interactions. As a result the viscosity of the formulation increases.

### Nature of the solvent

In solvents that are described as 'good solvents', the chains of a polymer will exist in the expanded state. Conversely, in the presence of a poor solvent, the polymer chains will exist in a non-expanded (coiled) state. The viscosity of a polymer solution is dependent on the expansion of the polymer chains. Therefore, the concentration of polymer that results in gel formation and the physicochemical (rheological) properties of the gel are dependent on the solvent system into which the hydrophilic polymer is dissolved. In poor solvents gelation does not occur.

### pH of the solvent

As previously discussed in this chapter, the pH of the solvent directly affects the ionisation of acidic or basic polymers which, in turn, affects the conformation (expansion) of the polymer chains. In the non-ionised state acidic and basic polymers exist in a coiled (non-expanded) state and gelation does not occur. The rheological properties of ionic polymers are optimal with a range of pH values at which maximum expansion of the polymer chains occurs. The rheological properties of non-ionic polymers are unaffected by the pH of the solvent, usually over a large pH range (circa 4–10).

### Ionic strength of the solvent phase

The rheological properties of both non-ionic and (in particular) ionic polymers are affected by the ionic strength of the solvent. At high concentrations of electrolytes (and hence large ionic strength), non-ionic polymers may be 'salted out' of solution due to desolvation of the polymer chains. Conversely, at lower concentrations of electrolyte, shielding of the charge on the pendant groups of the ionic polymer by a counterion will occur. This will therefore reduce the capacity of the polymer to interact with the solvent and hence the rheological properties of the gel will be compromised. If the concentration of electrolyte is sufficiently large, salting out of the ionic polymer will result.

### Temperature

Certain hydrophilic polymers may undergo a thermally induced transition that results in an increase in the rheological properties. Two examples of this are solutions of methylcellulose and hydroxypropylcellulose which have been reported to undergo gelation at elevated temperatures (circa 50–60°C). Whilst this transition has limited biological relevance, one polymer system, poly(oxyethylene)–poly(oxypropylene) block co-polymers (the Pluronic or Synperonic series) undergoes a thermal transition within a biologically useful temperature range (<37°C). At temperatures below this (sol–gel) transition temperature ($T_{sol/gel}$), solutions of this polymer exhibit Newtonian flow and low viscosity (the sol state). Conversely, above $T_{sol/gel}$ the polymer sol is converted into a gel with pronounced elasticity and viscosity. In solution at temperatures below $T_{sol/gel}$ and above the critical micelle concentration, the polymer exists in the micellar state. Elevation of the temperature (to above the $T_{sol/gel}$) results in the further production of micelles and (close) intermicellar aggregation. This results in a gel of pronounced rheological structure. Lowering the temperature of the system to below the $T_{sol/gel}$ will result in deaggregation of the micelles and the re-emergence of the sol (low-viscosity) state. The ability to modulate the rheological structure of these gels in the manner described has led to an interest in their use as drug delivery systems within the oral cavity and rectum.

### Ionic gelation

Certain hydrophilic polymers may undergo gelation in the presence of inorganic metal ions. Examples of these include:

- The gelation of polyhydroxypolymers, e.g. poly(vinyl alcohol) may occur in the presence of suitable anions, e.g. borate, permanganate. Poly(vinyl alcohol) is known to form structured gels in the presence of borate anions. The mechanism of the interaction between the polymer and borate anions is shown in Figure 4.13. The gels formed by this mechanism exhibit excellent mechanical strength, due to the borate anion-mediated cross-links. A non-pharmaceutical application of this interaction is the children's toy Kids Slime.
- As highlighted previously, gelation of alginic acid occurs in the presence of positively charged di/trivalent ions, e.g. $Mg^{2+}$, $Ca^{2+}$, $Al^{3+}$.

## Tips

- The main difference between type 1 and type 2 gels is the nature of the cross-links between adjacent polymer chains. In type 1 gels the cross-links are covalent, whereas in type 2 gels the crosslinks are non-covalent (secondary).
- The vast majority of gels used as dosage forms are type 2. The use of type 1 gels is reserved for wound dressings.
- The commonly used polymers for the formulation of type 2 gels are cellulose derivatives, alginates and poly(acrylic acid). Gels are normally formed by increasing polymer concentration and, if the polymer is ionic, by altering the pH.

## Formulation considerations for pharmaceutical gels

There are several formulation considerations open to the pharmaceutical scientist concerning the formulation of pharmaceutical gels. These include: (1) the choice of vehicle; (2) the inclusion of buffers; (3) preservatives; (4) antioxidants; (5) flavours/sweetening agents; and (6) colours.

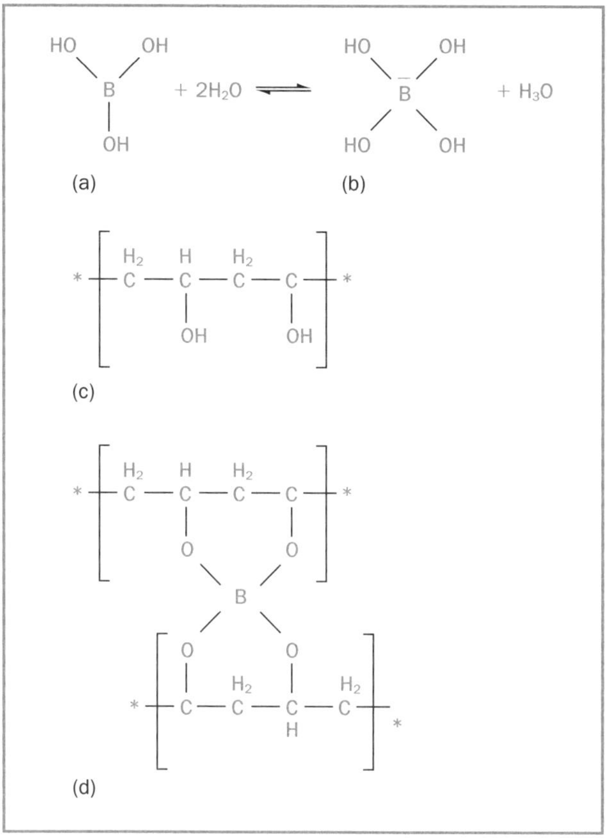

**Figure 4.13** Diagrammatic representation of the interaction between borate anions and poly(vinyl alcohol) that leads to gel formation: (a) boric acid; (b) borate anion; (c) poly(vinyl alcohol); (d) the interaction between poly(vinyl alcohol) and borate anions.

## The choice of vehicle

Purified water is the normal solvent/vehicle used in the formulation of pharmaceutical gels. However, co-solvents may be used, e.g. alcohol, propylene glycol, glycerol, polyethylene glycol (usually polyethylene glycol 400) to enhance the solubility of the therapeutic agent in the dosage form and/or (in the case of ethanol) to enhance drug permeation across the skin.

If the drug has poor chemical stability and/or poor solubility in water or water-based vehicles, pharmaceutical gels may be formulated using polyhydroxy solvents, e.g. propylene glycol, glycerol, polyethylene glycol 400 and polyacidic polymers, e.g. poly(acrylic acid). In these systems gelation is facilitated by hydrogen bonding between the hydroxyl and carboxylic acid groups and this results in: (1) expansion of the pendant groups on the polymer chain and (2) non-covalent cross-linking of adjacent polymer chains.

Non-aqueous gels may also be prepared using vegetable oils, e.g. maize oil, cottonseed oil, as the solvent system. In these systems gelation is achieved by the addition of aluminium salts of fatty acids (e.g. aluminium monostearate, aluminium distearate or aluminium tristearate) or trihydroxystearin. These viscosity modifying agents are added to a the oil which is then heated (circa 160°C for the aluminium salts and circa 60°C for trihydroxystearin) and allowed to cool. Gelation occurs whenever the oil cools to room temperature. If required lipophilic antioxidants may be included in these systems.

### The inclusion of buffers

As in other pharmaceutical formulations, buffers (e.g. phosphate, citrate) may be included in aqueous and hydroalcoholic-based gels to control the pH of the formulation. It should be noted that the solubility of buffer salts is decreased in hydroalcoholic-based vehicles.

### Preservatives

Pharmaceutical gels require the inclusion of preservatives and, in general, the choice of preservatives is similar to that described for solutions, suspensions, emulsions and ointments that have been described in the earlier chapters and in this chapter. It should be remembered that certain preservatives, e.g. parabens, phenolics, interact with the hydrophilic polymers used to prepare gels, thereby reducing the concentration of free (antimicrobially active) preservative in the formulation. Therefore, to compensate for this, the initial concentration of these preservatives should be increased.

### Antioxidants

As in other formulations, antioxidants may be included in the formulation to increase the chemical stability of therapeutic agents that are prone to oxidative degradation. The choice of antioxidants is based on the nature of the vehicle used to prepare the pharmaceutical gel. Therefore, as the majority of pharmaceutical gels are aqueous-based, water-soluble antioxidants, e.g. sodium metabisulphite, sodium formaldehyde sulphoxylate, are commonly used.

### Flavours/sweetening agents

Flavours and sweetening agents are only included in pharmaceutical gels that are designed for administration into the oral cavity, e.g. for the treatment of infection, inflammation or ulceration. As before, the choice of sweetener/flavouring agents is dependent on the required taste, the type and concentration being selected to mask the taste of the drug substance efficiently.

Examples of flavours/sweetening agents used for this purpose
have been detailed in Chapter 1.

### Colours

Colours, e.g. those described in Chapter 1, may be (but are not
usually) added into pharmaceutical gels.

Formulation considerations for pharmaceutical gels are shown
diagrammatically in Figure 4.14.

## Manufacture of pharmaceutical gels

In the manufacture of pharmaceutical gels, generally the
water-soluble components/excipients are initially dissolved in
the vehicle in a mixing vessel with mechanical stirring. The
hydrophilic polymer must be added to the stirred mixture slowly
to prevent aggregation and stirring is continued until dissolution
of the polymer has occurred. It should be noted that excessive
stirring of pharmaceutical gels results in entrapment of air.
Therefore, to prevent this the mixing rate must not be excessive or
a mixing vessel may be used to which a vacuum may be pulled,
thereby removing air.

## Quality control of ointments, pastes, lotions, gels and related formulations

Typically the quality control assessment of ointments, pastes,
lotions, gels and related formulations (that have been described in
this chapter) includes:

1. **Concentration of therapeutic agent**: As with all
   pharmaceutical products, post-manufacture the
   concentration of therapeutic agent must lie within 95–105%
   of the nominal concentration. Over the shelf-life of the
   product the concentration of drug must not fall below 90%
   of the nominal amount. If this occurs then the product has
   expired.
2. **Uniformity of content**: As defined previously, if this is a single
   dose product then uniformity of content is identical to that
   described in Chapters 1–3. For other preparations, evaluation
   of uniformity of content is performed and compared to product
   specifications.
3. **Concentration of preservative**: The concentration of
   therapeutic agent must lie within 95–105% of the nominal
   concentration following manufacture and also upon storage.
   Preservatives are only included whenever there is water in the
   formulation.

**Figure 4.14** Diagrammatic representation of formulation considerations for gels.

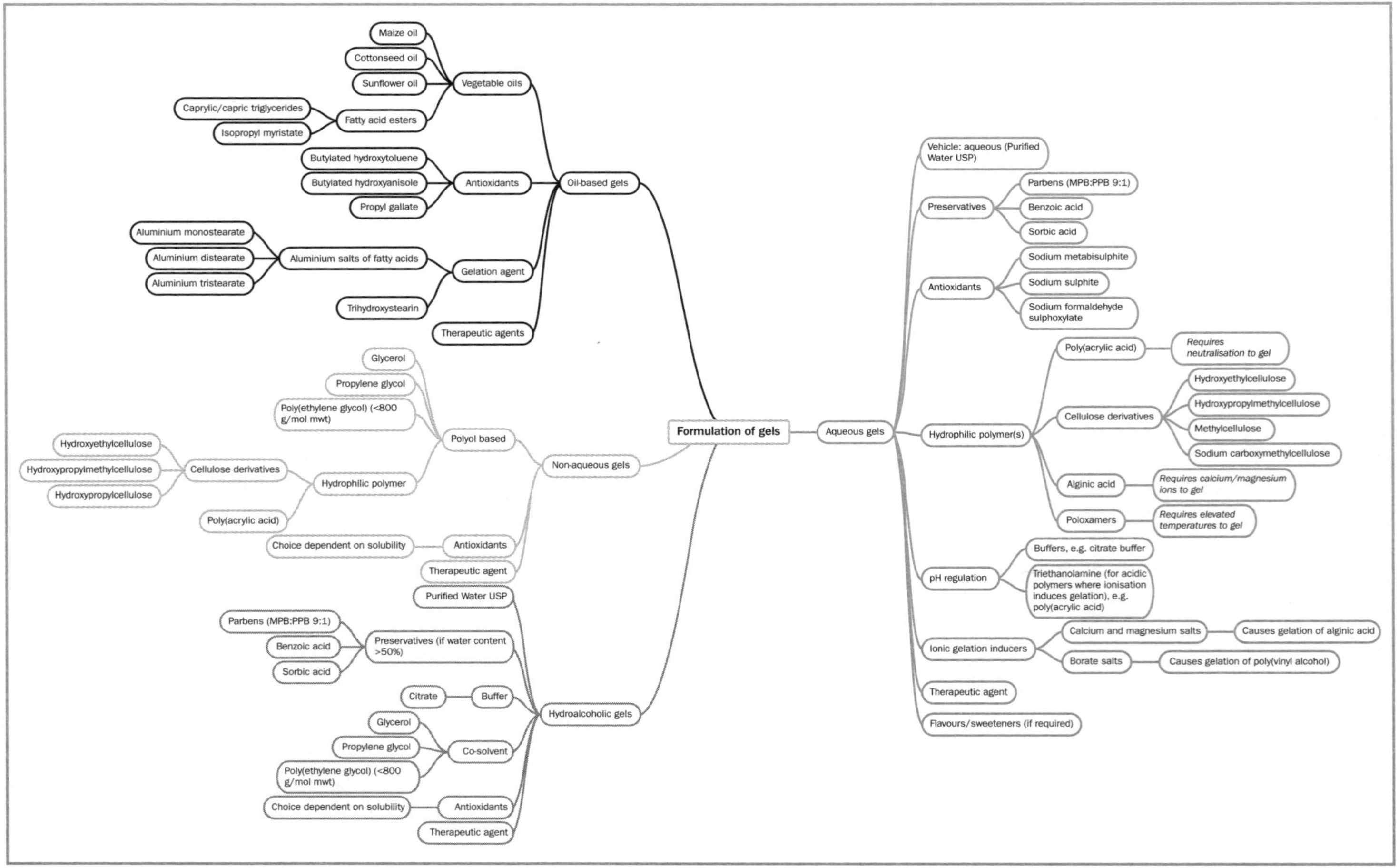

4. **Preservative efficacy testing**: The efficacy of the preservative(s) within the formulation, both following manufacture and during storage, must be assessed using the appropriate pharmacopoeial method. This test is not performed for formulations that are devoid of water.

5. **Appearance**: The products described in this chapter must be homogeneous in appearance (particularly w/o emulsion systems). It is rare for these products to be coloured. In addition, confirmation that the emulsion (w/o) is required to prove that phase inversion, either post-manufacture or during storage, has not occurred.

6. **pH**: The pH of the products described in this chapter is *not* performed generally due to the lack of water in the formulations and/or the high viscosity.

7. **Viscosity**: The viscosity of the products described in this chapter is measured following manufacture and during storage and compared to the specified range. It should be noted that these products will show a range of non-Newtonian flow properties and therefore analysis of the viscosity must take these properties into account.

8. **Fill volume/weight**: The fill volume/weight of the products described in this chapter is measured and compared to the specified range. Loss of volume over storage is usually due to loss of packaging integrity.

The reader should note that the list of methods is indicative of the quality control methods that may be employed. The reader should consult the appropriate pharmacopoeias for a more detailed description of the above methods and others that have not been explicitly covered in this chapter.

## Multiple choice questions

1. **With respect to pharmaceutical gels, which of the following are true?**
   a. Pharmaceutical gels reflect light, causing the product to have a 'creamy white' appearance.
   b. The structural properties of the gel are specifically due to the high water content of the formulation.
   c. Pharmaceutical gels are frequently formulated using long-chain hydrocarbons.
   d. They may contain surface-active agents to enhance the solubility of the therapeutic agent.

2. **With respect to ointments, which of the following are true?**
   a. The bases of hydrocarbon ointments are typically derived from petroleum.

   b.  Hydrophobic ointment bases are unsuitable for exuding
       lesions.
   c.  An ointment may be formed by fusion of silicone oil and
       a wax.
   d.  Drug solubility in an ointment base may be modified by the
       inclusion of a co-solvent such as propylene glycol.

3.  **Concerning pharmaceutical ointments, which of the
    following are true?**
   a.  Ointments may stain patients' clothes.
   b.  Ointments are generally highly viscous formulations.
   c.  Medicated ointments may be used to treat haemorrhoids.
   d.  Drugs that are prone to hydrolysis should not be formulated
       as ointments.

4.  **With respect to ointment formulations, which of the
    following are true?**
   a.  Ointment formulations may be water-miscible.
   b.  Ointments prepared using hydrocarbon bases require the
       inclusion of preservatives.
   c.  Ointments may be formulated as emulsions.
   d.  Ointments may require the addition of antioxidants to
       enhance the stability of the therapeutic agent.

5.  **Concerning ointments, which of the following are true?**
   a.  Ointments are structurally similar to creams.
   b.  An ointment base consists of two components which are
       liquids at room temperature.
   c.  Hydrocarbon ointments are mixtures of $C_{30}$–$C_{50}$ hydrocarbons
       and $C_{16}$–$C_{30}$ hydrocarbons.
   d.  Cosmetically, ointments are unsatisfactory, being greasy and
       difficult to remove.

6.  **With respect to paste formulations, which of the following
    are true?**
   a.  Pastes contain high drug loadings.
   b.  Pastes may not be applied to exuding wounds.
   c.  Pastes are opaque and therefore used as sunblock
       formulations.
   d.  Pastes are cosmetically acceptable formulations.

7.  **With respect to pharmaceutical lotions, which of the
    following are true?**
   a.  Pharmaceutical lotions may be formulated as solutions or
       suspensions, with solutions being preferred.
   b.  Pharmaceutical lotions are typically used for the treatment of
       local conditions.

   c. Pharmaceutical lotions are principally aqueous formulations.

   d. Pharmaceutical lotions do not require the inclusion of preservatives.

8. **With respect to pharmaceutical liniments, which of the following are true?**

   a. Pharmaceutical liniments are formulated as solutions or suspensions.

   b. Pharmaceutical liniments are typically used for the treatment of local conditions.

   c. Pharmaceutical lotions are aqueous formulations.

   d. Pharmaceutical lotions may be rubefacient, thereby increasing drug penetration across the skin.

9. **Which of the following statements are true concerning type 1 pharmaceutical gels?**

   a. The mechanical properties of the gels are primarily due to chemical cross-linking.

   b. Type 1 gels may be formulated to exhibit large swelling ratios in biological fluids.

   c. Type 1 gels are easily dispensed from an ointment tube.

   d. Type 1 gels are plasticised by water.

10. **Which of the following statements are true concerning type 2 pharmaceutical gels?**

   a. Type 2 pharmaceutical gels are pseudoplastic in nature.

   b. Type 2 gels require the inclusion of preservatives.

   c. Gelation in type 2 gels may be performed by increasing the concentration of hydrophilic polymer within the aqueous vehicle.

   d. Type 2 gels may only be formulated using water as the vehicle.

11. **Methylcellulose is a hydrophilic polymer that forms type 2 gels. Which of the following statements are true concerning a pharmaceutical gel containing 2% w/w methylcellulose as a lubricant for urethral catheters?**

   a. It is shear-thinning.

   b. It does not need the inclusion of preservative.

   c. It requires the addition of an antioxidant to prevent degradation of the polymer.

   d. It must contain a colouring agent.

12. **Zinc and Salicylic Acid paste BP is composed of zinc oxide 24% w/w, salicylic acid 2% w/w, starch 24% w/w and white soft paraffin 50% w/w. Which of the following statements are true?**

a. Zinc oxide is included to enhance the physical stability of the formulation.
b. The paste is hydrophobic and will not absorb exudate.
c. The formulation requires the addition of a preservative.
d. It is shear-thinning.

13. **The formulation of Calamine Lotion is provided in Table 4.1.**

**Table 4.1**    Formulation of Calamine Lotion

| Component | Concentration |
|---|---|
| Calamine | 15% w/w |
| Zinc oxide | 5% w/w |
| Glycerol | 5% w/w |
| Bentonite | 3% w/w |
| Sodium citrate | 0.5% w/w |
| Liquefied phenol | 0.5% w/w |
| Freshly boiled and cooled water | ad 100% w/w |

**Which of the following statements concerning Calamine Lotion are true?**
a. The formulation is a solution.
b. Glycerol is included to enhance the solubility of calamine.
c. The formulation requires the addition of a preservative.
d. Bentonite is included to enhance the physical stability of the formulation.

14. **Which of the following quality control methods should be applied to Calamine Lotion following manufacture?**
a. Preservative efficacy.
b. pH.
c. Assay of the concentration of bentonite.
d. Confirmation of colour against product specification.

15. **Salactol is a commercial product that is composed of salicylic acid (16.7% w/w), lactic acid (16.7%) in flexible collodion. Which of the following statements are true concerning Salactol?**
a. The pH of the formulation is acidic.
b. Following application a flexible film is formed at the site of application.
c. The formulation requires the addition of a preservative
d. Lactic acid is included in the formulation as an antioxidant.

16. **Which of the following statements are true concerning poly(acrylic acid) gels?**

a. The viscosity of the gel increases as the pH is elevated from pH 4 to pH 7.

b. The flow properties of poly(acrylic acid) gels are pseudo-plastic.

c. The viscosity of poly(acrylic acid) gels at pH 7 is unaffected by the presence of dissolved basic drugs.

d. The viscosity of poly(acrylic acid) gels at pH 7 is unaffected by the presence of dissolved acidic drugs.

# chapter 5
# Parenteral formulations

## Overview

**In this chapter we will:**

- examine the types of parenteral formulations
- provide an overview of the advantages and disadvantages of parenteral formulations
- describe the formulation considerations for parenteral formulations
- briefly describe manufacturing consideration of parenteral formulations.

## General description

Parenteral administration of drugs involves the injection of therapeutic agents, in the form of solutions, suspensions or emulsions, into the body. In so doing, one of the major barriers to drug entry (the skin) is breeched. Parenteral formulations have been officially recognised since the mid 19th century when morphine solution appeared in the 1874 addendum to the British Pharmacopoeia (1867). Currently many classes of drug are formulated as parenteral dosage forms and, indeed, the control of certain disease states is dependent on parenteral administration, e.g. type 1 diabetes mellitus. Parenteral products are therefore essential components of modern medicine.

## Routes of parenteral administration

There are several different routes by which parenteral products may be administered and indeed there are very few, if any, organs into which parenteral dosage forms may not be injected. However, there are three routes by which parenterals are most frequently administered: (1) intravenous (IV); (2) intramuscular (IM); and (3) subcutaneous (SC). These sites are located beneath the epidermis/dermis within the skin.

## KeyPoints

- Parenteral formulations are extensively used in the treatment and control of numerous disease states, e.g. diabetes, infection, pain.
- Parenteral formulations can be subdivided into three categories, namely solutions, suspensions and emulsions. Furthermore parenteral formulations may be defined as either large-volume or small-volume parenterals.
- Parenteral formulations are principally administered by three routes (intravenous, intramuscular and subcutaneous).
- The volume of injection and the requirements for onset of action primarily define the type of parenteral formulation (solution, suspension, emulsion) required.
- Parenteral formulations are *sterile* and therefore require specialist (and more expensive) manufacturing processes.
- Parenteral formulations are *pyrogen-free*.

A diagrammatic representation of these routes of administration is shown in Figure 5.1.

**Figure 5.1** A diagrammatic representation of the intravenous (IV), intramuscular (IM) and subcutaneous (SC) routes of parenteral administration.

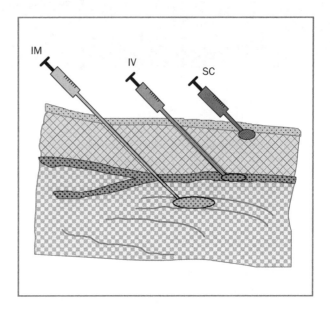

### Intravenous route

- This involves administration of the parenteral formulation into a vein, usually a large proximal vein. The veins are located beneath the subcutaneous tissue, embedded within the muscle.
- IV administration achieves a rapid and predictable response.
- It ensures 100% drug bioavailability.
- Both large- (up to 500 ml) and small- (up to 10 ml) volume formulations may be administered intravenously. Large volumes are infused into the vein at a controlled rate, e.g. total parenteral nutrition, infusion of solutions of electrolytes/nutrients either containing or devoid of drugs.
- Formulations are usually solutions or emulsions (in which the size of the disperse phase is small, <1 μm). Suspensions (or solutions that precipitate within the bloodstream) must not be administered IV due to disruption of blood flow.
- Due to the subsequent dilution of the injected dose and the relative insensitivity of walls of veins, IV administration may be employed for the administration of drugs that would normally be too irritating if administered by other routes.
- Care must be taken regarding the rate of administration of the parenteral formulation. If the administration is performed too quickly, an excessive concentration of drug at the target organ may result, leading to drug-induced shock.

■    Training is required to ensure that the dosage form is actually administered to the vein and that puncture of the vein is avoided.

## Intramuscular route

■    This involves administration into a muscle, usually the gluteal (buttocks), vastus lateralis (lateral thigh) or deltoid (upper arm) muscles. The musculature resides below the subcutaneous tissue (which itself lies beneath the epidermis and the dermis).
■    The volume of injection is small, usually 1–3 ml or up to 10 ml in divided doses.
■    Faulty injection technique may lead to local muscle damage.
■    IM injection results in relatively rapid absorption, second only to IV with respect to the time taken for the onset of action. The nature of the formulation directly affects the rate of absorption of drugs administered by the IM route. The rate of drug absorption from aqueous solutions is greater than from aqueous suspension or non-aqueous (oil-based) solutions of drugs.
■    IM injections are usually used for controlled-release formulations.

## Subcutaneous route

■    This involves administration into the subcutaneous tissue, a layer of fat located below the dermis.
■    There is a slower onset of action and sometimes less total absorption of therapeutic agents when compared to the IV or IM routes of administration. As before, the nature of the formulation directly affects the rate of drug absorption from this site: oily solutions or aqueous suspension of therapeutic agents exhibit slower drug absorption.
■    The volume of injection is typically circa 1 ml; however, large-volume parenteral solutions (electrolyte or dextrose, up to 1000 ml) may be infused subcutaneously. This technique is termed hypodermoclysis and is only employed when there is difficulty in accessing a vein. On some occasions hyaluronidase may be administered to increase the available volume at the site by catalysing the temporary breakdown of the connective tissue at the site (hyaluronic acid).
■    Viscous formulations are not generally administered subcutaneously.
■    Typical sites of SC injection include the arms, legs and abdomen.
■    SC administration is the route of choice for the administration of insulin.

## Miscellaneous routes

In addition to the primary routes of administration cited above, there are other parenteral routes of administration that are not used as frequently as the IM, IV and SC routes but do play an important role in medicine. These include: (1) intradermal (ID); (2) intra-arterial (IA); (3) intrathecal (IT); (4) intradural and extradural; (5) and intracardiac (IC) routes.

### Intradermal route
- This involves injection into the dermal layer of the skin.
- The ID route is generally used for diagnostic purposes, e.g. for the diagnosis of allergy and for the tuberculin test.
- Despite being a vascular site, absorption is slow and limited from this region.
- Only small volumes may be injected, circa 0.1 ml.

### Intra-arterial route
- The parenteral formulation is injected into an accessible artery.
- This route requires specialist training to administer therapeutic agents as if the artery is missed, possible damage to adjacent nerves may result.
- The IA route is used to administer radiopaque media to visualise organs, e.g. heart, kidney.
- It is used to administer anticancer drugs to ensure that the highest possible concentration of drug reaches the target organ.

### Intrathecal (IT) route
- This route is used to administer therapeutic agents to the cerebrospinal fluid to ensure that the appropriate concentration of drug is obtained at this site (e.g. for the treatment of infection).

### Intradural and extradural routes
- Intradural and extradural administration is employed to achieve spinal anaesthesia.
- Intradural administration involves injection of the therapeutic agent within the dural membrane surrounding the spinal cord.
- Extradural administration involves injection of the therapeutic agent outside the dural membrane and within the spinal caudal canals.

### Intracardiac (IC) route
- The intracardiac route involves injection of the formulation directly into the muscles of the heart.

- This route is normally used whenever there is a cardiac emergency.

Other more specialised routes are employed for the delivery of the therapeutic agent directly to the target site/region.

## Advantages and disadvantages of parenteral formulations

### Advantages

- An immediate physiological response may be achieved (usually by the IV route). This is important in acute medical situations, e.g. cardiac arrest, anaphylactic shock, asthma, hyperglycaemia, hypoglycaemia.
- Parenteral formulations are essential for drugs that offer poor bioavailability or those that are rapidly degraded within the gastrointestinal tract (e.g. insulin and other peptides).
- They offer a method to administer drugs to patients who are unconscious or uncooperative or for patients with nausea and vomiting (and additionally dysphagia).
- As trained medical staff primarily administer parenteral formulations, there is control of both the dosage and frequency of administration. The main exception to this is the administration of insulin, which, in the absence of complications (e.g. ketoacidosis), is performed exclusively by patients.
- Local effects may be achieved using parenteral formulations, e.g. local anaesthesia.
- Parenteral formulations provide a means by which serious imbalances in electrolytes may be corrected (using infusion solutions).
- Parenteral formulations may be readily formulated to offer a wide range of drug release profiles, including:
  - rapidly acting formulations (generally drug solutions that are administered IV)
  - long-acting formulations (generally drug suspensions, or solutions in which the drug is precipitated out of solution at the injection site, administered by the IM or SC routes). Examples of these include intermediate/long-acting insulin formulations and steroid injections.
- In patients who cannot consume food, total parenteral nutrition offers a means by which nutrition may be provided using specially formulated solutions that are infused into the patient.

> **Tip**
>
> The route of administration of the parenteral product directly affects the pharmacokinetic profile of the dosage form. Typically the intravenous route is employed for rapid onset of action whereas controlled-release formulations are administered by the intramuscular or subcutaneous routes.

### Disadvantages

- The manufacturing process is more complicated than for other formulations due to the requirement for aseptic technique. The level of training of staff involved in the manufacture of parenteral formulations is high and often specialist equipment is required to ensure that the finished product specification is achieved.
- Skill of administration is required to ensure that the dosage form is administered by the correct route. If a parenteral suspension, which is designed for administration by the IM or SC route, is incorrectly administered by the IV route, a pulmonary microcapillary blockage may occur leading to a blockage in the flow of blood at that site.
- Parenteral formulations are associated with pain on administration.
- If the patient is allergic to the formulation (the therapeutic agents and/or the excipients), parenteral administration will result in both rapid and intense allergic reactions.
- It is difficult to reverse the effects of drugs that have been administered parenterally, even immediately after administration. This is not strictly the case with other routes of administration, e.g. oral, transdermal.

## Formulation considerations for parenteral formulations

Parenteral formulations may be categorised as solutions (aqueous or oil-based), suspensions (aqueous or oil-based) or emulsions. For the most part, these formulations use similar (or indeed identical) excipients but, as may be expected, there are certain excipients that are unique to each category. Details of the key formulation excipients for each formulation type are provided below. In addition, the reader will observe a similarity in the excipients used for the formulation of solutions, suspensions and emulsions for parenteral administration and those used for non-parenteral use. The main formulation considerations for parenteral formulations are described below.

### Type of preparation

The initial choice that must be made concerns the type of parenteral formulation required, a choice that is informed by the physicochemical properties of the therapeutic agent, the intended route of administration of the formulation, the volume to be administered and the general preference for a particular formulation based on the perceived pharmacological effect or onset of pharmacological effect. In some cases, the formulation scientist may be asked to formulate a specific preparation, e.g. an aqueous suspension for IM administration. However, in the absence of this information, several issues, detailed below, must be addressed.

## Solubility of the therapeutic agent

With respect to the formulation of pharmaceutical products, therapeutic agents may be categorised into three groups:

1.  Good solubility, in which the therapeutic agent is freely soluble in the chosen solvent (either aqueous or oil-based) at the concentration required in the parenteral product. In this case a parenteral solution is a possible formulation option.
2.  Moderate solubility but insufficient to produce a solution in conventional solvents (e.g. water, oil). In this scenario, the use of co-solvents may sufficiently increase the solubility of the therapeutic agent in the vehicle to produce a parenteral solution containing the required drug concentration. This is generally the preferred strategy for therapeutic agents of moderate solubility. However, if required, the therapeutic agent may be formulated as a suspension, although one cautionary note regarding this approach is the potential recrystallisation of soluble drug during storage, a phenomenon that may affect the physical stability of the preparation.
3.  Low solubility in the chosen vehicle. The preferred choice for therapeutic agents that exhibit low solubility in the chosen vehicle is a parenteral suspension formulation.

## Preferred route of administration

If there is a preferred route of administration for the parenteral product, this will directly influence the nature of the parenteral product. In particular:

- IV products must be *aqueous solutions* and, furthermore, must not precipitate in the bloodstream following administration. Emulsions may also be administered by this route provided the particle size of the internal phase is sufficiently small.
- Parenteral suspensions (aqueous or oil-based) and oil-based parenteral solutions must be administered either subcutaneously or intramuscularly. Aqueous solutions may also be administered intramuscularly or subcutaneously.
- There are other restrictions for those less commonly used routes of parenteral administration that are specific for each route.

## Volume of dose to be administered

The volume of product to be parenterally administered will directly affect the type of formulated product.

- Large-volume parenterals (up to 500 ml) are administered intravenously (although the SC route of administration is infrequently used for this purpose).
- Small-volume parenterals may be administered by all routes (bearing in mind the restrictions of oil-based and suspension formulations).

## Onset/duration of action

A wide range of predictable plasma drug concentration–time profiles may be obtained with parenteral formulations that are dependent on both the type of formulation and route of administration. Indeed, this is one of the major rationales for the use of parenteral administration of therapeutic agents. Whilst the pharmacokinetics of parenteral dosage forms is not under direct consideration in this book, it is important to summarise the effects of both formulation type and route of administration on such properties.

- Formulations administered intravenously will have an immediate pharmacological effect. The rates of drug absorption from the other main routes of administration (SC and IM) are slower.
- The absorption of therapeutic agents from aqueous solutions when administered by the IM or SC routes is faster than from oil-based solutions, oil-based suspensions and aqueous suspensions. As a result, the slower absorption of therapeutic agents from oil-based solutions/suspensions and aqueous suspensions enables these formulations to offer a prolonged clinical effect whenever administered by the IM or SC routes. This may be illustrated by the following examples:
  - When injected subcutaneously, the onset of action of soluble insulin (aqueous solution) is rapid (circa 30 minutes), peaks between 2 and 4 hours and has a duration of action of up to 8 hours. Conversely, intermediate/long-acting insulins (aqueous suspensions), when administered subcutaneously, have an onset of action of 1–2 hours, a peak action between 4 and 12 hours and a duration between 16 and 35 hours.
  - Triamcinolone acetonide is administered intramuscularly or intra-articularly as an aqueous suspension to suppress inflammation. The duration of action of a single dose is approximately 21 days.

## Physicochemical properties of the therapeutic agent

The physicochemical properties are important determinants of the stability and absorption of the therapeutic agents when formulated as parenteral suspensions (aqueous or oil-based). Conversely, when formulated as a parenteral solution the effects of the physicochemical properties of the therapeutic agent on the above properties are limited. In particular, the following properties directly affect the rate of dissolution (and hence the rate of absorption) of poorly soluble therapeutic agents following IM or SC administration.

## Solid-state properties

In the solid state therapeutic agents may exist in either crystalline or amorphous states. In *crystalline* compounds the molecules are packed (bonded) in a defined, repeating order. Crystalline compounds exhibit a defined melting point, which occurs whenever there is sufficient energy applied to the crystal to overcome the attractive forces between the molecules in the lattice. As the strength of the interactive forces increases, so does the melting point of the crystal. *Polymorphism* refers to the ability of molecules to exist in more than one crystalline form. Usually one crystalline form is the stable form and the other(s) are less stable, being referred to as metastable. Over time the metastable form(s) will revert to the stable form. Importantly, different polymorphic forms of a particular therapeutic agent will exhibit different melting points and, as a result, will exhibit different dissolution rates. Polymorphic forms of a particular therapeutic agent will possess the same chemical structure; however, the different solubilities of the different polymorphs will lead to different rates of dissolution after IM or SC administration and hence different rates of drug absorption.

## Solubilities of insoluble salt forms

The reader will be aware that different salt forms of a particular therapeutic agent exhibit different aqueous solubilities. The rate of dissolution of poorly soluble therapeutic agents is directly proportional to the saturated solubility of the compound, as defined in the Noyes–Whitney equation:

$$\frac{dM}{dt} = \frac{DAC_s}{h}$$

where: $\dfrac{dM}{dt}$ refers to the rate of dissolution of the therapeutic agent; $h$ refers to the thickness of the unstirred diffusion layer that surrounds each particle that is undergoing dissolution;

$D$ is the diffusion coefficient of the dissolved drug molecule through the unstirred diffusion layer; $A$ is the surface area of the particle undergoing dissolution; $C_s$ is the saturated solubility of the drug, i.e. the concentration that exists in solution adjacent to the dissolving particle.

Therefore, by altering the solubility of the salt form, the rate of dissolution of the drug particles following administration (IM or SC) may be modified. This approach has been successfully used commercially in the formulation of intermediate and long-acting insulin preparations. For example, the solubility of protamine insulin is dramatically lower than soluble insulin and this is reflected by the greater duration of action of the former system. The duration of action of insulin may be further enhanced by forming a salt with zinc or with protamine and zinc, which has lower solubility and hence a lower rate of dissolution.

### Particle size

Particle size is a fundamental property that directly controls both the rate of dissolution and physical stability of parenteral suspensions. Referring to the Noyes–Whitney equation, it may be observed that the rate of dissolution of a poorly soluble drug increases as the surface area of the particle increases. In practice the surface area of the drug particle increases as the mean diameter of the particles decreases. Therefore, decreasing the particle size may increase the rate of dissolution of an insoluble therapeutic agent.

The role of particle size on the absorption of an insoluble drug may be illustrated in the following example, which relates to an aqueous testosterone suspension.

- A suspension of testosterone propionate (particle size range 40–100 μm) exhibits a duration of action of *8* days following IM administration.
- A suspension of testosterone propionate (particle size range 50–200 μm) exhibits a duration of action of *12* days following IM administration. Therefore, in this formulation the area of drug particle in contact with the biological fluids is less than in the previous example and, in accordance with the Noyes–Whitney equation, the longer duration may be explained by the slower rate of drug dissolution.
- A suspension of testosterone isobutyrate (particle size range 50–200 μm) exhibited a duration of action of *20* days following IM administration. The greater lipophilicity, and hence lower aqueous saturated solubility, of this testosterone ester would result in a slower rate of dissolution than for the more hydrophilic propionate ester of this drug.

In addition to the effect of particle size on the solubility and hence rate of dissolution of poorly soluble drugs, particle size plays an important role in the physical stability of parenteral suspensions. The reader will recall the Stokes' equation, in which the rate of particle sedimentation is related to particle size as follows:

$$\frac{dv}{dt} = \frac{2r^2 (\rho_s - \rho_l)g}{9\eta_l}$$

where: $\dfrac{dv}{dt}$ refers to the rate of particle sedimentation; $r$ refers to the radius of the dispersed particles; $(\rho_s - \rho_l)$ refers to the density difference between the solid phase and the liquid phase; $g$ refers to gravity; and $\eta_l$ refers to the viscosity of the liquid phase.

It is accepted that reducing the rate of sedimentation of the dispersed drug particles will enhance the physical stability of suspensions. One method by which this may be achieved is to reduce the particle size of the drug particles (Chapter 11). Increasing the particle size of the dispersed drug will both increase the rate of sedimentation (and possibly decrease the physical stability of the formulation) and decrease the rate of dissolution of the drug, the latter leading to a slower onset of activity but a prolonged duration of action following IM administration. The interplay between particle size and both the rate of dissolution (and hence absorption) and the physical stability of suspensions should be fully appreciated by students.

## Vehicle

All parenteral formulations may be formulated using an aqueous vehicle, an oil vehicle or a hydroalcoholic vehicle, the choice being determined (in part) by the required solubility of the active agent in the formulation and the desired type of formulation.

> **Tip**
>
> The nature of the formulation directly affects the onset and duration of action of parenteral products. When administered by the same route, the onset of action of parenteral suspensions is slower than for solutions but the duration of action is markedly greater.

### Aqueous vehicles

Water for injection is the major vehicle of choice for:

- freely soluble therapeutic agents (for the preparation of parenteral solutions)
- therapeutic agents of low aqueous solubility (for the preparation of parenteral suspensions)
- the external phase of parenteral emulsions.

Water for injection has specifications set regarding:

- appearance (clear, odourless and within a defined pH range, 5–7)
- purity (limits on the mass of ions, heavy metals and oxidisable compounds and also on the total amount of dissolved solids, <10 ppm)
- sterility:
  - Water for Injections USP is non-sterile and is used in the preparation of parenterals that will be terminally sterilised (i.e. during or after the manufacturing process).
  - Sterile Water for Injections USP is available. This is water for injections that has been sterilised and which has been packaged in single units (1 litre in volume). It may contain a greater mass of dissolved solids due to leaching of solid matter from the containers during sterilisation. It is intended to be used as a vehicle for products that have been packaged and sterilised, e.g. for the reconstitution of antibiotic powders as solutions or suspensions.
- pyrogens: Water for injection must be free of pyrogens (fever-producing compounds) that are primarily associated with Gram-negative bacteria. It is important to have knowledge of the physicochemical properties of pyrogens as these properties directly influence the choice of methods that may be used to ensure removal of these compounds. In particular pyrogens are:
  - thermostable, thereby invalidating their removal using simple heating cycles
  - water-soluble, thereby invalidating their removal using conventional filtration techniques
  - unaffected by bactericides.

In light of the above, pyrogens are effectively removed from water using either distillation or reverse osmosis. Following treatment, water for injection must be stored in *pyrogen-free containers* at a defined temperature (either 5°C or 60–90°C) if the period of storage exceeds 24 hours. Removal of pyrogens from the storage containers is typically performed by heating the container at either 250°C for 30–45 minutes or at 180°C for 3–4 hours.

As a variation on the above, *Bacteriostatic Water for Injections USP* is also available. Similar to Water for Injections USP, bacteriostatic water for injection is sterile and devoid of pyrogens. It additionally contains an antimicrobial agent, e.g. 0.9% w/v benzyl alcohol (a bacteriostatic preservative) and is commonly supplied in a multidose container (≤30 ml). Samples from this container may be repeatedly removed and used to dissolve or dilute therapeutic agents prior to injection. To prevent potential toxicity, Bacteriostatic Water for Injections USP is only used whenever the volume of formulation to be administered is less than 5 ml. Care must be given in the use of Bacteriostatic Water for Injections USP to ensure that the antimicrobial agent does not deleteriously interact with the therapeutic agent.

### Non-aqueous vehicles

- Non-aqueous vehicles are employed for the production of:
  - non-aqueous parenteral solutions of therapeutic agents that are water-insoluble
  - non-aqueous parenteral suspensions of therapeutic agents that are water-soluble and/or exhibit aqueous instability
  - the internal phase of parenteral emulsions.
- Fixed oils are predominantly used as non-aqueous vehicles (e.g. corn oil, cottonseed oil, peanut oil, sesame oil); however, non-aqueous esters may be used, e.g. ethyl ethanoate:
  - Sesame oil is generally the oil of choice as it is more stable.
  - Oils must be free from rancidity and must not contain mineral oils or solid paraffins.
- Two major problems associated with the use of non-aqueous pharmaceutical solutions are:
  - Pain/irritation on injection. It should be noted that the viscosity of fixed oils increases at lower storage temperatures. This will, in turn, affect the ease of administration by injection and the pain/irritation at the site of injection. It is essential to ensure that the viscosities of oil-based solutions and suspensions are minimised both to reduce pain on injection and to enhance the ease of administration (injection).
  - Patients may exhibit sensitivity to the oils and therefore the oil used in the formulation must be explicitly stated on the label/patient information.
- In some oily formulations, agents may be added to enhance the solubility of the therapeutic agent in the oil vehicle. Benzyl benzoate (itself a non-aqueous vehicle) may be used for this purpose.

## Inclusion of co-solvents

- As highlighted in Chapter 1, co-solvents are employed whenever the solubility of the drug in water (or occasionally in oil-based vehicles) alone is insufficient for the required application. The types and choices of co-solvent that may be employed in parenteral formulations are generally similar to those used to formulate pharmaceutical solutions; however, when used in parenteral formulations, the potentially greater toxicity of these agents when administered parenterally should be carefully considered. Furthermore, the toxicity of co-solvents is dependent on the route of administration; toxicity is greater whenever administered by the IV in comparison to the IM and SC routes.

- Examples of co-solvents used in parenteral formulations include:
  - glycerol
  - ethanol (high concentrations of ethanol are known to produce pain on injection)
  - propylene glycol
  - polyethylene glycol 400.
- In veterinary formulations other co-solvents may be used, including 2-pyrrolidone and dimethylacetamide; however, these co-solvents are not registered for use in parenteral formulations for humans.
- The concentration of co-solvent used should be sufficient to render the drug soluble within the formulation (over the shelf-life of the product) but should not be irritant or toxic to the patient.

## Surface-active agents

The use of surface-active agents in solutions and suspensions designed for oral administration has been addressed in Chapters 1 and 2. In general the incorporation of surface-active agents within parenteral formulations is conceptually identical. In this section some of the basic concepts are revisited within the context of the specific use of these agents in parenteral formulations. In particular:

- Surface-active agents may be incorporated into parenteral solutions to enhance the solubility of the therapeutic agent to the required concentration. In this scenario, the concentration of surface-active agent employed will exceed the critical micelle concentration (CMC) of the surface-active agent. Surface-active agents may be incorporated into aqueous or non-aqueous (oil-based) vehicles for this purpose.
- When included in parenteral suspension formulations, surface-active agents act to enhance the physical stability of the formulation by adsorbing to the surface of the dispersed therapeutic agent and preventing caking of the solid particles. For this purpose concentrations of surface-active agents that are below the CMC may be used. It is extremely common for non-ionic surface-active agents to be used in this manner. Examples include:

- polyoxyethylene sorbitan fatty acid esters (Tween series), within the concentration range 0.1–0.5% w/v
- poly(oxyethylene)-poly(oxypropylene) block co-polymers (Poloxamers), within the concentration range 0.01–5% w/v
- lecithin, within the concentration range 0.5–2.0% w/w.

■ The choices of surface-active agent and the concentration to be used are dependent on the nature of the vehicle and the type of parenteral formulation (i.e. solution or suspension). Accordingly surfactants with low and high hydrophile–lipophile balance values will be used to stabilise oil-based and aqueous drug suspensions, respectively. Similarly, these surfactants are used to solubilise drugs in oil-based and aqueous vehicles, respectively; in this scenario higher concentrations (greater than the critical micelle concentration) are required.

■ The use of surface-active agents to solubilise therapeutic agents is commercially employed. Examples of these formulated systems include:

- Steroids that have been solubilised for parenteral use using combinations of non-ionic surfactants, e.g. polyoxyethylene sorbitan fatty acid esters (Tween series) and sorbitan esters (Span series).
- The poorly water-soluble vitamins (A, D, E and K) may be solubilised using surface-active agents as parenteral solution formulations. For example, phytomenandione is formulated as a colloidal solution in a mixed-micelles vehicle and is designed for administration either by slow IV injection or by IV infusion (after incorporation within a 5% glucose solution).
- The poorly water-soluble antifungal agent amphotericin B is commercially available as a complex with sodium deoxycholate (Fungizone), L-α-dimyristoylphosphatidylcholine and L-α-dimyristoylphosphatidylglycerol (Abelcet), sodium cholesteryl sulphate (Amphocil). These powders are constituted with water for injection to produce colloidal solutions prior to use as IV infusions.

## Buffers

As highlighted in Chapter 1, buffers are commonly included in parenteral formulations to control the pH of the formulation. This is similarly the case for parenteral formulations where control of the pH of the formulation may:

■ Maintain the solubility of the drug in the vehicle over the shelf-life of the preparation. Importantly, if there is

# Tips

- The preferred vehicle of choice for parenteral formulations is water.
- In selecting a solubilisation strategy, consideration must be given to the toxicity of the solubilisation agent(s), e.g. co-solvent, surfactants, in the host (human or animal).

precipitation of the therapeutic agent within a parenteral solution during storage, the preparation can no longer be referred to as a solution and thus the shelf-life of the product has been reached. Furthermore, the IV administration of a parenteral solution in which there is precipitated drug may lead to a blockage within the capillaries, with the associated deleterious effects on the organ to which the blood would normally be transported.

- Enhance the chemical stability of the therapeutic agent by maintaining the pH of the formulation within the range of optimum chemical stability of the therapeutic agent.

Examples of commonly used buffers include acetic acid/sodium acetate, citric acid/sodium citrate and sodium phosphate/disodium phosphate (see Chapter 1).

## Polymers to modify formulation viscosity and/or drug solubility

The inclusion of polymers within parenteral suspensions occurs more frequently than in parenteral solutions. In parenteral solutions the inclusion of hydrophilic polymers will increase the viscosity of the formulations, which may in turn result in difficulties in administration. It must be remembered that during administration the formulation must flow through the narrow bore of the injection needle. Under these circumstances small changes in formulation viscosity will be amplified during the passage through the needle. In addition increased formulation viscosity may result in pain at the injection site. Whilst hydrophilic polymers may be added to solutions designed for oral administration to increase the viscosity, e.g. to aid the accurate measurement of a dose on the measuring spoon/cup, this requirement is unnecessary for parenteral solution formulations. Therefore, the inclusion of hydrophilic polymers in parenteral solution formulations is restricted to aqueous solutions to enhance the solubility of the therapeutic agent by complexation. Poly(vinylpyrrolidone) (PVP) is an example of a polymer that may be used for this purpose; it is present in aqueous tetracycline and aqueous oxytetracycline injections for veterinary applications. Importantly, the molecular weight of PVP that is employed in the formulation of parenteral solutions is low (circa 12 000) and this, in conjunction with the linear nature of this polymer in aqueous solutions, ensures that large concentrations (up to 18% w/v) may be used without the viscosity of the resultant formulation being excessive for clinical use.

Lipophilic polymers are rarely, if ever, used to solubilise therapeutic agent in oil-based vehicles for parenteral administration. As stated previously, if required the solubility of drugs in the oil-based vehicle may be enhanced by the incorporation of surfactants (e.g. sorbitan esters, Span) or co-solvents, e.g. benzyl benzoate.

When used in aqueous suspensions, hydrophilic polymers maintain the physical stability of the formulation by a number of mechanisms.

### Stearic stabilisation

In this the polymer chains adsorb on to the surface of the dispersed drug particles and, in so doing, the close approach of two particles is stearically inhibited. In terms of the DLVO theory (see Chapter 2), the presence of the adsorbed hydrophilic polymer is sufficient to prevent the two particles interacting at the primary minimum. In addition, if the polymer is a polyelectrolyte (e.g. sodium alginate), the polymer chains may effectively form a bridge between two particles when in the presence of an oppositely charged divalent or trivalent ion. The ion provides an effective charge on the surface of the particles, with which the charged polymer chain interacts. In so doing the two particles are maintained/stabilised at distances greater than the primary minimum.

### Enhancement of the viscosity of the formulation

As previously defined, the physical stability of suspensions is dependent, at least in part, on the rate of sedimentation of the suspended particles (Stokes' equation). In addition the rate of sedimentation is inversely related to the viscosity of the formulation. The presence of hydrophilic polymers in parenteral suspensions will increase the viscosity of the formulation and will therefore act to stabilise the formulation. However, it must be remembered that as the viscosity of the formulation is increased, the ease of administration decreases and likelihood of pain upon injection increases. Therefore, the concentration of hydrophilic polymer that may be used for this purpose (and hence the viscosity achieved) is limited by these potential clinical limitations.

The physical stabilisation of aqueous parenteral drug suspensions is usually achieved by the incorporation of surface-active agents and/or hydrophilic polymers. It should be noted that the range of surfactants and hydrophilic polymers that may be used in this fashion is lower than for comparator preparations designed for oral administration. This is primarily due to the potentially greater toxicity of excipients following parenteral administration.

The stabilisation of oil-based suspensions for parenteral administration is generally not performed using (lipophilic) polymers due to their limited availability. Instead the physical stability of oil-based systems may be effectively enhanced using salts of fatty acids or fatty acid esters, which primarily increase the viscosity of the formulation. Examples of these include:

■ Aluminium salts of stearic acid (e.g. aluminium stearate, aluminium distearate, aluminium tristearate). These are normally prepared by dissolving the required concentration of aluminium salt (up to 5%, depending on the salt type) into the oil vehicle at high temperatures (circa 165°C). Upon cooling the drug may be dispersed into the rheologically structured vehicle.

■ Trihydroxystearin (Thixcin-R). This may be dissolved in the oil-based vehicle without heating and, in a similar fashion to the example above, produces a rheologically structured vehicle into which the therapeutic agent may be dispersed.

### Preservatives

Preservatives are incorporated into parenteral formulations whenever:

■ The product is a multidose preparation. In this, several separate doses will be removed from the same container; the inclusion of preservatives is necessary to control microbial growth due to microbial introduction into the product.

■ The product has not been terminally sterilised, e.g. by irradiation or heat. In this situation the preservative is required to guard against any possible breakdown in the aseptic manufacturing process.

In all other situations the presence of a preservative is deemed unnecessary.

Examples of preservatives employed in parenteral formulations include:

■ Esters of parahydroxybenzoic acid, e.g. methyl and propyl parahydroxybenzoic acid are often used in combination in a ratio of 9:1. The overall concentration is usually circa 0.2% w/v.

■ Phenolic compounds, e.g. phenol (0.25–0.5% w/v) or chlorocresol (0.1–0.3% w/v).

Formulation considerations for the inclusion of preservatives into parenteral formulations include the following:

- In aqueous parenteral suspensions and in some aqueous parenteral solutions hydrophilic surfactants (included to enhance/maintain the solubility of the therapeutic agent or to ensure the physical stability of the formulation over the proposed shelf-life) may interact with esters of parahydroxybenzoate. In so doing the effective (free, unbound) concentration of preservative and hence the preservative efficacy are decreased. This problem is resolved by increasing the concentration of preservative (generally up to 0.25% w/v).
- Preservatives may similarly interact with the container and closure of the parenteral product, necessitating an increase in the concentration of preservative required or, preferably, a change in the type of container closure. For example, phenol has been shown to interact with rubber closures. In this situation, rather than increasing the concentration of phenol added, the rubber closure may be exchanged with a suitable replacement, e.g. nitrile closures.
- It is essential that the preservative does not adversely affect the chemical and physical stability of the parenteral product. For example, as insulin formulations are usually multidose preparations, preservatives are required to inhibit microbial contamination of the product. The physical stability of zinc insulin is compromised in the presence of phenol (but not methyl parahydroxybenzoic acid).
- In the preservation of parenteral emulsions, the formulation scientist must be aware of the ability of the preservative to distribute between the inner oil phase and the outer aqueous phase. The preservative is required in the aqueous phase of the emulsion to exert the antimicrobial effect. Distribution between the two phases will therefore decrease the concentration of preservative in the aqueous phase and, accordingly, reduce the preservative efficacy. To overcome this, the concentration of preservative in the dosage form should be increased. This may be easily calculated by consideration of the partition coefficient of the preservative between the oil and aqueous phases and the solubility of the preservative in the two phases (Chapter 3).
- Oil-based parenteral products (solutions and suspensions) do not generally require the inclusion of a preservative due to the low water activity of this medium.
- The potential toxicity of preservatives must be considered when formulating parenteral products in light of the greater potential toxicity of preservatives when administered parenterally. It is therefore desirable to avoid the inclusion of preservatives whenever possible.

## Agents to modify the osmolarity of parenteral products

Osmolarity refers to the mass of solute that, when dissolved in 1 litre of solution, will produce an osmotic pressure equivalent to that produced by a one-molar (1 mol) solution of ideal unionised substance. The units for osmolarity that are used in conjunction with parenteral preparations are mosmol/kg. An *isotonic* solution is one that exhibits the same effective osmotic pressure as blood serum, whereas hypotonic and hypertonic solutions refer to solutions in which the osmotic pressure exerted by the solution is less than and greater than blood serum, respectively. The tonicity of parenteral formulations is an important design criterion. In the presence of a hypotonic solution, red blood cells will swell (due to water entry into the cell) and eventually burst (termed haemolysis), whereas in the presence of a hypertonic solution, water will leave the red blood cells, leading to crenation.

Ideally the administration of parenterals and in particular the IV administration of parenteral should be isotonic (circa 291 mosmol/l) to avoid potential damage. However, many parenteral products designed for IV administration are hypertonic. Products within the osmolarity range 300–500 mosmol/l may be administered by the IV route rapidly (with care being taken with formulations at the higher extremes of this range to ensure that administration is slower). IV fluids may be profoundly hypertonic, e.g. the tonicity of Hyperamine 30 and Sodium Bicarbonate Intravenous Infusion BP (4.2% and 8.4% w/v) is 1450, 1000 and 2000 mosmol/l, respectively. Clinically this does not present any problems as these fluids will be administered to patients by a central venous line in which the infusion rate is slow and the infusion is rapidly diluted within the bloodstream.

It is recommended that parenteral formulations designed for IV administration should not be hypotonic. Therefore, hypotonic solutions should be rendered isotonic by the addition of compounds that will increase the osmotic pressure of the solution. Typically sodium chloride or dextrose is used for this purpose. There are two methods by which the mass of these compounds required to render the solution isotonic may be calculated: (1) consideration of the gram-molecular concentration; and (2) consideration of the freezing-point depression of the solution. These are individually addressed below.

### Gram-molecular concentration

- The gram-molecular concentration refers to the number of moles of substance in 100 grams of solvent.
- For example if 1 gram molecule (mole) of a non-ionic compound is dissolved in 100 grams of water, the gram-molecular concentration is 1%.

- As osmotic pressure is a colligative property, the number of moles of substance in solution is important. Therefore if 1 mol of NaCl is dissolved in water, 2 mol of ions are produced.
- A solution is isotonic whenever the gram-molecular concentration is 0.03%.
- The following three examples show how this may be used to produce isotonic solutions.

## Worked examples

### Example 5.1

Calculate the concentration (% w/v) of dextrose (molecular weight 180 g/mol) that should be added to water to produce an isotonic solution.

Dextrose is non-ionic and therefore 1 mol of dextrose when added to 100 grams of water will produce a gram-molecular concentration of 1%. Therefore:

$$0.03 \times 180 = 5.4\% \text{ w/v}$$

### Example 5.2

Calculate the concentration (% w/v) of sodium chloride (molecular weight 58.5 g/mol) that should be added to water to produce an isotonic solution.

Sodium chloride is ionic and therefore 1 mol of NaCl, when added to 100 grams of water, will produce a gram-molecular concentration of 2%. Therefore:

$$\left( \frac{0.03 \times 58.5}{2} \right) = 0.9\% \text{ w/v}$$

### Example 5.3

Calculate the concentration of sodium chloride that must be added to a 1% solution of lidocaine hydrochloride (molecular weight 270 g/mol) to render this isotonic.

Initially the gram-molecular concentration of the drug solution must be calculated (remembering that 1 mol of lidocaine hydrochloride dissociates to produce 2 mol of ions). The above equations must be rearranged in terms of the gram-molecular concentration. Thus:

$$\left( \frac{2 \times 1}{270} \right) = 0.007\% \text{ w/v}$$

As this is less than 0.03%, the reader will observe that a simple solution of lidocaine hydrochloride (1% w/v) is hypotonic.

The gram-molecular percentage that must be added to correct this imbalance is therefore:

$0.03 - 0.007 = 0.023\%$ w/v

Therefore, the concentration of sodium chloride required to render this solution isotonic is:

$$\left(\frac{0.023 \times 58.5}{2}\right) = 0.67\% \text{ w/v}$$

### *Freezing-point depression*

- The inclusion of ions in a solvent will lower the freezing point of that solvent: the extent of the depression of the freezing point is dependent on the number of ions in solution. This is a basic colligative property.
- An isotonic solution exhibits a freezing-point depression of 0.52°C.
- Therefore the solution of drug should be adjusted to produce a freezing-point depression of 0.52°C to render the solution isotonic.
- Tables are available which provide the freezing-point depressions for various compounds.
- The following are two examples of the use of this technique.

## Example 5.4

Calculate the concentration (% w/v) of sodium chloride that should be added to water to produce an isotonic solution. The freezing-point depression of a 1% solution of NaCl is 0.576°C (derived from tables).

Therefore, the concentration of sodium chloride required to render a solution isotonic is:

$$\left(\frac{0.52}{0.576}\right) \times 1 = 0.9\% \text{ w/v}$$

## Example 5.5

Calculate the concentration of sodium chloride that must be added to a 1% solution of lidocaine hydrochloride (freezing-point depression of a 1% solution is 0.130°C) to render this isotonic.

Initially the freezing-point depression for the 1% solution of drug is calculated:

$1 \times 0.130 = 0.130°C$

Therefore the difference in freezing-point depressions of an isotonic solution and the described drug solution is:

$$0.52 - 0.13 = 0.39\,°C$$

The freezing-point depression of a 1% w/v solution of sodium chloride is 0.576°C. This allows the calculation of the required mass of sodium chloride to be added:

$$\frac{0.39}{0.576} = 0.677g$$

## Antioxidants

As detailed in Chapter 1, many drugs are susceptible to degradation by oxidation, a process involving the addition of an electronegative atom or radical or the removal of an electropositive atom, radical or electron. Oxidation may occur due to the action of molecular oxygen; however, this is a slow process, especially in aqueous solution in which the concentration of dissolved oxygen is low. Alternatively oxidation may be facilitated by free radicals, with breakdown occurring via a chain reaction process. Radicals are formed due to the action of light, heat or transition metals (e.g. iron, copper) that are present in the formulation. Several important classes of therapeutic agents may undergo oxidative degradation, including phenothiazines, polyene antimicrobial agents, steroids, morphine and tetracyclines.

Antioxidants are included in parenteral formulations to slow down or inhibit oxidative degradation of therapeutic agents. These agents either act to prevent the formation of free radicals (e.g. butylated hydroxyanisole, butylated hydroxytoluene) or alternatively are strong reducing agents and are therefore oxidised in preference to the therapeutic agent (e.g. sodium metabisulphite, sodium formaldehyde sulphoxylate). Furthermore, chelating agents, e.g. ethylenediamine diacetic acid, may be added to extract dissolved transition metals, thereby reducing their ability to generate free radicals or to be involved in electron transfer reactions.

A further strategy that may be used to enhance the stability of the therapeutic agent is to flush the injection container/vial with nitrogen prior to closure. In so doing oxygen is removed from the headspace within the packaged product. Although successful, the limitation of this approach is the availability of specialised filling equipment that will provide satisfactory gas purging prior to closure of the product.

Diagrammatic representations of the formulation considerations for aqueous and non-aqueous parenterals are shown in Figures 5.2 and 5.3.

**Figure 5.2** Diagrammatic representation of formulation considerations for aqueous parenterals.

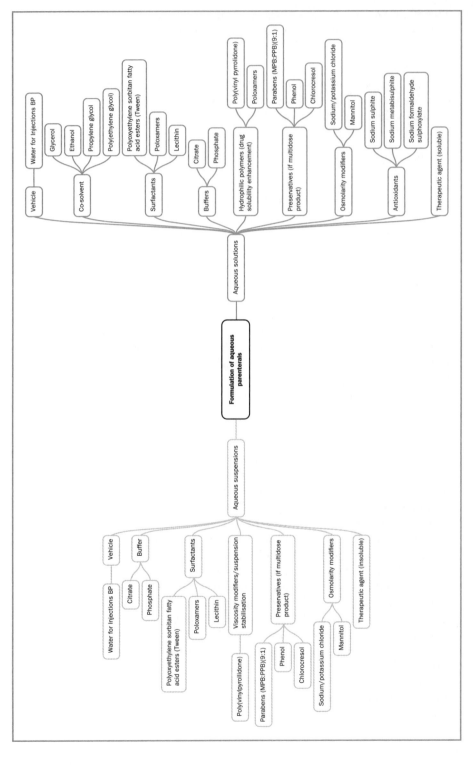

**Figure 5.3** Diagrammatic representation of formulation considerations for non-aqueous parenterals.

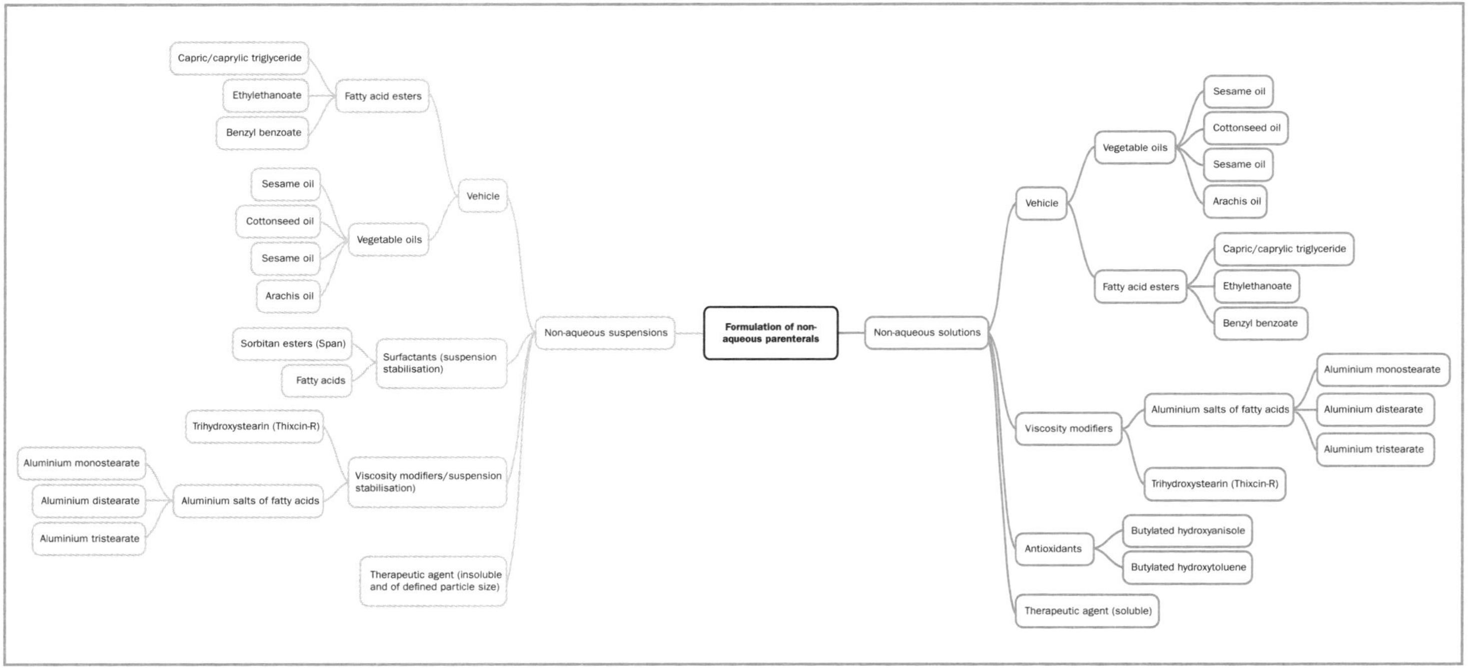

# Tips

- The control of the osmotic properties of parenteral formulations for human use is an important formulation consideration, which should not be overlooked by the formulation scientist.
- Parenteral formulations are sterile and therefore preservatives are only usually required whenever the preparation is a multidose formulation, e.g. veterinary parenterals.
- Possible interactions between the preservative and viscosity-modifying agent must be considered in the selection of the type and concentration of preservative for parenteral formulations.

## Parenteral emulsions

The reader will be aware that the primary focus of this chapter has been the formulation of parenteral solutions and suspensions, due to the overwhelming majority of parenteral formulations being formulated as these dosage form types. However, under certain circumstances, emulsions are employed as parenteral formulations. As the reader will be aware, emulsions are disperse systems in which one immiscible liquid is dispersed in another liquid. Whilst two forms of emulsion exist, water in oil (w/o, in which water droplets are dispersed in an oil phase) and oil in water (o/w, in which oil droplets are dispersed in an aqueous phase), it is the latter type (o/w) that is usually administered parenterally. Examples of the parenteral use (past and present) of emulsions include:

- The subcutaneous administration of allergenic compounds in a w/o emulsion was performed by Freund & McDermott (1942) to enhance the resultant antibody response.
- The IM administration of o/w emulsions to provide controlled drug release.
- The IV administration of o/w emulsions as total parenteral nutrition emulsions. In these, 10–20% oil is emulsified within an aqueous phase using phospholipids and lecithin surfactants to stabilise the emulsion. The emulsion is rendered isotonic by the addition of glycerol and glucose. The oils are subsequently broken down to triglycerides that provide essential fatty acids and act as a source of calories for patients who cannot consume food orally.

There are several problems associated with the use of parenteral emulsions that have restricted their pharmaceutical use:

- There is a limited list of surface-active agents that may be employed to stabilise parenteral emulsions (due to toxicity concerns).
- When administered intravenously it is essential that the droplet size is less than 1 µm to prevent blockage of blood flow within the capillaries. The physical instability of emulsions, which normally causes the droplets of the internal phase to coalesce, is therefore a potentially dangerous consequence of a poorly formulated emulsion.
- Emulsions are difficult to sterilise. Normal sterilisation methods, e.g. heat and filtration, are generally inappropriate.

## Manufacture of parenteral formulations

As detailed previously, two essential requirements of parenteral formulations are sterility and the absence of pyrogens. It is of no surprise to the reader that these two requirements directly influence the methods by which parenteral formulations are manufactured. All raw materials must be of sufficient (injectable) grade and therefore are assured to be pyrogen-free. Furthermore, the methods used to remove pyrogens from both equipment (storage and manufacture) and water when used as a vehicle for parenterals have been detailed previously.

Before describing different strategies for the manufacture of sterile parenteral formulations, it is worth while briefly discussing the concept of sterility and the different methods by which the sterilisation of parenteral formulations may be achieved. Knowledge of these processes is important in the understanding of sterile product manufacture.

Sterilisation may be defined as the absence of viable microorganisms (either through the destruction of all living microorganisms or by their removal) in pharmaceutical preparations. There are five established methods by which pharmaceutical raw materials and pharmaceutical preparations may be sterilised: (1) moist-heat sterilisation; (2) dry-heat sterilisation; (3) filtration sterilisation; (4) sterilisation by exposure to ionising radiation; and (5) gas sterilisation.

### Moist-heat sterilisation

The key features of moist-heat sterilisation include:

- It is performed in an autoclave and employs steam under pressure.
- It offers efficient sterilisation at lower temperatures than dry-heat sterilisation, due to the presence of moisture.
- The mode of action is thought to be due to the denaturation/coagulation of microbial proteins.
- At normal pressure the temperature of water cannot exceed 100°C. In moist-heat sterilisation the pressure in the autoclave is increased to enable an increase in the processing temperature.
- Exposure of the pharmaceutical product at the required temperature for the required time will result in efficient sterilisation. For example, at 103.4 kPa (i.e. 15 pounds per square inch) and 121°C sterilisation is achieved in 20 minutes whereas at 68.91 kPa (10 pounds per square inch) sterilisation is achieved in 30 minutes. It must be remembered that the time of sterilisation must include a lag period, i.e. to ensure the required temperature is reached within the interior of the product.
- Moist-heat sterilisation is principally used to sterilise materials that are both thermostable (within the conditions

of the sterilisation cycle) and through which moisture can perfuse. These include:

- glassware
- dressings
- closures
- aqueous solutions: The aqueous nature of these systems ensures that moist-heat sterilisation is an ideal method for terminal sterilisation (assuming that the therapeutic agent is thermostable). In this process the parenteral solution is either presented to the autoclave sealed in vials or sealed using a closure and aluminium cap. The temperature within the container is elevated to the designated sterilisation temperature at which it is held for the appropriate period.

### Dry-heat sterilisation

The key features of dry-heat sterilisation include:

- In this process microorganisms are destroyed following cellular dehydration and then pyrolysis/oxidation.
- Dry-heat sterilisation is performed in ovens.
- Due to the lower microbicidal efficiency of dry-heat sterilisation (in comparison to moist-heat sterilisation), dry-heat sterilisation is performed at higher temperatures and requires longer times of exposure of the microorganism to this temperature. Examples of dry-heat sterilisation cycles include:
  - 170°C for 1 hour
  - 160°C for 2 hours
  - 140°C for 4 hours.
- As with moist-heat sterilisation, the described sterilisation conditions refer to the time of residence of the product/article in the oven following attainment of the defined temperature. A lag time is therefore required to ensure that the article/container has achieved this temperature.
- Dry-heat sterilisation is employed to sterilise materials/products that cannot be readily sterilised by moist heat (and which are thermostable following exposure to the sterilisation cycle), e.g.:
  - oils and other aqueous vehicles (e.g. glycerin, propylene glycol)
  - heat-stable therapeutic agents/excipients
  - glassware (e.g. bottles).

### Filtration sterilisation

The key features of filtration sterilisation include:

- In this method microorganisms are removed (not destroyed) from solutions using sterilising filters of pore diameter

0.22 μm. This pore diameter is sufficient to entrap/retain bacteria and fungi. Following use, the filters (containing the entrapped/retained microorganisms) are then safely discarded.

■ To maximise the lifetime of the filter (i.e. the volume that may be passed through the filter with blockage of fluid flow), the solution is passed through a series of clarifying filters of defined diameter, e.g. 1 μm, 0.45 μm, prior to passage through the sterilising filter.

■ Filtration sterilisation is used to sterilise solutions of therapeutic agents that are thermolabile.

■ It is an efficient, inexpensive technique.

■ As the product is filtered prior to filling into the final container (e.g. vial/bottle), one concern with filtration sterilisation is the possibility of the product being non-sterile due to a flaw in the manufacture (and hence function) of the sterilisation filter.

### Sterilisation by exposure to ionising radiation

The key features of this sterilisation method include:

■ It involves exposure of the raw material/product to a defined dose of ionising radiation. Typically gamma radiation is employed, sterilisation occurring following exposure to 25–40 kGy.

■ It requires specialist equipment and is therefore expensive for routine use.

■ It is used to sterilise therapeutic agents/excipients or the production of parenteral formulations that are manufactured and packaged under aseptic conditions but are neither terminally sterilised nor sterilised by filtration.

■ It must be noted that certain therapeutic agents and excipients are unstable in the presence of sterilising doses of ionising radiation. Therefore the effects of ionising radiation on the stability of formulation ingredients must be individually examined.

### Gas sterilisation

The key features of gas sterilisation include:

■ Gas sterilisation involves exposure of materials/products to mixtures of ethylene oxide or propylene oxide and an inert gas, e.g. carbon dioxide, within a specially designed apparatus.

■ Sterilisation efficiency increases in the presence of moisture (up to 60%) and elevated temperature (circa 55°C).
  • A reduction in the operating temperature will result in an increased time required for sterilisation.

■ Due to the highly penetrative nature of the gas medium, this technique is frequently used to sterilise medical devices (e.g. packaged catheters) and porous surgical accessories (e.g. blankets). However, this technique may also be employed for the sterilisation of therapeutic agents/excipients.

- Due to the toxicity of the gas mixture, sufficient time must be allowed after sterilisation to enable the sterilising gas to be desorbed from the product/ingredient.

## Specific manufacturing requirements for parenteral products

The manufacture of parenteral products occurs under aseptic conditions, in an area that is only used for the preparation of sterile products. The large-scale manufacturing equipment within the facility must be capable of sterilisation in situ. For example, the manufacturing vessels in which the solutions are mixed are generally jacketed and may be sealed. Temperature can then be applied to these to render them sterile. In addition small-scale equipment (which has been previously sterilised and sealed) and raw materials (including the therapeutic agent) are entered into the aseptic manufacturing area through a special portal (with a positive airflow to prevent the ingress of microorganisms). The air supply to the manufacturing area is filtered. Operators within the manufacturing area must wear special (sterile) work clothing, ensuring that there is no operator contamination of the product and environment.

Whilst the manufacture of parenteral formulations is similar to that of non-sterile comparator formulations, the main difference (in addition to the points described in the paragraph above) is the means by which the product is rendered sterile. In this context the manufacture of parenteral solutions (aqueous and oil-based) and parenteral suspensions (aqueous and oil-based) will be briefly discussed separately.

### Aqueous parenteral solutions

The main steps in the manufacture of aqueous parenteral solutions are as follows:

- The formulation components (including the therapeutic agent) are dissolved in the main mixing vessel within the manufacturing suite.
- If the active ingredient is thermostable, the formulation is filled into the final containers and sealed. Sterilisation is then performed using moist-heat sterilisation. The inclusion of a preservative is unnecessary unless the preparation is a multidose product.
- If the active ingredient is thermolabile, the product is sterilised using filtration under sterile conditions and collected into a second mixing vessel, from which filling into the final container is performed. In some cases it may be possible to fill the product into the final container immediately after

filtration, obviating the need for a second mixing vessel. In this case preservatives are normally included.

### Oil-based parenteral solutions

Moist-heat sterilisation cannot be used to sterilise oil-based solutions and therefore the manufacturing procedure must accommodate this limitation.

The main steps involved in the manufacture of oil-based parenteral solutions are as follows:

- Sterilisation of the oil-based vehicle (containing the various soluble excipients) within the mixing vessel in the aseptic manufacturing suite (performed using dry-heat sterilisation). If the therapeutic agent is stable under the conditions of dry-heat sterilisation, the drug may be incorporated (dissolved) at this stage. This product is then ready for filling into the final container (which is then sealed). Alternatively the product may be manufactured and terminally sterilised using dry-heat sterilisation.
- If the therapeutic agent is not stable under the above sterilisation conditions, sterile drug should be dissolved in the sterile oil-based vehicle using normal mixing facilities under sterile conditions. This product is then ready for filling into the final container, which is subsequently sealed.

### Aqueous parenteral suspensions

As suspensions contain drug that has been dispersed in the chosen vehicle, filtration cannot be performed to render the product sterile. Therefore, sterile drug must be added to the chosen vehicle and suitably dispersed under aseptic conditions. Typically, the main steps in the manufacture of aqueous parenteral suspensions are:

- Dissolution of the formulation excipients in an aqueous (or hydroalcoholic) vehicle within the main vessel.
- Sterilisation of the vehicle by filtration and collection into a second vehicle, under aseptic conditions.
- Dispersal of the (sterile) therapeutic agent into the sterile vehicle. If required the particle size of the disperse phase may be reduced by, for example, passage through a ball mill (whose contents and packing material are both sterile).
- The formulation may then be filled into the final container, followed by sealing.
- Drug suspensions are often physically unstable if exposed to moist-heat sterilisation; however, if the suspension is stable under these conditions, then the formulation may be manufactured and filled into the final container, followed by terminal sterilisation.

### Oil-based parenteral suspensions

The manufacture of oil-based suspensions combines the manufacturing considerations of both oil-based solutions and aqueous suspensions and includes:

- Sterilisation of the oil-based vehicle (containing the various soluble excipients) within the mixing vessel in the aseptic manufacturing suite (performed using dry-heat sterilisation).
- Mixing of the sterile therapeutic agent into the sterile vehicle, prior to filling into and sealing the final container.
- If the formulation can physically and chemically withstand the conditions for dry-heat sterilisation, the product may be manufactured and filled in the final container. This may then be sterilised using dry heat.

## Further reading

Freund J, McDermott K. (1942) Sensitization to horse serum by means of adjuvants. *Proc Soc Exp Biol Med* 49: 548–553.

## Quality control of parenterals

As the reader will have discerned, the main product types that are administered parenterally are (1) aqueous solutions, (2) aqueous suspensions, (3) oily solutions, (4) oily suspensions and (5) emulsions. The quality control methods that are applied to these formulation platforms are identical to the comparator non-parenteral formulations. However, it is important to note that, unlike these comparators, all parenteral formulations must be sterile and therefore quality control methods must include an assessment of sterility. Therefore, following manufacture and over the designed period of the shelf-life, the following analyses are applied to parenteral formulations:

1. **Concentration of therapeutic agent**: Following manufacture the concentration of therapeutic agent must lie within 95–105% of the nominal concentration. This range offers sufficient flexibility for the product to be successfully manufactured. Products whose drug concentration lies outside this range cannot be released for sale. Over the shelf-life of the product the concentration of drug must not fall below 90% of the nominal amount.
2. **Uniformity of content**: As with other preparations, the individual contents of parenteral formulations are examined and compared to the average masses. Limits are then set

regarding the numbers of units that deviate from a defined percentage of the average mass.

3. **Concentration of preservative**: Preservatives are only included in multidose parenteral formulations. As before, the concentration of therapeutic agent must lie within 95–105% of the nominal concentration following manufacture and also upon storage.

4. **Preservative efficacy testing**: The efficacy of the preservative(s) within multidose parenterals must be assessed using the appropriate pharmacopoeial method. These tests evaluate the resistance of the product to microbial challenge and are performed both on products post-manufacture and also during storage.

5. **Bacterial toxins and pyrogens**: Examination of the presence of bacterial endotoxins is performed using the amoebocyte lysate test whereas the testing of pyrogens is performed by examination of the body temperature of rabbits following administration. These tests are applied to all parenteral preparations in which the injection volume is 15 ml or more or if there is a claim on the label that the preparation is free from bacterial endotoxins/pyrogens.

6. **Appearance**: All parenteral products (solutions, suspensions and emulsions) will have specifications for appearance. It is unusual for parenteral products for human use to contain a colouring agent. However, parenteral solutions may be coloured due to the presence of the drug substance (e.g. oxytetracycline, tetracycline). Parenteral solutions designed for veterinary application may contain a colouring agent.

7. **pH**: The pH of aqueous solutions, aqueous suspensions and emulsions is measured and compared to the specified range. Oily platforms do not have a pH specification. Changes in the pH of parenteral products upon storage may be due to decomposition of the therapeutic agent, leaching of alkalis from the glass container or extraction of acids or alkalis from the closure.

8. **Viscosity**: The viscosity of parenteral products is measured and compared to the specified range. Viscosity is particularly important for oily parenterals, as these products may be associated with pain upon injection if there is an increase in viscosity. In addition pharmaceutical companies may apply their own in-house assessment of redispersability of parenteral suspensions.

9. **Uniformity of mass**: For single dose parenteral preparations, the masses of twenty individual units are measured and the average mass calculated. To pass this test the masses of not

more than two of the individual units may deviate by greater than 10% of the average mass and no units should deviate by more than 20% of the average mass.

Alternatively, the fill volume of the product is measured and compared to the specified range. Loss of volume over storage is usually due to loss of packaging integrity.

10. **Dispersed phase size analysis**: The particle size distributions of the dispersed drug in both aqueous and oily suspensions and, in addition, droplet size in parenteral emulsions are characterised. As before changes (increases) in the particle size distribution provide evidence of possible instability due to particle–particle interactions and/or crystal growth. Similarly, increased droplet size within parenteral emulsions is indicative of possible instability and, if the droplet size is sufficiently large, may result in clinical complications.

11. **Freeze–thaw storage**
    As described previously, it is common to examine the physical properties, in particular the particle size distribution and redispersibility of parenteral suspensions, following exposure to a series of cycles of freezing and thawing. This storage regimen is useful for two reasons. Firstly, exposure to extremes of temperature may occur as the product is transported from one country to another and consequently it is important to understand the effects of this process on the physicochemical properties of the product. Secondly freeze–thaw cycling provides evidence of possible instability of the product upon storage.

12. **Sterility testing**
    Parenteral products are required to be sterile, both post-manufacture and over the shelf-life of the product. The assessment of sterility of parenteral products is performed using pharmacopoeial methods (e.g. BP/USP/EP). These methods describe the protocol for the determination of the microbial content within the product, including the composition of the growth media, validation tests and the observation and interpretation of results.

The reader should note that other tests, similar to those described in Chapters 1 and 2, are applied to parenteral systems. Furthermore, the list of methods is indicative of the quality control methods that may be employed. The reader should consult the appropriate pharmacopoeias for a more detailed description of the above methods and others that have not been explicitly covered in this chapter.

## Multiple choice questions

1. **Regarding parenteral formulations, which of the following are true?**
a. Parenteral formulations are sterile products.
b. Parenteral formulations must be pyrogen-free.
c. The majority of parenteral formulations for human administration are multidose formulations.
d. The vehicle of choice for human parenteral formulations is water.

2. **Regarding the intravenous route of parenteral administration, which of the following are true?**
a. Large-volume parenteral formulations may be administered intravenously.
b. Intravenous administration results in a rapid onset of drug action.
c. Emulsion parenteral formulations must never be administered intravenously.
d. The intravenous route of parenteral administration provides limited drug bioavailability.

3. **Regarding the intramuscular route of parenteral administration, which of the following are true?**
a. Large-volume parenteral formulations are routinely administered intravenously.
b. Solution parenteral formulations may be administered intramuscularly.
c. Emulsion parenteral formulations may be administered intramuscularly.
d. Intramuscular administration of parenteral formulations provides a rapid onset of drug action.

4. **With respect to parenteral formulations, which of the following are true?**
a. The particle size of the suspended therapeutic agent affects the subsequent onset of drug action.
b. The majority of veterinary parenteral formulations are formulated as suspensions.
c. The majority of insulin formulations are administered intramuscularly.
d. Increasing the particle size of the suspended therapeutic agent will decrease the duration of action following intramuscular administration.

5. **Regarding the osmotic pressure of parenteral solutions, which of the following are true?**
a. A solution which has a molecular concentration of 0.9% is isotonic with blood plasma.
b. 1 mol of a substance in 100 g water has a gram-molecular concentration of 1%.
c. Sodium chloride ionises in solution to yield 2 gram ions.
d. Dextrose, boric acid and phenol are regarded as practically non-ionising in aqueous solutions.

6. **Regarding parenteral solutions, which of the following are true?**
a. Parenteral solutions must contain a preservative.
b. All parenteral solutions must be buffered.
c. Glycerol is frequently chosen as a co-solvent in small-volume parenteral solutions.
d. Parenteral solutions may be administered by the intradermal route of administration.

7 **Regarding the use of surface-active agents in parenteral formulations, which of the following are true?**
a. Surface-active agents are employed to lower the interfacial free energy of water, thereby reducing the viscosity of the injection.
b. Cationic surface-active agents are commonly used in parenteral solution formulations to enhance the solubility of therapeutic agents.
c. Surface-active agents are used to enhance the solubility of therapeutic agents at concentrations lower than the critical micelle concentration.
d. Lecithin is commonly used to stabilise emulsions designed for parenteral administration.

8. **Concerning Water for Injections, which of the following are true?**
a. Water for Injections must be sterile.
b. Water for Injections has defined limits on the total amount of dissolved solids.
c. Water for Injections is pyrogen-free.
d. Water for Injections contains buffers to control the pH.

9. **Regarding oily solution parenterals, which of the following are true?**
a. The viscosity of the formulation may increase as the storage temperature is decreased.

b. Examples of oils that may be used as the vehicle include arachis oil and liquid paraffin.

c. The rheological properties of oily injections may be modified by the addition of aluminium salts of stearic acid.

d. Oily injections require the inclusion of a preservative.

10. **Regarding the manufacture of parenteral formulations, which of the following are true?**

a. Parenteral suspensions may be terminally sterilised by filtration.

b. Commonly used sterilisation cycles destroy pyrogens.

c. Aqueous parenterals may be sterilised using dry-heat sterilisation.

d. Oil-based parenterals are commonly sterilised using moist-heat sterilisation.

11. **A veterinary parenteral formulation has been designed containing 20% w/w therapeutic agent that has been physically dispersed within a non-aqueous vehicle. Which of the following statements are true concerning this formulation?**

a. The physical stability of the formulation will be improved by the addition of aluminium monostearate.

b. The formulation should be sterilised by filtration.

c. The physical stability of the formulation will be improved by decreasing the particle size of the suspended drug.

d. Hydrolysis of the therapeutic agent within this formulation will be lower if aluminium monostearate is included in the formulation.

12. **Insulin is delivered parenterally to patients who have type 1 diabetes mellitus. Which of the following statements are true regarding parenteral insulin formulations?**

a. Moist heat sterilisation is the preferred method of sterilisation of soluble insulin solution.

b. Insulin products contain preservatives.

c. The onset of action and duration of action of parenteral insulin suspensions is greater than that for soluble insulin when both preparations are administered subcutaneously.

d. The route of parenteral administration affects the bioavailability of insulin.

13. **Which of the following statements are true regarding the formulation and use of total parenteral nutrition systems?**

a. Total parenteral nutrition formulations are o/w emulsions.

b. Total parenteral nutrition formulations are hypo-osmotic

c. Total parenteral nutrition formulations include a preservative
d. Total parenteral nutrition formulations are administered by IV.

14. **Procaine penicillin is an insoluble penicillin salt that is used in veterinary medicine for the treatment of infection. Which of the following statements are true regarding procaine multidose penicillin parenteral formulations?**
a. The onset of action would be expected to be short (<30 minutes) and the duration of action would be expected to be short (<24 hours).
b. The onset of action would be expected to be short (<30 minutes) and the duration of action would be expected to be long (several weeks).
c. The onset of action would be expected to be long (>60 minutes) and the duration of action would be expected to be long (several weeks).
d. The onset of action would be expected to be long (>60 minutes) and the duration of action would be expected to be short (<30 minutes).

15. **Which of the following statements are true regarding single dose IM aqueous parenteral suspension (e.g. suspensions containing steroids)?**
a. Increasing the particle size of the suspended steroid will increase the rate of dissolution at the site of injection.
b. The formulation does not contain a preservative.
c. The formulation is normally administered by IM but may be administered by IV.
d. The viscosity of aqueous steroid suspensions is unaffected by temperature.

# chapter 6
# Ocular, nasal and otic dosage forms

## Overview

**In this chapter the following points will be discussed:**

- an overview/description of the uses of ocular dosage forms
- formulation strategies for ocular dosage forms, including consideration of the excipients used
- the advantages and disadvantages and uses of ocular dosage forms
- considerations for the manufacture of ocular dosage forms
- an overview into the formulation of nasal and otic dosage forms.

## Introduction

This chapter predominantly deals with the formulation of dosage forms that are designed to be administered topically to the eye (principally the conjunctiva or the eyelid) for the treatment of primarily local disorders: glaucoma, a disease of the interior of the eye, may also be treated using these dosage forms. There are three main types of dosage form that are clinically used: (1) solutions; (2) suspensions; and (3) ointments. The administration of these to the eye is usually performed using a dropper (or a container with a dropper nozzle) or a tube with a nozzle. Access of principally solution or suspension formulations to the remote anatomical regions of the eye may also be achieved parenterally (termed intraocular injections) to the anterior, posterior and vitreous chambers (Figure 6.1). Ocular formulations (akin to parenteral formulations) must be sterile. Formulation considerations for nasal and otic formulations will be discussed later in this chapter.

## KeyPoints

- Ocular dosage forms are commonly used to treat local ocular disorders, e.g. infection and inflammation; however, intraocular disorders, notably glaucoma, may also be successfully treated.
- Ocular dosage forms are principally solutions, ointments and suspensions. Intraocular injections are available for the treatment of more serious disorders.
- These formulations exhibit similar concerns as comparator formulations regarding physical and chemical stability.
- Ocular dosage forms must be sterile.
- Nasal and otic dosage forms are non-sterile dosage forms that are inserted into the nasal cavity and ear canal (respectively) for the treatment of local disorders.

**Figure 6.1** Cross-section of the anatomy of the eye.

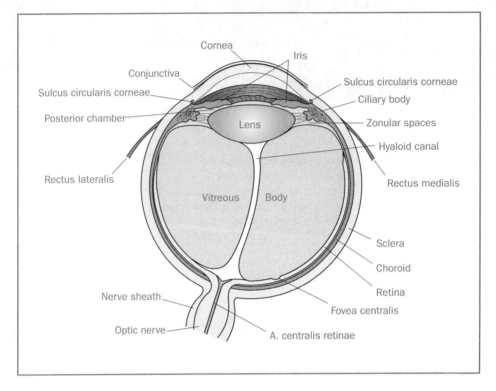

## Advantages and disadvantages of ocular dosage forms

### Advantages

- The application of the therapeutic agents directly to the site of action ensures that the therapeutic agent is available at higher concentrations than may be achieved following oral administration.
- Administration of the therapeutic agent locally ensures that the incidence of side-effects is minimised.
- Following training, the administration of the dosage form locally to the eye may be easily performed by the patient.

### Disadvantages

- The retention of the drug at the site of action is relatively poor, due principally to the low tear volume (7 μl for the blinking eye, 30 μl for the non-blinking eye). The typical volume of two drops of a solution formulation is circa 100 μl and therefore the majority of the applied dose is lost either through spillage on to the face or via the lacrimal duct.

- In addition, the retention time of applied solutions on the surface of the eye is poor. For example, it has been reported that, following the administration of a pilocarpine solution to the eye, removal of the solution from the precorneal region of the eye occurs in less than 2 minutes, resulting in absorption of approximately 1% of the originally applied dose of drug.
- Therefore, to overcome these deficiencies in practice, the patient is required to administer the ocular solution formulations (containing high concentrations of therapeutic agent) frequently, which is inconvenient and may lead to patient non-compliance. These inadequacies have inspired pharmaceutical scientists to devise strategies by which the retention of the drug within the precorneal region may be enhanced.
- Ocular formulations are sterile and therefore specialist facilities are required for the manufacture of these dosage forms.
- Local side-effects may be experienced to ocular dosage forms (to either the high concentration of therapeutic agent ($\leq 5\%$ w/w) or excipients used in the formulation). Typically pain and irritation are the major side-effects encountered by patients.
- The application of ointment formulations to the eye may result in a temporary blurring of vision.

## Administration of therapeutic agents to the eye

The main anatomical features of the eye are illustrated in Figure 6.1.

The main features of the eye that are of interest for ocular drug delivery and hence the formulation of ocular dosage forms are: (1) conjunctiva; (2) cornea; and (3) lacrimal fluid.

### Conjunctiva
- The conjunctiva is located at the side of the eye and joins on to the cornea and eyelids.
- The surface area of the conjunctiva is large (circa 18 cm).
- The conjunctiva helps produce and maintain the tear film.
- The permeability of the conjunctiva to the diffusion of therapeutic agents is greater than that of the cornea.

### Cornea
- The cornea is composed of three layers:
  - epithelium (adjacent to the conjunctiva): a multilayered epithelium that is rich in lipids
  - stroma (central region): this is an aqueous matrix composed of collagen and keratocytes
  - endothelium: a lipid-rich, single-cellular epithelium that maintains corneal hydration.

- The diffusion of drugs into the inner chambers of the eye is controlled by the cornea; diffusion occurs via paracellular routes.
- The lipid outer and inner layers (epithelium/endothelium) of the cornea and the predominantly aqueous stroma control drug diffusion into the internal regions of the eye. To be effectively absorbed, therapeutic agents must exhibit intermediate solubility in both these lipid and aqueous phases and must be of low molecular weight.
- The cornea is non-vascular and negatively charged.

### Lacrimal fluid

- Lacrimal fluid is secreted from glands and is located on the surface of the eye.
- The pH of the lacrimal fluid is 7.4 and this fluid possesses a good buffer capacity (due to the presence of carbonic acid, weak organic acids and protein), being able to neutralise *unbuffered* formulations effectively over a wide range of pH values (3.5–10.0).
- Lacrimal fluid is isotonic with blood. Typically ocular aqueous dosage forms are not specifically formulated to be isotonic (0.9% w/w NaCl equivalent) and may be formulated within a range of tonicity values equivalent to between 0.7% and 1.5% w/w NaCl.
- The rate of turnover of lacrimal fluid is approximately 1 µl/min and the blinking frequency in humans is circa 15–20 times per minute. These physiological functions act to remove the therapeutic agent/formulation from the surface of the eye.

### Formulation considerations for aqueous ocular dosage forms

There are two categories of aqueous ocular dosage forms: (1) ocular solutions; and (2) ocular suspensions. Whilst the vast majority of aqueous ocular dosage forms are solutions, suspensions may be required whenever the therapeutic agent exhibits problems regarding chemical stability or, in the case of steroids (e.g. dexamethasone, prednisolone), the potency of the lipophilic drugs is greater than that of the water-soluble salts. The majority of formulations considerations for ocular dosage forms are similar to those described for pharmaceutical solutions in general. Accordingly the main considerations for the formulation of ocular solutions and suspensions are as follows.

### Choice of drug salt for use in ocular solutions

One of the main determinants surrounding the choice of salt type for inclusion in solution formulations is the solubility – the salt form being chosen to achieve the required solubility. To compensate for the poor retention of therapeutic agents within the precorneal region (and therefore to achieve the maximum pharmacological effect), the concentration of therapeutic agents in ocular solutions is relatively high. Under these circumstances the choice of drug salt is important as the local ocular application of certain drug salts results in greater pain/irritation (e.g. stinging). This may be exemplified by considering three salt forms of adrenaline (epinephrine) (the hydrochloride, borate and bitartrate salts), their physicochemical properties in solution and the associated pain when applied to the eye. Adrenaline bitartrate is the most acidic salt in solution (pH 3–4) and, due to the strong buffer capacity, has been reported to produce moderate to severe stinging following application. Conversely, only mild stinging is associated with the topical application of the borate salt due to the low buffer capacity and lower acidity of this salt (pH 5.5–7.5). Adrenaline hydrochloride, although relatively acidic (pH 2.5–4.5), exhibits medium buffer capacity and is therefore more effectively neutralised than the bitartrate salt. This results in reduced stinging in comparison to the bitartrate salt. Importantly, as the discomfort of the salt form increases, there will be a greater tendency for the drug to be washed off the surface of the eye due to the production of lacrimal fluid.

### Physical properties of the dispersed therapeutic agent

The physical properties of the therapeutic agent are important design features of ophthalmic suspensions and indeed in all types of suspension dosage forms. In addition to the physical properties previously outlined in Chapter 2, the particle size used in ocular suspensions is of primary importance due to the relationship between particle size and ocular irritation. Typically 95% of dispersed particles should have an average particle diameter that is less than 10 µm.

### Solution pH/inclusion of buffers

The pH and the control of the pH of ocular formulations are important determinants of the stability of the therapeutic agent, the ocular acceptability of the formulation and the absorption of the drug across the cornea. Ideally the pH of the formulation should be that which maximises the chemical stability (and, if required, absorption) of the therapeutic agent. This issue is particularly important due to the effect of pH on the stability of alkaloid drugs, e.g. atropine, pilocarpine, carbochol. As highlighted in the previous section, the pH and the buffer capacity directly affect the subsequent discomfort of the formulation.

**Table 6.1**   The effect of temperature and pH on the stability (half-life) of pilocarpine and atropine

| Drug | Temperature (°C) | pH | Half-life |
|------|------------------|-----|-----------|
| Atropine | 121 | 6.8 | 1 h |
| | 121 | 5 | 60 h |
| | 25 | 6.8 | 2 years |
| | 25 | 5 | 130 years |
| Pilocarpine | 121 | 6.8 | 34 min |
| | 121 | 5 | 24 h |
| | 25 | 6.8 | 66 days |
| | 25 | 5 | Several years |

Ideally the pH of the ocular solution should be controlled at 7.4 as this is the pH of tear fluid. However, the choice of pH of the formulation is also dictated by the stability of the therapeutic agent at that pH (which in turn serves to define the shelf-life of the formulation) and whether (or not) absorption of the active agent across the cornea is required.

## Chemical stability

pH is an important contributor to the degradation of therapeutic agents that undergo hydrolysis, e.g. alkaloids. The effect of pH on the stability of two alkaloids at two different temperatures is shown in Table 6.1.

As may be observed, the stability of the two alkaloids at room temperature increases as the pH is increased. Therefore, solutions of these alkaloids are prepared at the lower pH to ensure that the product has a commercially viable shelf-life. Interestingly, exposure to high temperatures, such as those encountered during moist-heat sterilisation, also reduces the stability of these therapeutic agents (although this is significantly less within a formulation of lower pH). Therefore, the choice of pH may also affect the method by which the formulation is sterilised. Importantly, if the pH of alkaloidal solutions is buffered to pH 6.8 (e.g. using a phosphate buffer), the formulations must not be sterilised using thermal methods. For this purpose aseptic preparation involving sterile filtration is used.

## Drug absorption across the cornea

The successful treatment of glaucoma using ocular formulations requires that there is sufficient drug absorption across the cornea. To be effectively absorbed the drug must exhibit differential

**Figure 6.2**  Diagrammatic representation of the effect of ionisation (and hence pH) on drug absorption across the cornea.

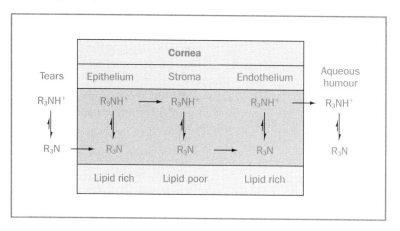

solubility, i.e. the ionised and non-ionised forms coexist. The effects of pH and ionisation on the absorption of drugs across the cornea are represented in Figure 6.2.

As may be observed, sufficient concentration of the non-ionised form is required to partition into and diffuse across the lipid-rich outer layer of the cornea (the epithelium). The inner layer of the cornea (the stroma) is predominantly aqueous and therefore ionisation of the drug must occur to enable partitioning into this phase. Following diffusion to the interface of between the stroma and the endothelial (lipid-rich) layer, absorption of the non-ionised (but not the ionised) form occurs. The non-ionised drug then diffuses to the endothelium/aqueous humour interface where ionisation and dissolution into the aqueous humour occur. The $pK_a$ of the therapeutic agent determines the ionisation of the therapeutic agent at defined pH values. The role of pH on the stability of alkaloids was highlighted in the previous section. Therefore, whilst the chemical stability of alkaloids is enhanced in formulations of lower pH (5), under these conditions the therapeutic agent will be predominantly ionised (circa 99%) and drug absorption will be poor. Increasing the pH of the formulation will decrease the percentage ionisation of the therapeutic agent (and hence increase absorption); however, the chemical stability may be compromised.

To overcome this problem alkaloidal drugs are frequently used in the form of the acid salt, where the pH of the solution will be acidic and stability is optimised. No buffers are added and therefore whenever the formulation is instilled into the eye, the lacrimal fluid adjusts the pH to physiological conditions, thereby facilitating absorption.

## Tip

Most eye drop formulations are solutions due to the less complex nature of the formulation (as compared to suspensions) and greater comfort in use.

If pH control is required, e.g. for acidic drugs, a borate buffer may be used to produce solutions of acidic pH (circa 5). Stabilisation of this pH within the lacrimal fluid will, if required, enhance the absorption of acidic drugs. Phosphate buffers may be used to control the pH of eye drops (at physiological pH). Finally it should be remembered that control of the pH is often used to prevent precipitation of the therapeutic agent in the formulation.

### Vehicle

The vehicle that is predominantly used for the formulation of aqueous ocular dosage forms is Purified Water USP. Water for injections is not a specific formulation requirement. Occasionally oil is used if the therapeutic agent is profoundly unstable within an aqueous vehicle. The choice of oils for ocular use is similar to that for parenteral use.

### Viscosity-modifying agents

Viscosity-modifying (enhancing) agents are hydrophilic polymers that are added to ocular solutions for two main reasons: (1) to control the rate at which the drop flows out of the container (and thus enhance ease of application); and, more importantly, (2) to control the residence time of the solution within the precorneal environment. For example, it has been shown that the retention of an aqueous solution within the precorneal region is short (frequently less than 1 minute); however, if the viscosity is increased, the retention may be enhanced. Furthermore, it has been reported that there is a critical formulation viscosity threshold (circa 55 mPa/s) above which no further increase in contact time between the dosage form and the eye occurs. It must be remembered that there is an upper limit below which the viscosity of eye drop formulations must be maintained as this may lead to blockage of the lacrimal ducts. The viscosities of commercially available products are frequently less than 30 mPa/s. The enhancement of the viscosity of ocular suspensions will serve to enhance the physical stability of ocular suspensions.

Ideally the viscosity-modifying agent should exhibit the following properties:

■ *Easily filtered:* all eye drop solutions are filtered during the manufacturing process. This may be simply to remove any particles (e.g. clarification using a 0.8-μm filter) or clarification in conjunction with filtration sterilisation.

- *Easily sterilised:* sterilisation of eye drop solution is usually performed by filtration (see above) or by heat. If heat sterilisation is performed, the viscosity-modifying agent must be chemically and physically stable under these conditions.
- *Compatible with other components:* the interaction of hydrophilic polymers with certain preservatives is a well-known phenomenon and is usually resolved by increasing the concentration of preservative. However, the pharmaceutical scientist must ensure that there is no or only limited interaction between the therapeutic agent and the viscosity-modifying agent, e.g. basic therapeutic agents and polyacidic polymers.

Examples of polymers that are used as viscosity-modifying agents include:

- *Hydroxypropylmethylcellulose (Hypromellose USP).* Hydroxypropylmethylcellulose (HPMC) is a partially methylated and *O*-(2-hydroxypropylated) cellulose derivative (Figure 6.3). In aqueous ocular formulations HPMC is used in the concentration of 0.45–1.0% w/w.

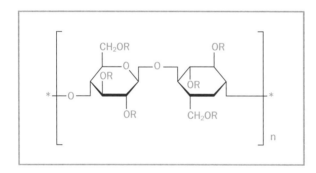

**Figure 6.3** Structural formula of hydroxypropylmethylcellulose (R is H, $CH_3$ or $CH_3CH(OH)CH_2$).

- *Poly(vinyl alcohol)* (Figure 6.4). This is a water-soluble vinyl polymer that is available in three grades: (1) high-viscosity (average molecular weight 200 000 g/mol); (2) medium-viscosity (average molecular weight 130 000 g/mol); and (3) low-viscosity (average molecular weight 20 000 g/mol). It is used to enhance the viscosity of ocular formulations in concentrations ranging from 0.25% to 3.00% w/w (the actual concentration being dependent on the molecular weight of polymer used).
- *Poly(acrylic acid)* This is a water-soluble acrylate polymer that is cross-linked with either allyl sucrose or allyl ethers of pentaerythritol. It is predominantly used in ocular aqueous formulations for the treatment of dry-eye syndrome. However, it may be used to increase the viscosity of ocular formulations that contain a therapeutic agent.

**Figure 6.4**    Structural formula of poly(vinyl alcohol).

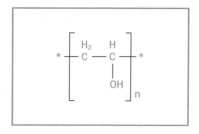

## Preservatives

Ocular formulations are sterile; however, as they are designed as multidose systems, the addition of a preservative is required. The main features of preservatives and preservation in general have been described in previous chapters and therefore this section will specifically address issues regarding the preservation of ophthalmic formulations, including a description of the key preservatives that are used in this class of pharmaceutical product.

### Cationic preservatives

The two main cationic preservatives that are commonly used in ocular solutions and suspensions are *benzalkonium chloride* and *benzethonium chloride*. It should be noted that cationic preservatives are incompatible with anionic therapeutic agents and anionic agents that are used for the control (e.g. pilocarpine nitrate, physostigmine) or diagnosis (e.g. sodium fluorescein) of ocular conditions. Furthermore, interaction of these compounds may occur with non-ionic hydrophilic polymers (used to modify the viscosity of ocular formulations).

### Benzalkonium chloride

This is a mixture of alkylbenzyldimethylammonium chlorides that is used in ocular solutions/suspensions at a concentration between 0.002% and 0.02% w/v (typically 0.01% w/v). The resistance of certain microorganisms that are ocular pathogens to benzalkonium chloride (most notably *Pseudomonas aeruginosa*) has been observed. Therefore, it is customary to include 0.1% w/v disodium edetate (disodium EDTA) in ocular formulations in which benzalkonium chloride is used. This agent acts to enhance the antimicrobial activity of benzalkonium chloride by chelating divalent cations in the outer membrane of the bacterial cell, thereby rendering the bacteria more permeable to the diffusion of the antimicrobial agent. The antimicrobial properties of benzalkonium chloride decrease whenever the pH of the formulation falls below 5.

## Benzethonium chloride

Benzethonium chloride (unlike benzalkonium chloride) is a pure compound (and not a mixture of compounds) (Figure 6.5). It is commonly used in ophthalmic formulations within the concentration range 0.01–0.02% w/v (although it has been reported to exhibit lower antimicrobial activity than benzalkonium chloride).

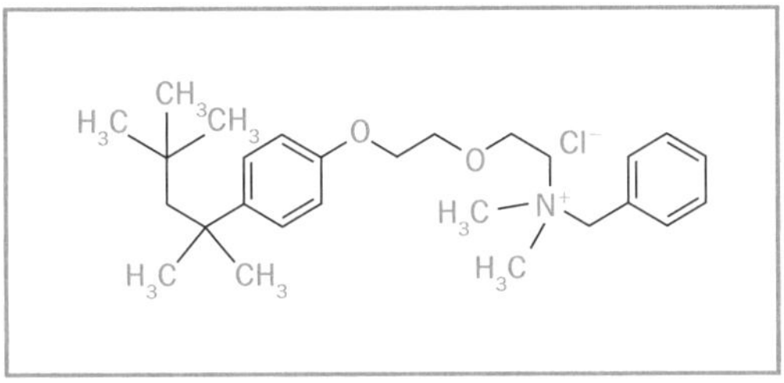

**Figure 6.5**   Structural formula of benzethonium chloride.

### Esters of parahydroxybenzoates (parabens)

Mixtures of methyl and propyl esters of parahydroxybenzoic acid are used in ophthalmic formulations (typically at a combined concentration of 0.2% w/w). There is a concern regarding the ocular irritancy of the parabens, which limits their use in ophthalmic preparations. This problem is augmented by the need to increase the concentration of parabens in ocular formulations that contain hydrophilic polymers (due to the interaction between these two species).

### Organic mercurial compounds

These are antimicrobial agents that contain mercury and, due to environmental and toxicity concerns, are not commonly used in ocular formulations nowadays. The main examples of these compounds that have found pharmaceutical use are phenylmercuric acetate (Figure 6.6a), phenylmercuric nitrate (Figure 6.6b) (which is sometimes supplied as a mixture with phenylmercuric hydroxide, Figure 6.6c) and thimerosal (Figure 6.6d). The typical concentration ranges of the antimicrobial agents used in ocular formulations are: 0.001–0.002% w/v for phenylmercuric acetate, 0.002% w/v for phenylmercuric nitrate and 0.001–0.15% w/v and 0.001–0.004% w/v thimerosal (when used in ophthalmic solutions and suspensions, respectively).

The phenylmercuric salts have been reported to be deposited in the lens of the eye (termed mercuria lentis) when formulated in preparations that are designed for chronic usage, e.g. for the treatment of glaucoma. Thimerosal is not associated with this problem; however, it has been associated with ocular

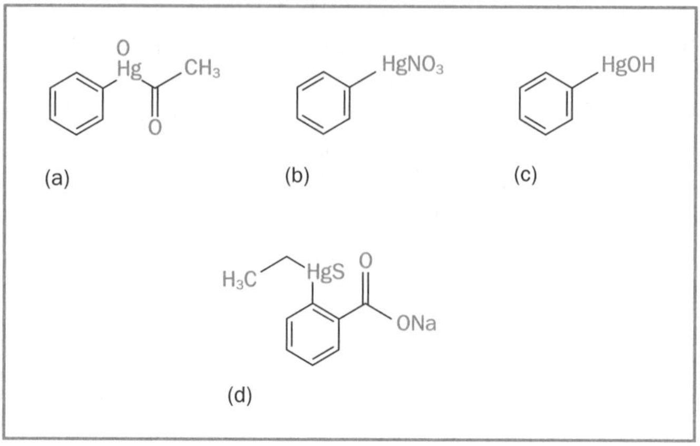

**Figure 6.6**   Structural formulae of (a) phenylmercuric acetate, (b) phenylmercuric nitrate, (c) phenylmercuric hydroxide and (d) thimerosal.

**Figure 6.6**   Structural formulae of (a) phenylmercuric acetate, (b) phenylmercuric nitrate, (c) phenylmercuric hydroxide and (d) thimerosal.

sensitisation. As a result, these preservatives are only used in ocular formulations whenever there is no suitable option.

### Organic alcohols

Two main organic alcohols are used as preservatives for ophthalmic formulations: (1) chlorobutanol; and (2) phenylethylalcohol. The main features of these are described below.

### Chlorobutanol

Chlorobutanol (Figure 6.7) is used in ocular formulations at a concentration of circa 0.5% w/v. Under alkaline conditions hydrolysis of chlorobutanol occurs, liberating HCl as a by-product (the rate of reaction increases with increasing temperature, e.g. during autoclaving).

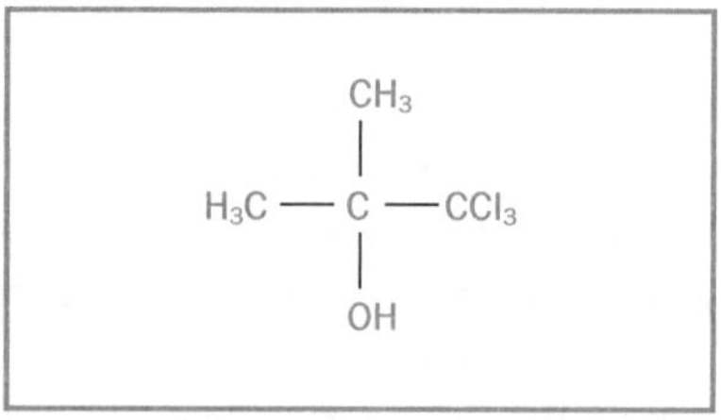

**Figure 6.7**   Structural formula of chlorobutanol.

Therefore, the use of chlorobutanol is reserved for acidic ophthalmic preparations. In addition, chlorobutanol is volatile and lost from solution if stored in polyolefin containers (due to partitioning). Accordingly, preparations that employ this preservative must be stored in glass containers. One final problem associated with the use of chlorobutanol is its limited solubility. The concentration of chlorobutanol used in ophthalmic preparations

(0.5% w/v) is close to the saturation solubility (0.7% w/v at room temperature). Therefore, if the formulation is stored below room temperature, precipitation of the preservative may occur in situ.

### Phenylethylalcohol

Phenylethylalcohol (Figure 6.8) has similar properties to chlorobutanol and indeed shares similar problems, e.g. poor solubility, volatility and partitioning into plastic containers. The typical concentration used in ophthalmic preparations is 0.25–0.50% v/v.

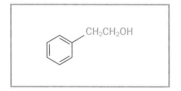

**Figure 6.8**   Structural formula of phenylethylalcohol.

### Antioxidants

As in other formulations, antioxidants may be added to ocular solutions/suspensions to optimise the stability of therapeutic agents that degrade by oxidation, e.g. epinephrine. Sodium metabisulphite (circa 0.3%) is an example of an antioxidant that is commonly used for this purpose. Other examples have been cited in previous chapters.

### Surface-active agents

Surface-active agents are predominantly employed in aqueous suspension to enhance the physical stability of the dispersed particles; however, there have been reports concerning the use of these agents to solubilise therapeutic agents in aqueous ocular solutions. One of the primary concerns regarding the use of surface-active agents in ocular dosage forms is the potential toxicity/irritancy. Accordingly, non-ionic surfactants are preferentially (and predominantly) used whereas the use of anionic surfactants on ocular solution/suspension dosage forms is avoided.

A summary of the formulation considerations for aqueous ocular solutions and ocular gels is provided in Figure 6.9.

## Tips

- The use of mercuric preservatives in eye drop formulations is less common than for other preservatives due to possible concerns regarding toxicity.
- Cationic surfactants are commonly used as preservatives in eye drop formulations.
- Eye drop formulations are sterile. The inclusion of preservatives is required due to the multidose nature of this type of dosage form.

## Tip

The formulation of eye drop dosage forms involves the use of similar components to those used in the formulation of conventional solutions; however, it should be remembered that ocular dosage forms are sterile.

**Figure 6.9** Schematic of formulation considerations for aqueous ocular solutions and gels.

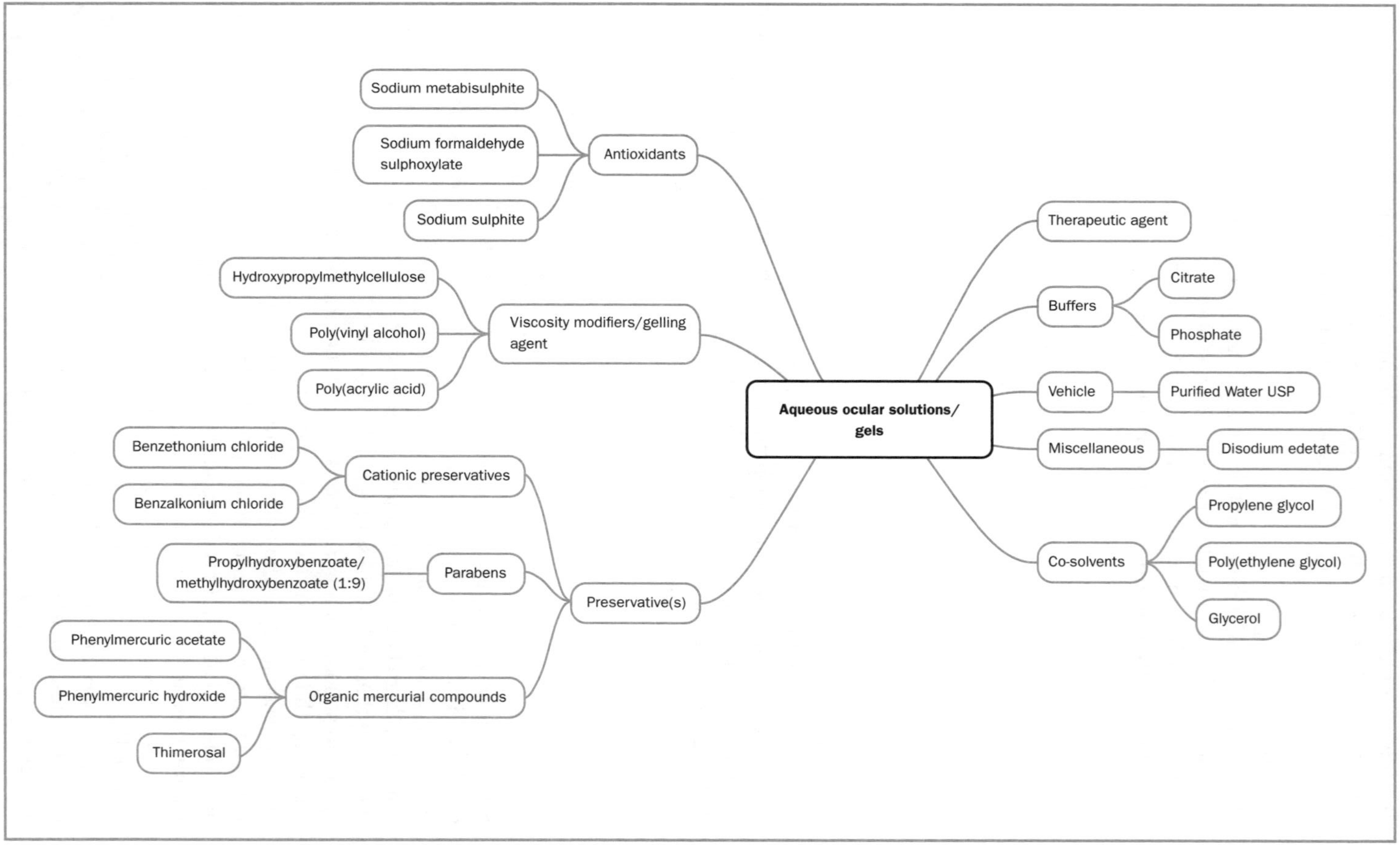

## Manufacture of aqueous ophthalmic formulations

The manufacture of aqueous ophthalmic solutions/suspensions is similar to that described for the manufacture of parenteral solutions and suspensions (Chapter 5). Akin to parenteral formulations, ophthalmic solutions/suspensions must be sterile. The manufacture of these systems may involve at least one of the following steps.

### Production under clean conditions followed by sterilisation by autoclaving

Typically ophthalmic solutions may be manufactured and packaged in the final container under clean conditions. Sterilisation may then be performed using moist-heat sterilisation (assuming the therapeutic agent is chemically stabile under these conditions).

### Production under clean or aseptic conditions followed by aseptic sterilisation by filtration

If the therapeutic agent is thermolabile then sterilisation by heat should be avoided. The production of ocular solutions is then performed by initially manufacturing the dosage form under clean or aseptic conditions. Clarification of the solution is performed by filtration through an appropriate filter (circa 1 µm), followed by sterilisation filtration (0.22-µm filter) and packaging (both under aseptic conditions).

### Production under aseptic conditions

Ophthalmic suspensions may not be sterilised by filtration. Therefore, the manufacture of these systems usually involves the dispersion of the sterile therapeutic agent into the sterile vehicle (with added excipients) and subsequent packaging, both under aseptic conditions.

## Formulation considerations for ophthalmic ointments

Ophthalmic ointments are commonly used for the treatment of ocular infection and/or inflammation. One major disadvantage associated with the use of ophthalmic ointments is their general greasiness and the associated (temporary) blurring of vision. For these reasons patients may prefer aqueous ocular formulations or, alternatively, may prefer to reserve the use of ophthalmic ointments to night-time application. Due to their greater retention within the precorneal region (approximately 2–4 times greater than for aqueous solutions), the frequency of administration of ointments is generally lower than their aqueous-solution counterparts.

As is the case for all ointment formulations, ophthalmic ointments are prepared by dispersion of the therapeutic agent in the pre-prepared ointment base. The physicochemical properties and pharmaceutical strategies for the formulation of ointment bases have been detailed in Chapter 4 and, accordingly, these details will not be repeated in this chapter. This chapter will, however, define the particular formulation requirements of ophthalmic ointments and redirect the reader to the aforementioned chapter for further details.

## Ointment bases for ocular administration

Various types of ointment bases may be employed for ocular administration, as follows.

### Hydrocarbon bases

Hydrocarbon bases are widely used in ophthalmic ointments and are generally composed of a mixture of paraffins (to achieve the correct viscosity). In the formulation of these systems, yellow soft paraffin is preferred to white soft paraffin: white soft paraffin is a bleached form of yellow soft paraffin, thereby being associated with a lower potential for ocular irritancy. Examples and properties of the various excipients that are used to formulate these systems have been provided in Chapter 4.

### Non-emulsified absorption bases

As detailed in Chapter 4, these bases are composed of one or more paraffins (e.g. liquid and yellow soft paraffin) and a sterol containing emulsifying agent (e.g. lanolin derivatives). These bases are less greasy than those prepared using hydrocarbons alone and, in addition, aqueous solutions of drug may be incorporated into non-emulsified absorption bases.

### Water-soluble bases/aqueous gels

Water-soluble bases may be successfully administered to the eye and are associated with a lower incidence of blurred vision (due to their water-miscibility). These bases are primarily formulated using polyethylene glycols. More recently aqueous gels containing a dissolved or dispersed therapeutic agent have been successfully used for the treatment of infection and/or inflammation. Aqueous gels composed of poly(acrylic acid) are commonly used for this purpose and, in addition, have been successfully employed for the treatment of glaucoma (requiring fewer administrations per day in comparison to aqueous solutions containing this drug). Furthermore, poly(acrylic acid) gels (devoid of therapeutic agent) have also been used for the treatment of keratoconjunctivitis sicca (dry-eye syndrome).

## Excipients in ocular ointments

The nature of the excipients required in ophthalmic ointments is dependent on the nature of the ointment base used in the formulation. Typically hydrocarbon-based ointment bases and bases that are either devoid of or contain lower concentrations of water require no further excipients. Exceptionally the inclusion of non-aqueous antioxidants, e.g. butylated hydroxytoluene, butylated hydroxyanisole, may be required if the therapeutic agent is prone to degradation by oxidation. Conversely, aqueous systems, e.g. poly(acrylic acid) gels (which may contain >95% w/w water), will require the same types of excipients that are commonly employed in aqueous solutions, e.g. preservatives, buffers and water-soluble antioxidants.

## Manufacture of ophthalmic ointments

The manufacture of ophthalmic ointments is very similar to that used for the formulation of ointment formulations; however, one important difference is the requirement that ophthalmic ointments must be sterile. Due to their relatively high viscosity, terminal sterilisation of ointments is a difficult operation and may only be performed (in limited examples) at elevated processing temperatures. Therefore, ointments are typically manufactured and packaged under aseptic conditions. The components of the base are customarily prepared in an enclosed mixing vat to which heat may be applied both to aid dissolution and mixing of the ingredients and, importantly, to sterilise the ointment base. The sterile therapeutic agent may then be added to the sterile base and mixed until homogeneous. Filling of the ointment into the final container is performed under aseptic conditions.

A summary of the formulation considerations for ocular ointments is provided in Figure 6.10.

## Nasal formulations

### Introduction

Nasal formulations are aqueous-based systems that are instilled within or sprayed into the nasal cavity and are predominantly employed for the treatment of congestion (ephedrine hydrochloride, pseudoephedrine hydrochloride), allergic rhinitis (e.g. beclomethasone dipropionate, sodium cromoglicate, levocabastine) and infection (mupirocin, chlorhexidine). Importantly, under the appropriate conditions, the nasal

### Tips

- The formulation of ocular ointments involves the use of similar components to those used in the formulation of conventional ointments; however, it should be remembered that ocular dosage forms are sterile.
- Eye drop formulations are generally preferred to ocular ointments by patients due to the ease of administration and non-blurring of vision associated with eye drops.

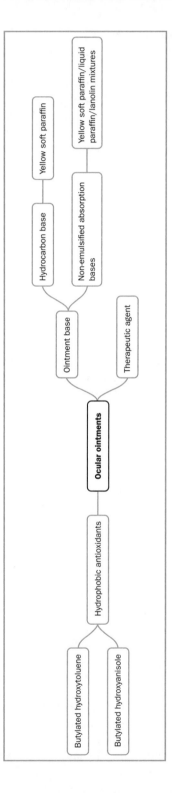

**Figure 6.10** Schematic of formulation considerations for ocular ointments.

route of administration provides a rapid systemic delivery of therapeutic agents (whilst avoiding first-pass metabolism). Examples of this approach include the treatment of migraine (ergot alkaloids, sumatriptin), diabetes insipidus (desmopressin), prostate cancer/endometriosis (gonadotrophin analogues) and smoking cessation (nicotine).

## Formulation considerations for nasal preparations

In addition to providing the required pharmacological activity, nasal formulations must not interfere with the cleansing action of the cilia on the nasal mucosa. This requirement has led to the avoidance of the use of non-aqueous vehicles for nasal formulations. Many of the formulation requirements of nasal solutions/sprays are based on the preservation of ciliary function and include the following.

### Control of the pH of the formulation

Nasal formulations are usually buffered within the pH range 5.5–6.5 to preserve nasal function. For this purpose buffers (e.g. citrate, phosphate) are included in these formulations. The reader should recall that the pH of the formulation affects the ionisation of therapeutic agents and this will, in turn, affect the rate of systemic absorption of the therapeutic agent following nasal absorption. This phenomenon does not affect the activity of drugs within nasal formulations that act locally within the nasal cavity, e.g. antimicrobial agents.

### Tonicity of the formulation

To maintain ciliary function nasal formulations are formulated to be isotonic. The approach taken to achieve this is identical to that described for parenteral formulations, i.e. the inclusion of a suitable salt (e.g. sodium chloride).

### Choice of the vehicle

As detailed above, nasal formulations are aqueous and therefore use purified water as the vehicle. Non-aqueous solvents are not used to formulate nasal dosage forms as these may interfere with ciliary function. Small concentrations of co-solvents (e.g. glycerol, polyethylene glycol, propylene glycol) may be employed to enhance the solubility of the therapeutic agent in the formulation. In addition glycerol acts as a humectant and may therefore reduce or minimise irritation to the nasal mucosa.

### Addition of viscosity-modifying agents

As with other formulation types, control of the viscosity of nasal formulations is important as it directly affects both the ease of administration to, and the retention of, the formulation on the nasal mucosa. The viscosity of nasal formulations may be easily modified by the inclusion of a suitable hydrophilic polymer, e.g. methylcellulose, hydroxyethylcellulose, sodium carboxymethylcellulose, poly(acrylic acid). The viscosity of the formulation is usually modified to be similar to that of nasal mucus (to preserve nasal ciliary function). It must be remembered that the viscosity of the formulation must be suitable to the method of application. For example, the viscosity of nasal formulations delivered using a spray application is lower than those delivered using a bottle/dropper, due to the requirement of the former system for the formulation to be aerosolised as fine droplets.

### Preservatives

Similar to ocular formulations, nasal formulations are multidose dosage forms and therefore require the addition of a preservative. The preservatives used for this purpose include:

- chlorobutanol (0.5% w/w)
- combinations of parabens (0.2% w/w total)
- benzalkonium chloride (0.002–0.02% w/v)
- thimerosal (0.002–0.005% w/v), frequently in combination with benzalkonium chloride.

The reader should be aware that there have been reports that the presence of preservatives in nasal decongestant formulations may be irritant to the nasal mucosa, resulting in rebound nasal congestion. In spite of these concerns preservatives are necessary and are therefore included in nasal formulations.

### Antioxidants

These are included to enhance the chemical stability of therapeutic agents that are prone to oxidative degradation, e.g. adrenaline. As before, water-soluble antioxidants are preferred, e.g. sodium metabisulphite, sodium sulphite.

A summary of the formulation considerations for nasal solutions and gels is shown in Figure 6.11.

## Tip

Nasal preparations must be formulated to ensure that there are no detrimental effects on ciliary function.

### Manufacture of nasal preparations

The manufacture of nasal formulations is similar to that for other non-sterile solution dosage forms. The reader should consult earlier chapters, in particular Chapter 1, to acquire further details on this process.

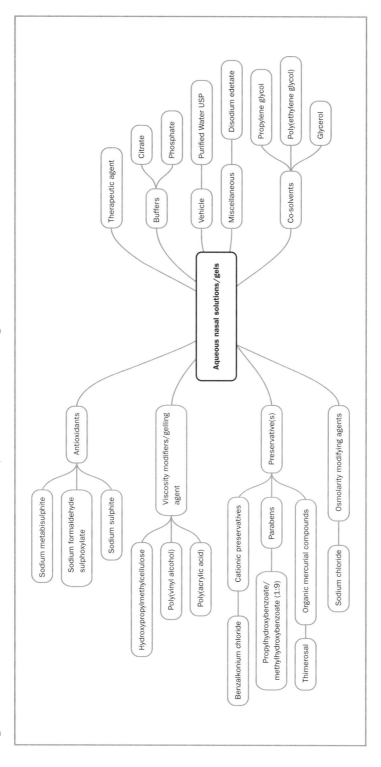

**Figure 6.11** Schematic of formulation considerations for aqueous nasal solutions and gels.

## Otic formulations

### Introduction

Otic dosage forms are applied to the ear canal and are used for the treatment of infection, inflammation and/or pain. Whilst suspensions and ointments may be applied to the ear canal, the vast majority of nasal dosage forms are solutions and therefore only this category of dosage form will be described. Again, to prevent repetition, only the main formulation considerations that are directly relevant for otic formulations will be described and, when a similarity exists with previously discussed formulation types, reference will be made to the appropriate chapter(s). It will be of interest for the reader to note that the considerations for the formulation of otic ointments and suspensions are similar to those previously described in Chapters 2 and 4, respectively.

### Formulation considerations for otic preparations

In common with other pharmaceutical formulations, otic formulations may be formulated as aqueous or non-aqueous solutions and contain one (or more than one) category of therapeutic agent, e.g. local anaesthetics (benzocaine), steroids (prednisolone sodium phosphate, hydrocortisone), antimicrobial agents (polymixin B sulphate, chloramphenicol), in addition to agents that soften ear wax (e.g. carbamide peroxide). The following factors should be considered in the formulation of solutions for application to the ear.

### Vehicle

The vehicle in otic formulations may be:

- aqueous, e.g. purified water
- non-aqueous, e.g. mineral oil (e.g. liquid paraffin), vegetable oil (e.g. arachis oil)
- non-aqueous but miscible with water, e.g. glycols (propylene glycol, glycerol).

Non-aqueous vehicles are predominantly used as vehicles in preparations that are used to remove ear wax, as their lipophilic character assists in the solubilisation of the wax. The choice of aqueous solvent or non-aqueous solvent as the vehicle is primarily dependent on the solubility of the therapeutic agent(s) in each of these. Accordingly, preparations containing water-soluble therapeutic agent(s), e.g. prednisolone sodium phosphate, may be prepared using water as the vehicle, whereas drugs that exhibit lower aqueous solubility may be formulated using mixtures of water and glycols, e.g. hydrocortisone or, alternatively, using the glycol solvent alone, e.g. chloramphenicol. In some cases, the drug may be formulated as a suspension using the same solvents/solvent mixtures as the vehicle. Interestingly, the hygroscopic nature of glycerol and propylene glycol is employed in otic formulations to reduce the swelling of, and remove exudate from, inflamed tissues.

### Preservatives

Otic formulations are multidose and frequently aqueous systems and therefore require the inclusion of a preservative to inhibit microbial growth. The preservatives used in otic formulations are similar to those used in ocular/nasal formulations and include:

- chlorobutanol (0.5% w/w)
- combinations of parabens (0.2% w/w total)
- benzalkonium chloride (0.002–0.02% w/v)
- thimerosal (0.002–0.005% w/v).

Benzalkonium chloride and thimerosal are frequently combined in otic preparations.

Preservatives are not required in otic formulations of low water content.

### Viscosity-modifying agents

As in other topical formulations, the control of product viscosity is important in the design of otic formulations, affecting both the ease of administration to, and retention at, the site of application. Modification of the viscosity of aqueous-based formulations may be achieved using hydrophilic polymers (as described previously in this chapter). The viscosity of glycol-based (and especially glycerol-based) formulations will be greater than their aqueous counterparts and therefore viscosity modification by the inclusion of polymeric components may not be necessary.

### Antioxidants

Antioxidants may be included in otic formulations to improve the stability of drugs that are prone to oxidative degradation. The choice of antioxidant is based on the solubility characteristics of the formulation.

A summary of the formulation considerations for otic preparations is shown in Figure 6.12.

## Quality control of ocular, nasal and otic products

The quality control methods that are applied to ocular, nasal and otic products are identical to other product types that have been described in previous chapters, e.g. solutions, gels, ointments, suspensions. Therefore, the quality control methods associated

**Figure 6.12** Schematic of formulation considerations for otic formulations.

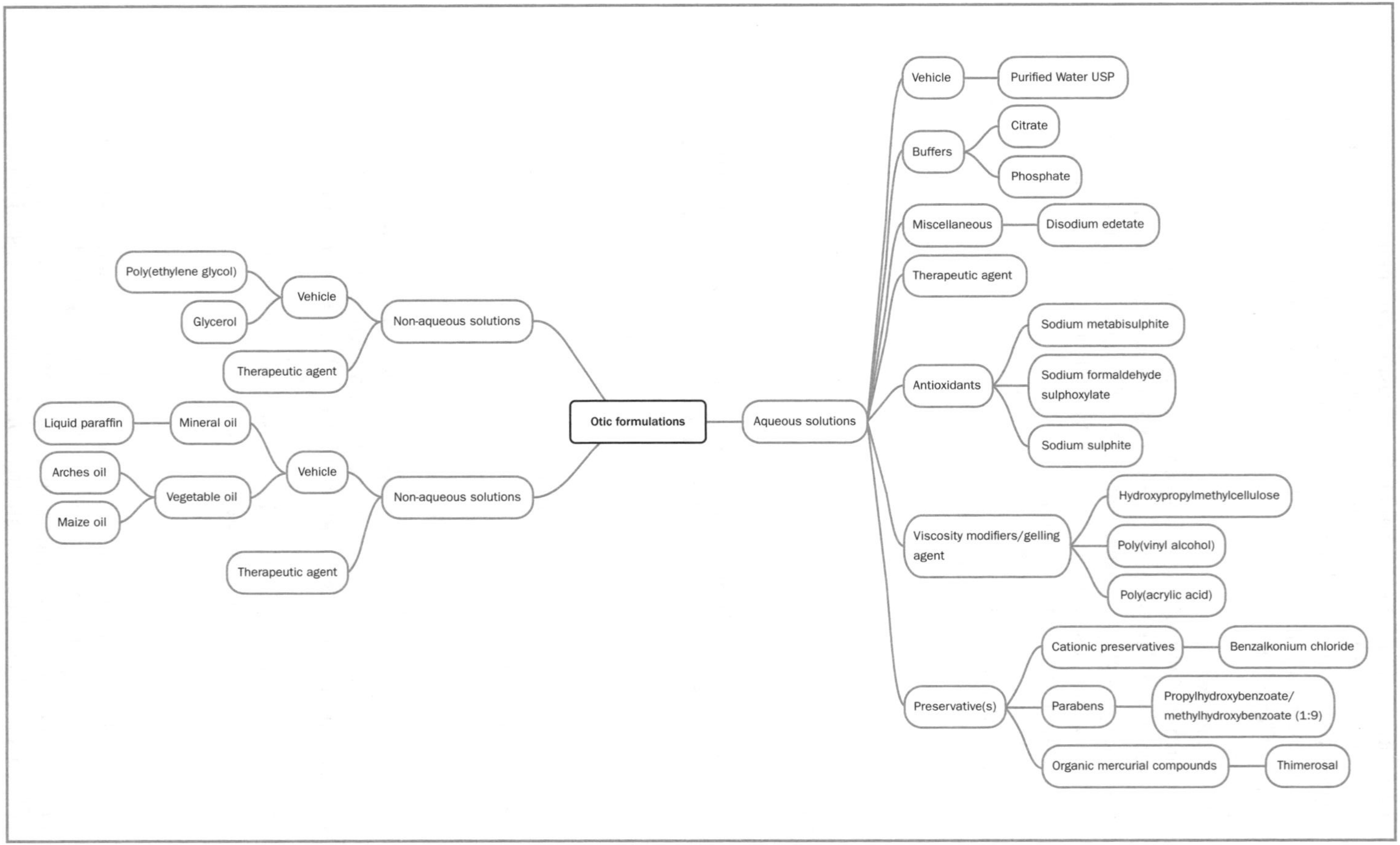

with these products have been previously defined. Akin to parenteral formulations, ocular products must be sterile and therefore quality control methods must include an assessment of sterility. Therefore, following manufacture and over the designed period of the shelf-life, the following analyses are applied to ocular, nasal and otic products:

1. **Concentration of therapeutic agent**: Following manufacture the concentration of therapeutic agent must lie within 95–105% of the nominal concentration. Over the shelf-life of the product the concentration of drug must not fall below 90% of the nominal amount.

2. **Concentration of preservative**: Analysis of preservative concentration within these dosage forms is performed. As before, the concentration of therapeutic agent must lie within 95–105% of the nominal concentration following manufacture and also upon storage.

3. **Preservative efficacy testing**: The efficacy of the preservative(s) within the formulations must be assessed using the appropriate pharmacopoeial method.

4. **Appearance**: All ocular, nasal and otic products will have specifications for appearance.

5. **pH**: The pH of aqueous based formulations is measured and compared to the specified range.

6. **Viscosity**: The viscosity of parenteral products is measured and compared to the range defined in the product specifications.

7. **Uniformity of mass/fill volume**: In a similar fashion the uniformity of mass/fill volume is examined in a defined number of units and compared to the product specifications.

8. **Dispersed phase size analysis**: The particle size distributions of the dispersed drug within suspension formulations are determined using particle size analysis methodology. For example Betaxolol Eye Drops Suspension has limits associated with particle size ranges (not less than 99.5% of particles are less than 25 µm, not less than 99.95% are less than 50 µm and none exceeds 75 µm.

9. **Freeze–thaw storage**: If required, freeze–thaw cycling is performed on the products defined in this chapter.

10. **Sterility testing**: As described in the introduction to this section, ocular products are required to be sterile, both post-manufacture and over the shelf-life of the product. The assessment of sterility of parenteral products is performed using pharmacopoeial methods (e.g. BP/USP/EP). These methods describe the protocol for the determination of the microbial content within the product, including the composition of the growth media, validation tests and the observation and interpretation of results.

The reader should note that other tests, similar to those described in previous chapters, are applied to parenteral systems. Furthermore, the list of methods is indicative of the quality control methods that may be employed. The reader should consult the appropriate pharmacopoeias for a more detailed description of the above methods and others that have not been explicitly covered in this chapter.

## Multiple choice questions

1. **Regarding ocular dosage forms, which of the following statements are true?**
   a. Ocular dosage forms are always sterile.
   b. Ocular dosage forms are principally employed to treat local conditions, e.g. infection, inflammation.
   c. Ocular dosage forms may be employed to treat intraocular disorders.
   d. Ocular dosage forms may be used to deliver therapeutic agents systemically.

2. **Regarding nasal dosage forms, which of the following statements are true?**
   a. Nasal dosage forms are always sterile.
   b. Nasal dosage forms are principally employed to treat local conditions, e.g. infection, inflammation.
   c. Nasal dosage forms are principally aqueous formulations.
   d. Nasal dosage forms may be used to deliver therapeutic agents systemically.

3. **Concerning ocular dosage forms, which of the following statements are true?**
   a. The retention of eye drop formulations at the site of application is poor.
   b. Elimination of eye drop formulations from the eye principally occurs via blinking and drainage into the lacrimal duct.
   c. The retention of ointments at the site of application is similar to that of eye drop formulations.
   d. Drug absorption following the administration of eye drop formulations frequently exceeds 5% of the originally applied dose.

4. **Regarding the cornea, conjunctiva and lacrimal fluid, which of the following statements are true?**
   a. The permeability of the conjunctiva to the diffusion of therapeutic agents is similar to that of the cornea.
   b. The cornea is non-vascular.

c. The cornea is multilayered in structure.
d. The lacrimal fluid is isotonic with blood.

5. **The following statements concern important factors when developing medicines for use in the eye. Indicate if each is true or false.**
a. Lacrimal fluid is usually normal in the pH range 6.0–7.4.
b. Drugs in ocular formulations are mainly absorbed through the cornea.
c. Without blinking, the eye can typically hold a volume of approximately 7 µl.
d. Lipophilic drugs are typically more readily absorbed than hydrophilic ones.

6. **Concerning the absorption of drugs following administration, which of the following statements are true?**
a. The choice of drug salt influences the subsequent absorption.
b. The pH of the formulation influences drug absorption.
c. Increasing the viscosity of the formulation results in a greater extent of drug absorption from eye drop formulations.
d. The rate of drug absorption from ointment formulations is greater than from eye drop formulations.

7. **Concerning viscosity modifiers for eye drop formulations, which of the following statements are true?**
a. Increasing the viscosity of eye drop formulations increases their retention at the site of application.
b. The viscosity of commercially available eye drop formulations is typically greater than 100 mPa/s.
c. The incorporation of viscosity-modifying agents may adversely affect the efficacy of preservatives.
d. Examples of polymers used to enhance the viscosity of eye drop formulations include poly(acrylic acid) and ethylcellulose.

8. **Concerning the use of preservatives in eye drop formulations, which of the following statements are true?**
a. Preservatives are not required in single-use eye drop formulations.
b. Anionic surfactants are commonly used as preservatives in eye drop formulations.
c. Cationic surfactants are commonly used as preservatives in eye drop formulations.
d. The use of organic mercurial preservatives may result in the deposition of the preservative within the lens of the eye.

9. **With respect to the formulation of ophthalmic ointments, which of the following statements are true?**
   a. The therapeutic agent must be dissolved in the ophthalmic ointment formulation.
   b. Ophthalmic ointments prepared using hydrocarbon bases require the addition of preservatives.
   c. Ophthalmic ointments prepared using hydrocarbon bases require the addition of buffer salts to control the pH of the formulation.
   d. The use of ophthalmic ointments may result in temporary blurring of vision.

10. **With respect to the formulation of nasal dosage forms, which of the following statements are true?**
    a. Nasal dosage forms are isotonic.
    b. Nasal dosage forms require the addition of preservatives.
    c. Co-solvents, e.g. propylene glycol, may be added to nasal formulations to enhance the solubility of the therapeutic agent.
    d. The pH of nasal formulations is typically alkaline (>pH 7).

11. **You have been asked to formulate an aqueous gel containing a basic antimicrobial agent designed for the treatment of conjunctivitis. Which of the following statements are true?**
    a. The product must be sterile.
    b. Preferably the drug should be soluble in the formulation.
    c. Poly(acrylic acid) is preferably used to control the viscosity of the formulation.
    d. The pH of the formulations should be approximately neutral.

12. **The viscosity of nasal formulations is an important formulation consideration to ensure optimum clinical activity. Which of the following statements are true regarding the rheological properties of nasal formulations?**
    a. All nasal formulations exhibit Newtonian flow.
    b. The rheological properties of nasal formulations may be modified by the inclusion of non-ionic polymers, e.g. methylcellulose.
    c. The rheological properties of nasal formulations containing non-ionic polymers, e.g. methylcellulose are dependent on the pH of the formulation.
    d. The rheological properties of nasal formulations containing non-ionic polymers, e.g. methylcellulose affect the retention of the product on the conjunctiva.

13. **Which of the following statements is/are true regarding ocular drug delivery?**
   a. The corneal surface is negatively charged.
   b. The clinical efficacy of ocular solutions is dependent on the volume of drops applied.
   c. Charged molecules permeate across the cornea easily.
   d. Large molecular weight drugs (>700 g/mol) do not readily permeate across the cornea.

14. **Which of the following statements is/are true regarding nasal solutions?**
   a. Successful systemic absorption of a therapeutic agent across the nasal mucosa requires the therapeutic agent to be unionised.
   b. Nasal solutions should exhibit high viscosity to aid application and retention.
   c. Oily nasal solutions require a preservative.
   d. Nasal solutions should be coloured to enable the patient to visualise the spreading of the formulation within the nasal cavity.

15. **Which of the following statements is/are true regarding ocular dosage forms?**
   a. The use of ointments is preferred to that of solutions.
   b. High viscosity solutions may decrease lachrymal drainage.
   c. The particle size of therapeutic agents in ocular suspensions is less than 10 µm.
   d. Ocular dosage forms are sterile products.

# chapter 7
# Vaginal and rectal dosage forms

## Overview

**In this chapter the following points will be discussed/described:**
- an overview/description of rectal and vaginal dosage forms and the rationale for their use
- formulation strategies for suppositories, pessaries and related products that are specifically designed for administration to the rectum or vagina
- the advantages and disadvantages of rectal and vaginal dosage forms
- considerations for the manufacture of rectal and vaginal dosage forms.

## Introduction

This chapter is divided into two sections that individually describe the formulation of rectal and vaginal products. The first of these sections describes the general physiology of the rectum, the advantages and disadvantages of rectal drug delivery and strategies for the formulation of rectal dosage forms; the main emphasis is on the formulation of suppositories. In the second section the general physiology of the vagina, the advantages and disadvantages of vaginal drug delivery and strategies for the formulation of vaginal dosage forms are addressed. In a similar fashion to the section on rectal dosage forms, an emphasis is placed on pessaries, although other unique dosage forms are described, albeit briefly.

## Rectal dosage forms

### Advantages and disadvantages of rectal dosage forms

#### Advantages
- Rectal dosage forms may be successfully employed to provide a local effect for the treatment of infection and inflammation, e.g. haemorrhoids, proctitis.

## KeyPoints

- Administration of drugs to the rectum and vagina may be performed to treat local disorders, e.g. infection and inflammation, or to achieve systemic absorption of the therapeutic agent in situations where alternative routes of drug administration are inappropriate.
- Suppositories are the main class of rectal dosage forms; however, creams, gels and foams are also used for this purpose.
- Pessaries, creams and gels are commonly used for the delivery of therapeutic agents to the vagina.
- The strategies for the formulation of gels and creams for rectal or vaginal administration are identical to those described in previous chapters.
- Suppositories and pessaries are semisolid or solid dosage forms that are inserted into the rectum and vagina, respectively. Advice must be given to patients regarding their administration.
- Rectal and vaginal dosage forms are non-sterile formulations.

- Rectal dosage forms are used to promote evacuation of the bowel (by irritating the rectum), to relieve constipation or to cleanse the bowel prior to surgery.
- Rectal dosage forms may be employed to provide systemic drug absorption in situations where oral drug absorption is not recommended. Examples of such applications include:
  - patients who are unconscious, e.g. in intensive care or who are postoperative
  - patients who are vomiting, e.g. gastrointestinal infection, migraine
  - gastroirritant drugs, e.g. non-steroidal anti-inflammatory agents, particularly in chronic usage
  - drugs that are prone to degradation in the stomach
  - drugs that are erratically absorbed from the upper gastrointestinal tract
  - administration of drugs that are extensively first-pass metabolised. If administered (inserted) correctly, the therapeutic agent is absorbed directly into the systemic circulation, thereby avoiding direct entry into the liver. NOTE: Any drug that is absorbed following oral administration will be systemically absorbed when administered rectally.
- Rectal dosage forms may be employed to provide local treatment of diseases of the colon, e.g. Crohn's disease, ulcerative colitis.
- Following advice from the pharmacist, the administration of the rectal and vaginal dosage forms may be easily performed by the patient.

### Disadvantages
- In certain countries, especially the USA and the UK, the rectal dosage forms are generally unpopular, especially for systemic administration of therapeutic agents, whereas the opposite is true in some European countries.
- Specialist advice is required concerning the administration of dosage forms.
- The absorption of therapeutic agents from the rectum is slow and prone to large intrasubject and intersubject variability. The presence of faeces within the rectum considerably affects both the rate and extent of drug absorption.
- Rectal administration of therapeutic agents may result in the development of local side-effects, in particular proctitis.
- The industrial manufacture of suppositories is more difficult than for other common dosage forms.

### Physiology of the rectum
A diagrammatic representation of the gastrointestinal tract, featuring the rectum, is shown in Figure 7.1.

**Figure 7.1** Diagrammatic representation of the gastrointestinal tract, with particular emphasis on the rectum (see insert).

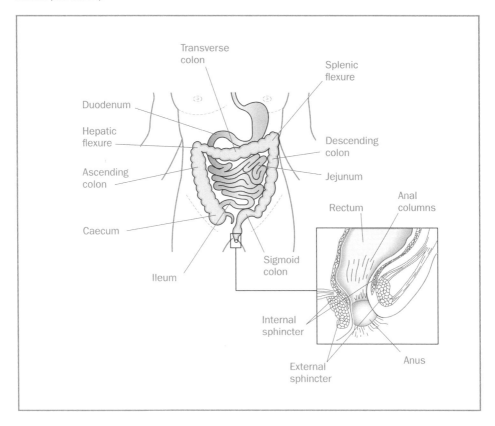

The main physiological features of the rectum that are related to drug delivery and hence to the formulation of rectal products are as follows:

- The length of the rectum is circa 15–20 cm. The rectum is joined to the sigmoid colon at the top and to the anus.
- The rectum is divided into two sections: (1) the anal canal; and (2) the ampulla. The ampulla is the larger of the two sections (approximately four times larger than the anal canal). Faeces are stored in the ampulla and excreted through the anus (a circular muscle) via the anal canal.
- Following absorption from the rectum, the therapeutic agent enters the haemorrhoidal veins. Blood from the upper haemorrhoidal vein enters the portal vein, which flows into the liver, where drug metabolism occurs. Conversely, blood in the middle and lower haemorrhoidal veins enters the general circulation (Figure 7.2).
- The wall of the rectum is composed of an epithelial layer that is one cell thick. Two cellular types exist: (1) cylindrical

**Figure 7.2**   Diagrammatic representation of blood flow into and from the rectum.

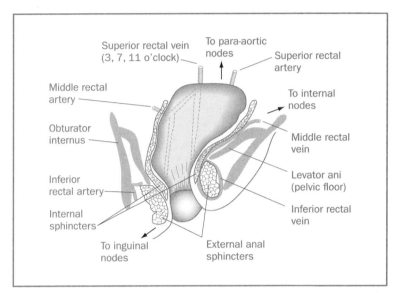

cells; and (2) goblet cells – the latter are responsible for the secretion of mucus. There are no villi (or microvilli).

- When empty the rectum contains circa 3 ml of mucus, spread over a rectal surface area of approximately 300 cm².
- The pH within the rectum is essentially neutral with minimal buffering capacity. Therefore, due to the inability of the fluids within the rectum to alter the degree of ionisation, the salt form of the drug is an important determinant of the resulting local efficacy and/or systemic absorption. The presence of faecal matter will markedly affect both the dissolution of the drug in the rectal fluids and the subsequent absorption of the drug into the systemic circulation.
- As detailed above, the fate of the absorbed drug is dependent on the area of the rectum from which absorption has occurred. Drugs that are absorbed into the *inferior* and *middle haemorrhoidal veins* will enter the circulation via the *inferior vena cava* and will subsequently avoid direct exposure of the drug to, and hence metabolism by, the liver. Absorption into the *upper (superior) haemorrhoidal vein* will result in entry into the liver (and subsequent metabolism) via the portal vein.
- Examples of the classes of drugs that are administered by the rectal route for systemic absorption include:

- antiemetics (e.g. prochlorperazine)
- analgesics (e.g. oxymorphone hydrochloride)
- anti-inflammatory agents (e.g. indometacin)
- ergot alkaloids.

■ There are no esterases or peptidases in the rectal fluid.
■ Local muscle activity within the rectal wall may influence the rate of dissolution of solid dosage forms within the rectum, i.e. suppositories.

## Factors affecting the rectal absorption of therapeutic agents

The process by which drugs are absorbed into the systemic circulation involves the dissolution of the drug in the rectal fluids (which is often preceded by the dissolution/melting of the dosage form), diffusion of the drug through the rectal fluids to the rectal mucosa and then absorption. There are various factors that affect the rate and extent of drug absorption, as follows.

### Site of absorption within the rectum

As highlighted previously, the site of absorption will affect the fate of the therapeutic agent within the bloodstream. Therefore the metabolism of therapeutic agents will depend on the location of the dosage form. It should be noted that it is difficult to ensure that drug absorption occurs exclusively via the inferior and middle haemorrhoidal veins and, therefore, first-pass metabolism of therapeutic agents will occur following rectal administration (although the extent of this is markedly lower when compared to drug absorption following oral administration).

### The partition coefficient and degree of ionisation of the therapeutic agent

Some rectal products are lipophilic (see later in this chapter) and therefore if the therapeutic agent is also lipophilic, the release of the drug will be slow and the solubility of the drug in the rectal fluids will be low (a factor that is compounded by the low volume of the latter). Conversely, the use of (dispersed) hydrophilic drugs in lipophilic bases is preferred, as these would exhibit a greater thermodynamic tendency to dissolve in the rectal fluid. It should be noted that the limited volume of fluid in the rectum affects both the rate and extent of drug dissolution and, therefore, increasing the drug loading in suppositories does not enhance drug absorption.

The reader will be aware that the degree of ionisation of therapeutic agents affects the resultant absorption by biological membranes, as absorption requires the presence of the unionised form of the drug. Commonly the pH and buffer capacity of biological fluids will control the degree of ionisation and hence the absorption of the therapeutic agent. However, the poor buffer capacity of the rectal fluid results in the inability of this fluid to control the pH and hence the ionisation. Therefore, the salt form of the therapeutic agent effectively controls the degree of ionisation.

### The particle size of the dispersed active agent

If the formulation is composed of an active agent that has been dispersed in the appropriate formulation base/vehicle, e.g. a hydrophilic drug dispersed in a lipophilic base or vice versa, the rate of dissolution of the drug is inversely proportional to the particle size of the dispersed active agent. A reduction in particle size of the dispersed therapeutic agent will also affect the physical stability of the formulation, as previously discussed for suspensions (Chapter 2).

### The physicochemical properties of the formulation base

One of the main criteria for rectal formulations is the ability to release the drug. Any interactions between the formulation and dosage form effectively delay release and hence local action/absorption. This interaction may be specific, e.g. resulting from a specific interaction between the drug and the vehicle, or may be non-specific, in which lipophilic drugs are dissolved in lipophilic bases. The latter concern may be obviated by choice of drug salt. In addition, it is important to ensure that the formulation is non-irritant to the rectal mucosa as irritation frequently results in defecation and hence removal of the dosage form before the required mass of drug has been released.

## Tips

- The location of the suppository within the rectum directly affects the route of entry of drug into the systemic circulation and hence the extent of metabolism of the drug. This is particularly relevant whenever suppositories are employed for the systemic delivery of therapeutic agents.
- The formulation of the suppository directly affects the resultant onset and duration of action.

### Formulation of rectal dosage forms

There are several formulation types that are used rectally, including creams, ointments, gels, solutions and suppositories. The formulation considerations for semisolid preparations (creams, ointments, gels) and solutions that are designed for rectal use are similar to those that are formulated for non-rectal use. Therefore, for more information on these aspects the reader should consult the earlier chapters. As a result the remainder of this section will address the formulation of suppositories.

## Formulation of suppositories

Suppositories are solid-dosage forms that are inserted into the rectum where they undergo softening, melting or dissolution to liberate the therapeutic agent. Suppositories may also be inserted into the vagina and other accessible body cavities, e.g. the urethra and, as such, are available in a wide range of sizes and shapes, as depicted in Figure 7.3.

**Figure 7.3**   Examples of the different shapes and sizes of suppositories. (Taken from Allen L V Jr (2008) *Suppositories*. London: Pharmaceutical Press.)

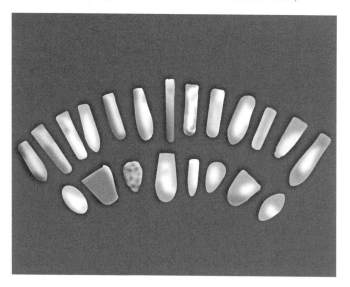

The typical weight range for suppositories is 1–4 grams, with the 2-gram suppository being the commonly used size. The smallest suppositories are mainly reserved for use in children, whereas the largest size may be administered to adults, e.g. glycerin suppositories that are used to relieve constipation in adults. Suppositories are tapered at one end (to aid insertion) and are frequently wider in the middle before tapering towards the other end (thereby aiding retention in the rectum and enabling the suppository to be pressed forward by the anal sphincter). The drug loading of suppositories ranges from circa 0.1% to 40% w/w. In general, suppositories are composed of an inert base into which the therapeutic agent is incorporated (dissolved/dispersed). The physicochemical properties of both the therapeutic agent and suppository base are important determinants of the clinical and non-clinical performance of suppositories. The role of the solubility, salt type and particle size on the dissolution and hence pharmacological activity of rectal dosage forms, e.g. suppositories, has been addressed earlier in this chapter. The next subsection describes the types and physicochemical properties of suppository bases.

## Suppository bases

There are a number of key properties that an ideal suppository base should exhibit. These are as follows:

- The base should be solid at the storage temperature of the formulation but should soften, melt or dissolve in the rectal fluid following insertion into the rectum, thereby releasing the therapeutic agent.
- The base should be non-irritant to the rectal mucosa.
- The base should not chemically interact with the therapeutic agent.
- The base should be chemically and physically stable over the period of storage (shelf-life).

There are two main categories of suppository base: (1) fatty (oleaginous) bases; and (2) water-miscible bases. A third (sub-) category may also be used that consists of lipophilic and hydrophilic excipients and therefore shares properties common to both fatty and water-miscible bases.

### Fatty (oleaginous) bases

Fatty bases are predominantly composed of naturally occurring or semisynthetic/synthetic fatty acid esters of glycerol. These systems are designed to melt within the rectum thereby facilitating drug release and subsequent dissolution. There are several types of fatty base.

#### Cocoa butter (theobroma oil)

This is a natural material that consists of a mixture of fatty acid esters of glycerol, e.g. stearic, palmitic and oleic, predominantly triesters, e.g. glyceryl tripalmitate. The presence of unsaturated (e.g. oleic acid) esters contributes to the low melting point of cocoa butter (30–36°C), thereby facilitating cocoa butter melting following insertion within the rectum. The incorporation of lipophilic drugs into cocoa butter has been reported to lower the melting-point range of suppositories produced using this base, which may lead to stability problems and may result in suppositories that are too soft to insert. To overcome this problem, beeswax (circa 4% w/w) or cetyl esters wax (circa 20% w/w) may be added to the suppository (thereby increasing the melting temperature). Cocoa butter is safe, non-toxic and non-irritating.

One major problem with the use of cocoa butter as a base for suppositories is polymorphism, i.e. the ability of this material to exist in different crystalline forms; this is accredited to the high content of triglycerides. Cocoa butter exists in four polymorphic forms: (1) alpha ($\alpha$, melting point circa 20°C); (2) beta ($\beta$, melting point circa 34–35°C); (3) beta prime ($\beta'$, melting point circa 28°C); and (4) gamma ($\gamma$, melting point circa 34–35°C). If cocoa butter is melted at circa 36°C and allowed to solidify slowly, the stable ($\beta$) polymorph will form. However, if the temperature is markedly elevated above 36°C and allowed to cool, the other polymorphic forms will be produced. Being unstable, the $\alpha$, $\beta$ and $\beta'$ polymorphs will slowly revert to the stable $\beta$ polymorph. However, as this conversion may require several days, there may be associated stability issues (notably poor setting properties or remelting of the suppositories following manufacture).

It should be noted that the melting/softening point of cocoa butter may be inappropriate (too low) for storage at room temperature and therefore either storage under controlled temperature or, as detailed previously, the inclusion of excipients that raise the melting/softening temperature of cocoa butter (e.g. beeswax, cetyl esters wax) may be required.

### Semisynthetic/synthetic triglycerides (Hard Fat BP, US Pharmacopeia National Formulary (USPNF) or Adeps solidus PhEur (European Pharmacopeia))

The main problems associated with cocoa butter (polymorphism and variation in composition) led to the development and use of suppository bases that were based on synthetic/semisynthetic glycerides. Typically these systems are composed of mixtures of triglycerides of higher *saturated* fatty acids (ranging from $C_8H_{17}COOH$ to $C_{18}H_{37}COOH$) and di/monoglycerides. Semisynthetic/synthetic triglyceride bases are safe, non-toxic and non-irritating and, unlike cocoa butter, can be reheated during processing (whilst not affecting the solidification temperature) and exhibit low batch-to-batch variability.

## Choice of synthetic/semisynthetic base for use in the formulation of suppositories

There are several different grades of synthetic/semisynthetic suppository bases that are commercially available and are currently used in the formulation of suppositories. These bases differ in a number of physicochemical properties, including the following.

### Hydroxyl number

This refers to the presence of hydroxyl groups, which in turn is a measure of the presence of mono- and diglycerides in the suppository base. The available suppository bases range in hydroxyl numbers from circa 5 to 15, e.g. Witespol H15, H175, H185, E75, E85, to >55, e.g. Witespol S51 (55–70), S58 (60–70). The use of bases exhibiting high hydroxyl numbers should be reserved for therapeutic agents that exhibit low reactivity towards this chemical group. It should also be noted that as the hydroxyl number of the base increases so does the relative hydrophilicity. This may affect drug release and subsequent absorption.

### Melting properties

There is a wide range in the melting points of commercially available synthetic/semisynthetic suppository bases. For example, Suppocire AIP and Witespol H32 exhibit low melting points (30–33°C and 31–33°C, respectively). Conversely, the melting points of other bases in these product ranges may exceed 40°C, e.g. Suppocire D, ND and DM (42–45°C) and Witespol E85 (42–44°C). Low melting bases (<37°C) are generally used for the formulation of suppositories in which systemic absorption of the therapeutic agent is desired, whereas bases of higher melting points are frequently used in formulations in which a local effect is desired.

The melting properties of these bases and, in addition, the rheological properties of the molten bases are important considerations for the successful formulation of suppositories. The importance of each of these topics is presented below.

### Melting point

In the formulation of suppositories that are intended to provide systemic administration of the therapeutic agent, the selected base should melt to facilitate drug release and hence absorption. On the other hand, the bases of suppository formulations that are designed to provide a local effect do not necessarily have to melt: it is sufficient for them to soften for the intended use.

The melting properties of the base should be considered in conjunction with the solubility of the chosen therapeutic agent. If the active agent is soluble in the base, this will lead to a reduction in the melting point of the base. Under these circumstances a suppository base may be used that offers a higher melting point, e.g. Witespol E75 or E85.

Consideration of the solubility of the therapeutic agent within both the molten and cooled base is important. If the drug is more soluble in the base during the manufacturing process (in which heat is applied to melt the base), the drug may precipitate out following cooling (particularly during storage). This may produce different polymorphic forms of the drug or crystals of a different size fraction, both of which will result in a change in the performance of the product.

## Viscosity of the melted base

The viscosity of the melted base can affect the performance of suppositories in two ways:

1.  During the manufacture of suppositories the base is melted and into this the drug may be dispersed. The molten base (containing the dispersed drug) is then poured into the appropriate mould prior to cooling. The viscosity of the base will therefore affect both the mixing of the drug with the molten base and, in addition, the flow of the molten dispersion into the moulds. Furthermore, the dispersion of solid drug particles into the molten base will dramatically affect the viscosity of the product. Therefore, in selecting a suppository base, the viscosity of the molten base is a major consideration.

2.  The viscosity of the melted base also affects the spreading of the formulation on the rectal mucosa and the subsequent drug release. Bases with high viscosity will exhibit poorer spreading properties and slower rates of drug release when compared to bases of lower melt viscosity.

## Water-soluble and water-miscible bases

There are two main categories of suppository base in this classification: (1) glycerol–gelatin base (which dissolves in the rectal fluids); and (2) water-miscible bases, composed of polyethylene glycols (PEGs).

### Glycerol–gelatin

This suppository base is prepared by dissolving gelatin (circa 20% w/w) in glycerol (70% w/w) with the aid of heating (circa 100°C). The required drug is generally dissolved/dispersed in an aqueous phase (<10% w/w) and then combined with the glycerol phase with stirring prior to pouring into the suppository mould. As an alternative method the gelatin may be dissolved in the heated aqueous phase, into which the drug is dispersed/dissolved and then added to the heated glycerol. The mechanical properties of the formed suppository may be manipulated by increasing or decreasing the mass of added gelatin. For example, as the mass of added aqueous phase is increased, the suppositories will become softer, thereby leading to potential problems regarding their insertion by the patient. Increasing the concentration of gelatin will negate this effect, e.g. in Gelato-glycero suppositories BPC the concentrations of aqueous phase, gelatin and glycerol are 27.5% w/w, 32.5% w/w and 40% w/w, respectively. In urethral suppositories the ratios of gelatin to glycerol to water (containing the active agent) is 3:1:1; the enhanced mechanical properties facilitate insertion into the urethra.

Due to the presence of water in glycerol–gelatin suppositories it is necessary to include preservatives, e.g. mixtures of methyl and propylparabens acid (total concentration 0.2% w/w). These are dissolved within the aqueous phase during the manufacturing process.

Glycerol–gelatin bases may be used for the formulation of suppositories that contain a water-soluble therapeutic agent. However, the use of this type of base is restricted by several disadvantages, including:

- *An associated physiological effect.* Glycerol–gelatin suppositories will induce defecation and, indeed, are used for this purpose to relieve constipation or to facilitate bowel evacuation prior to surgery.
- *Difficult to manufacture.* Suppositories prepared using glycerol–gelatin bases are generally more difficult to manufacture than other types of base, e.g. triglyceride systems.
- *Hygroscopic.* Glycerol–gelatin bases will absorb moisture from the atmosphere and therefore must be carefully packaged to prevent moisture uptake and, ultimately, to maintain both the shape and mechanical properties of the suppository. This ability of glycerol–gelatin bases to absorb water will also occur within the rectum, leading to dehydration and irritation of the rectal mucosa. This action prompts bowel evacuation. To minimise this phenomenon, the suppository may be moistened with water prior to insertion.
- *Potential interactions with therapeutic agents.* Gelatin is a protein that is obtained following partial acid hydrolysis (termed type A) or partial alkaline hydrolysis (termed type B) of animal collagen. The isoelectric point of type A ranges between 7 and 9 whereas the isoelectric point of type B lies between 4.7 and 5.3. Accordingly the pH of aqueous solutions of each gelatin type will affect ionisation: type A acts as a base if formulated at a pH below 4.7 whereas type B will act as an acid if the pH is raised above 5.3. As a result of this a possibility arises for gelatin to interact with therapeutic agents of opposite charge in a typical acid–base interaction. This will lower the solubility of gelatin and may lead to precipitation. Therefore it is important to select the pH of the aqueous phase and gelatin type carefully when formulating suppositories containing acidic or basic drugs.

### Water-miscible bases

Water-miscible bases are composed of PEGs possessing a molecular weight greater than 1000 g/mol. The melting point of these higher grades of PEGs increases as the molecular weight increases, e.g. the melting points of PEG 1000 and PEG 8000 are

37–40°C and 60–63°C, respectively. Typically the melting point of PEG suppository bases is circa 42°C; this is generally achieved and controlled using the appropriate mixtures of grade of this polymer. The higher melting point of these systems obviates the need for storage under cold conditions. In addition to controlling the melting point, different molecular weights of this polymer may also be blended to control the mechanical properties of PEG-based suppositories. In this scenario, the lower-molecular-weight PEGs, e.g. PEG 400, will act to reduce the brittle behaviour of these suppository bases.

Following insertion into the rectum, these suppositories will not melt but, due to their hygroscopic properties, will instead gradually dissolve (the volume of rectal fluid is too small to allow rapid dissolution) and, in so doing, will enable drug dissolution to occur. This ability to absorb moisture may lead to patient discomfort due to the extraction of water from the rectal mucosa into the suppository; however, this may be minimised by the inclusion of water (>20% w/w) and by moistening the suppository prior to insertion. As before, PEG-based suppositories will require storage in moisture-resistant packaging. There are two final concerns regarding the use of PEG-based suppositories. PEG is known to enhance the solubility of therapeutic agents and therefore this interaction between the drug and polymer may affect the subsequent release of the drug from the liquefied base. Secondly, the solubility of the drug in the solid base may change as functions of both storage conditions and time and this may result in crystal growth within the suppository. This in turn may affect the mechanical properties of, and drug release from, the suppository.

## Formulation excipients that may be used in the formulation of suppositories

As with other formulation types, the formulation of successful suppositories may necessitate the inclusion of other pharmaceutical excipients. The list of possible excipients is more limited for suppositories than for other dosage form types that have been discussed in this textbook to date. Examples include: (1) surface-active agents; (2) agents to reduce hygroscopicity; and (3) agents to control the melting point of the base.

> **Tip**
>
> Semisynthetic suppository bases are generally used for the formulation of suppositories.

### Surface-active agents

These are included to enhance the wetting properties of the suppository base with the rectal fluid. This in turn will enhance drug release/dissolution. The use of surfactants is mainly reserved for formulations composed of a lipophilic suppository base and/or a lipophilic drug. Examples of excipients that are included in suppository formulations include sorbitan esters and polyoxyethylene sorbitan fatty acid esters.

### Agents to reduce hygroscopicity

Agents that reduce hygroscopicity, e.g. colloidal silicon dioxide, may be included in fatty suppository bases to reduce the uptake of water from the atmosphere during storage and, in so doing, the physical and chemical stability of the dosage form/therapeutic agent may be enhanced.

The hygroscopicity of water-soluble or water-miscible bases is large and is indeed partly responsible for the in vivo performance of formulations base. Water uptake during storage will result in changes to the mechanical properties (softening) and shape of these dosage forms. Accordingly, protection against water uptake during storage is afforded by the use of moisture-resistant packaging.

### Agents to control the melting point of the base

As previously mentioned, the melting point of the base may be manipulated to enhance the mechanical properties and physical stability of the suppository in response to the deleterious effects of storage at higher temperature and/or the presence of a therapeutic agent that is soluble in the suppository base. Examples of excipients that are commonly used to increase the melting point of suppositories prepared using fatty bases include:

■  beeswax (white or yellow wax)
■  Cetyl esters wax USP-NF
■  stearic acid
■  stearyl alcohol
■  aluminium mono- or distearate
■  colloidal silicon dioxide
■  magnesium stearate
■  bentonite.

Conversely there may be a requirement to reduce the melting point of the fatty suppository base, e.g. to enable melting within the rectum. Examples of excipients that may be used for this purpose include:

■  glyceryl monostearate
■  myristyl alcohol
■  Polysorbate 80
■  propylene glycol.

As detailed previously, the melting point of PEG-based suppositories may be controlled by altering the grade of PEG used. In this approach the melting point may be increased by the inclusion of higher-molecular-weight PEG. Conversely, the melting point of water-miscible suppository bases may be lowered by incorporating the required concentration of low-molecular-weight PEG (PEG 400).

## Manufacture of suppositories

Suppositories are manufactured principally using a moulding method, i.e. a method in which the base is heated to above the melting temperature, the drug dispersed (or dissolved) in the heated liquid prior to dispensing the molten product into suppository moulds (Figure 7.4).

**Figure 7.4**   Armstrong suppository mold. (Photo courtesy of Professional Compounding Centers of America, Houston, TX. Taken from Allen L V Jr (2008) *Suppositories*. London: Pharmaceutical Press.)

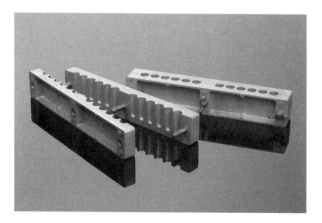

In the laboratory the suppository formulation is heated in a small vessel (typically an evaporating basin) and then manually poured into the mould. This is then allowed to cool (usually in a refrigerator) prior to removal and packaging. Suppositories prepared using cocoa butter or glycerol–gelatin bases are prone to sticking in the metallic suppository mould after cooling. To prevent this, the moulds are usually precoated (e.g. arachis oil may be used for glycerol–gelatin suppositories whereas an alcohol/surfactant mixture is commonly used for suppositories prepared using cocoa butter). The composition of the lubricant must differ from that of the base to prevent any interaction.

Alternatively, plastic moulds may be used that do not require lubrication (Figure 7.5). In large-scale manufacture, the preparation is heated within a vessel and the molten dosage form is automatically dosed into the mould, cooled and mechanically removed. The suppositories are then automatically packaged into the final container. Alternatively the molten formulation may be injected directly into the final packaging (termed mould strips), which are then sealed.

**Figure 7.5** Examples of plastic moulds that are used for the production of suppositories: (a) Plastic mould filled with the suppository formulation. (Courtesy of Prof Ryan Donnelly, School of Pharmacy, Queen's University Belfast.) (b) Plastic-filled suppository mould and formed suppositories. (Courtesy of Prof Ryan Donnelly, School of Pharmacy, Queen's University Belfast.)

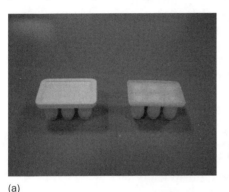

(a)

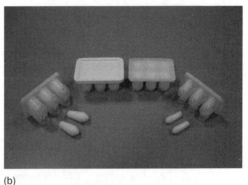

(b)

## Calculation of the mass of base required

One concern regarding the manufacture of suppositories is the calculation of the mass of base that is required. The volume of each suppository mould is known (and has been calibrated) and therefore the molten formulations are dispensed into the moulds according to volume. However, if the drug is *dispersed* in the molten formulation, the volume of the formulation will be dependent on the mass of drug present (remembering that solids displace an equal volume of base). To ensure that the correct volume of base is used, a calculation is performed based on the *displacement value*, i.e. the ratio of the weight of the drug to the weight of base displaced by the drug. The displacement factor may be visualised as the weight of drug required to displace unit weight of base. For example, the displacement value of cinchocaine hydrochloride is 1.5 for cocoa butter (and indeed other triglyceride bases). Therefore 1.5 grams of this therapeutic agent will displace 1 gram of cocoa butter.

In practice the displacement value is calculated as follows:

- The average weight of the suppository mould is calculated using the blank suppository base (the molten base is added to the correct volume, allowed to cool and then weighed).

- The weight of drug needed for the total number of suppositories is calculated as the product of the weight of drug per suppository and the number of suppositories manufactured.
- Suppositories are then prepared by adding the mass of drug to the notional mass of suppository base, melting and then dispensing into the suppository moulds. The weight of the cooled suppositories is then determined.

Examples of pertinent calculations are shown below.

## Worked examples

### Example 7.1
Determine the displacement value for a therapeutic agent in a triglyceride base. The concentration of drug in each suppository should be 10%.

*Step 1*
Prepare (e.g. six) blank suppositories and weigh.
   Total weight determined as 6 grams (1 gram per suppository).

*Step 2*
Prepare six suppositories containing 10% drug in each suppository (based on the volume of the suppository mould) and weigh.
   Total weight determined as 6.11 grams.

*Step 3*
Calculate the masses of base and drug in the six suppositories (remembering that the drug loading of the suppositories is 10%).

$$\text{Base} = \frac{90}{100} \times 6.11 = 5.50\,\text{g}$$

$$\text{Drug} = \frac{10}{100} \times 6.11 = 0.61\,\text{g}$$

*Step 4*
Calculate the mass of base displaced by the 0.61 gram of drug.

$$6.0 - 5.5 = 0.5\,\text{g}$$

*Step 5*
Calculate the displacement value for the drug.

$$\text{DV} = \frac{0.61}{0.50} = 1.22$$

A list of displacement values for a range of drugs is available and these are used to calculate the required weight of base. The use of these values is illustrated in the next example.

## Example 7.2

You have been asked to prepare six (2-gram) suppositories containing 5% cinchocaine hydrochloride (displacement value of 1.5). Calculate the final formulae for the requested dosage form.

As this is a small batch, an overage must be included in the calculation. Therefore, the weights required for nine suppositories will be determined:

- Calculate the weight of drug required for nine suppositories:

$$Wt(g) = \left(\frac{5}{100} \times 2\right) \times 9 = 0.90g$$

- Calculate the weight of base displaced by the desired weight of drug (900 mg):

$$DV = \frac{Wt_{Drug}}{Wt_{Base\,displaced}} \quad 1.5 = \frac{0.90}{Wt_{Base\,displaced}}$$

$$Wt_{Base\,displaced} = \frac{0.9}{1.5} = 0.60g$$

- Determine the amount of base required:
- Weight of base required = (theoretical weight – weight of base displaced)

$$Wt_{Required} = 18.0g - 0.6g = 17.4g$$

- Define final formulation (Table 7.1).

**Table 7.1** Calculated composition of the suppository formulation described in Example 7.2.

| Formulation component | Weight (g) for 9 suppositories |
|---|---|
| Cinchocaine hydrochloride | 0.90 |
| Suppository base | 17.40 |

Note: The final weight of the formulation does not necessarily add up to 9 g. Thus, in the above formulation the weight of each suppository is 2.03 g, i.e. (17.40 + 0.90)/9.

Although the above calculations are primarily directed to therapeutic agents that are dispersed within suppositories prepared using fatty bases (e.g. cocoa butter, semisynthetic/synthetic triglycerides), this argument may be extended to water-soluble and water-miscible systems in which the drug has been dispersed. Furthermore, the density of glycerol–gelatin suppositories is 1.2 g/cm, and therefore, after consideration of the displacement values (as described earlier in this section), the calculated amount of base is increased by a factor of 1.2. Displacement values are not considered in formulations in which the drug is soluble in the base.

## Vaginal dosage forms

The administration of drugs to the vagina is principally performed to achieve a local effect, e.g. for the treatment of infection. However, it is also known that, due to the highly vascular nature of the vaginal wall, systemic absorption of therapeutic agents may occur following administration by this route. Furthermore, absorption into the systemic circulation by this route avoids first-pass metabolism in the liver. Akin to the rectum, the fluid content within the vagina is low under normal physiological conditions.

### Advantages and disadvantages of vaginal dosage forms
#### Advantages
- Vaginal dosage forms may be successfully employed to provide a local effect for the treatment of infection (fungal and bacterial) and for the treatment of hormone deficiency (topical hormone replacement therapy for the treatment of vaginal atrophy).
- Vaginal dosage forms may be employed to provide systemic drug absorption of certain therapeutic agents. Examples of these applications include:
  - steroid hormones (e.g. oestradiol acetate) for the treatment of the systemic symptoms of the menopause and for long-term contraception
  - prostaglandins (for cervical ripening).

#### Disadvantages
- Whilst the use of topical preparations for the treatment of local disorders is accepted, the use of vaginal dosage forms for systemic drug administration is not as popular.
- Specialist advice is required concerning the administration of dosage forms to ensure correct placement and hence retention in the vagina. This may often require the use of special applicators.
- The industrial manufacture of vaginal pessaries (suppositories) is more difficult than for other common vaginal dosage forms.

## Vaginal physiology

The key aspects of vaginal physiology (Figure 7.6) that are related to drug delivery (and hence formulation) are as follows:

**Figure 7.6** Diagram of the vaginal physiology. (Adapted from Hyde J S, DeLamater J (2003) *Understanding Human Sexuality*, 8th edn. Boston: McGraw-Hill.)

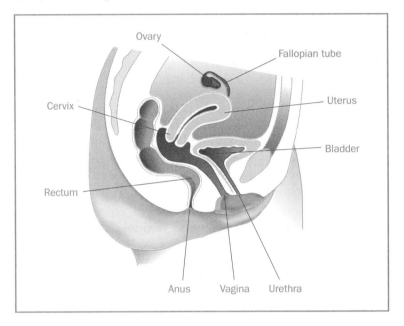

- There are ridges (folds) in the vagina, termed vaginal rugae, that increase the surface area of, and hence drug absorption from, the vagina. In addition the vaginal rugae will enhance the retention of the dosage form in the vagina (in particular solid and semisolid systems), thereby facilitating prolonged drug release to this site.
- The vaginal epithelium is composed of stratified squamous cells. The upper section of the vagina is a mucous membrane. Vaginal mucus is produced within the vagina, principally by the cervix, which, under normal physiological conditions, results in a limited volume of fluid within the vagina. Both the volume and consistency of vaginal mucus change during the menstrual cycle. The limited volume of fluid within the vagina will limit the dissolution of moderate to highly lipophilic drugs.
- The vagina is highly vascular; a plexus of arteries and veins is located round the vagina. The venous blood supply from the vagina does not enter the portal system and therefore first-pass metabolism does not occur. This renders the vagina a useful site for the systemic administration of therapeutic agent.

## Types of vaginal dosage forms

There are several types of dosage forms that are administered to the vagina. The formulation considerations for many of these types are similar to those described previously (and subsequently in more specialised chapters). Therefore only a brief overview of the specific formulation considerations that are relevant to vaginal dosage forms will be considered in this chapter. The main types of dosage forms that are administered via the vagina to achieve a local or systemic effect are: (1) semisolid formulations; (2) tablets and capsules; (3) pessaries (vaginal suppositories); and (4) vaginal implants.

### Semisolid formulations

Semisolid formulations (creams, ointments and gels) are commonly administered to the vagina (and vulva) and typically contain antimicrobial agents, hormones or contraceptive agents. The administration of these to the vagina may require a special applicator (nozzle). The formulation of creams, ointments and gels for vaginal administration is generally the same as previously described; however, in vaginal products the pH may be modified to match that of the vaginal secretions (circa 4.5). In addition, foams may be used as a drug delivery system for the vagina. These are aerosol products based on oil in water creams (aerosol products will be discussed in Chapter 8).

### Tablets and capsules

Vaginal tablets and capsules are commonly used dosage forms for the delivery of antimicrobial agents and steroid hormones. Due to the relatively low content of aqueous fluid in the vagina under normal physiological conditions, disintegrants may be incorporated into tablets to assist in tablet disintegration and subsequent drug dissolution. The formulation of tablets and capsules is described in Chapters 9 and 10. The administration of tablets and capsules to the vagina requires the use of an applicator. The patient should be advised by the pharmacist on the correct method to insert these dosage forms into the vagina (usually towards the back of the vagina).

### Pessaries (vaginal suppositories)

Pessaries (often termed vaginal suppositories) are visibly similar to rectal suppositories and are used for the treatment of infection (yeast and bacterial), vaginal atrophy and for contraceptive purposes. In addition, pessaries may be used to achieve systemic absorption of therapeutic agents. Pessaries are formulated using glycerol–gelatin or PEG bases, into which the drug is incorporated (as previously described in this chapter). The pH of pessaries that contain an aqueous phase may be buffered to pH 4.5. The administration of these dosage forms generally requires an applicator and training (akin to tablets and capsules).

### Vaginal implants

Vaginal implants are solid formulations that provide a controlled release of therapeutic agent into the vaginal fluids and may be used to achieve either a local effect or systemic absorption. One example of this is Femring, a doughnut-shaped implant that is located adjacent to the cervix and which provides controlled systemic absorption of oestradiol. The mechanical properties of the product ensure excellent retention of the dosage form at the site of application.

A diagrammatic representation of formulation considerations for rectal and vaginal preparations is shown in Figure 7.7.

## Quality control of rectal and vaginal products

As the reader will have observed the nature of many rectal and vaginal formulations (e.g. creams, ointments, enemas, solutions) is identical and therefore, the quality control methods that are relevant to these types of dosage forms have been previously described. This section will therefore define the quality control methods that are relevant for suppositories and pessaries.

1. **Concentration of therapeutic agent**: Following manufacture the concentration of therapeutic agent must lie within 95–105% of the nominal concentration. Over the shelf-life of the product the concentration of drug must not fall below 90% of the nominal amount.
2. **Appearance**: All rectal and vaginal products will have specifications for appearance.
3. **Uniformity of content**: Typically, the individual mass of the therapeutic agent in 10 suppositories/pessaries is determined. Passing the test requires the content of not more than one unit being outside the range 85–115% of the average and not more than three units being outside the range 75–125% of the average.

   If the mass of therapeutic agent in two or three units lies outside the 85–115% range but within 75–125% of the average content, the drug contents of 20 further units is determined. The batch will pass uniformity of content then if not more than three of the individual contents of the 30 units is outside the range 85–115% and none is outside 75–125% of the average content range.

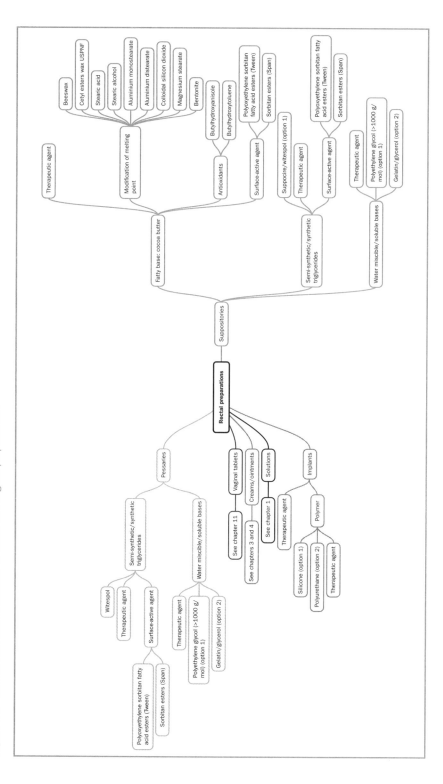

**Figure 7.7** Formulation considerations for rectal and vaginal preparations.

4. **Uniformity of mass (weight)**: In this method the individual weights of twenty suppositories/pessaries are measured and the average weight is determined. To successfully pass this test the weights of no more than two units should deviate from the average weight by more than 5% and none should deviate by more than 10%.

The reader should note that the list of methods is indicative of the quality control methods that may be employed. The reader should consult the appropriate pharmacopoeias for a more detailed description of the above methods and others that have not been explicitly covered in this chapter.

## Multiple choice questions

1. **Regarding suppositories, which of the following statements are true?**
   a. Suppositories are employed for the treatment of local disorders, e.g. haemorrhoids.
   b. Suppositories may be used to deliver therapeutic agents systemically.
   c. Suppositories are sterile products.
   d. Suppositories are normally formulated with the inclusion of colourants.

2. **Regarding pessaries, which of the following statements are true?**
   a. Pessaries are employed for the treatment of local disorders, e.g. infection.
   b. Pessaries may be used to deliver therapeutic agents systemically.
   c. Pessaries are sterile products.
   d. Pessaries are normally formulated with the inclusion of preservatives.

3. **Regarding the physiology of the rectum, which of the following statements are true?**
   a. The rectum is joined to the bottom of the sigmoid colon.
   b. The rectum is divided into two sections termed the anal canal and the ampulla – the ampulla is the smaller of the two sections.
   c. The rectal surface area is relatively small, i.e. <100 cm².
   d. Following absorption, drugs enter the systemic circulation via the haemorrhoidal vein.

4. **Factors affecting drug absorption from the rectum include which of the following?**
a. The pH of the rectal fluid.
b. The presence of faecal matter.
c. The presence of esterases in the rectal fluid.
d. The location of the suppository within the rectum.

5. **Regarding the formulation of suppositories, which of the following statements are true?**
a. Suppositories should melt or dissolve within the rectum.
b. The typical weight range for suppositories is between 1 and 4 grams.
c. The concentration of drug within suppositories typically ranges from 0.1% to 40% w/w.
d. Suppositories formulated using oleaginous bases require the addition of preservatives.

6. **Concerning the use of theobroma oil in the formulation of suppositories, which of the following statements are true?**
a. It is a natural product consisting of a mixture of fatty acid esters of glycerol.
b. It is a pure substance.
c. It exists in different polymorphic states, the number and type of which are dependent on the processing temperature.
d. The melting temperature of suppositories that have been formulated using theobroma oil may be enhanced by the addition of beeswax.

7. **In the formulation of suppositories which of the following excipients may be used for the following reasons?**
a. Surface-active agents – to facilitate manufacture of suppositories.
b. Propylene glycol – to reduce the melting point of suppository bases.
c. Colloidal silicon dioxide – to reduce the hygroscopicity of fatty suppository bases.
d. Ethanol – to enhance the solubility of the therapeutic agent within the suppository base.

8. **Concerning vaginal physiology, which of the following statements are true?**
a. The vagina is highly vascular.
b. Following absorption from the vagina, the drug enters the portal system and hence first-pass metabolism occurs.

   c. There is a limited volume of fluid within the vagina into which drug dissolution occurs.

   d. There is a high available surface area within the vagina for drug absorption.

9. **Concerning vaginal dosage forms, which of the following statements are true?**

   a. Creams and ointments may be used for vaginal administration.

   b. Pessaries are frequently formulated using oleaginous bases.

   c. Vaginal tablets are generally formulated to include disintegrants.

   d. Type 1 gels are commonly used as vaginal dosage forms.

10. **Which of the following agents may be used as the basis of suppository formulations?**

   a. Triglycerides of higher saturated fatty acids.

   b. Poly(acrylic acid).

   c. Poly(ethylene glycols).

   d. Glucose.

11. **Which of the following statements are true regarding glycerol–gelatin suppositories?**

   a. They may exert a physiological effect.

   b. They are hygroscopic.

   c. They do not need a preservative.

   d. They require a long preparation time.

12. **Which of the following statements are true regarding suppositories prepared using poly(ethylene glycol) as the suppository base?**

   a. They are prepared using low-molecular-weight grades.

   b. These bases do not melt within the rectum.

   c. They may exert a physiological effect.

   d. They do not need a preservative.

13. **Which of the following statements are true regarding drug release from glycerol–gelatin pessaries?**

   a. Increasing glycerol content will increase drug release rate.

   b. Drugs that are soluble within the pessary base will be released faster than those that are not soluble within the pessary base.

   c. Decreasing the particle size of insoluble drugs will decrease the rate of release.

   d. Ethanol may be added to enhance drug solubility within the pessary base.

14. **Which of the following statements are true regarding suppositories that have been formulated by incorporating a water-soluble drug within a lipophilic suppository base (theobroma oil)**
   a. The drug will be suspended in the base.
   b. The suppository would be expected to be hygroscopic.
   c. The presence of the drug would be expected to lower the melting point of the base.
   d. In production a lubricant is required to aid removal of the suppository from the mould.

15. **You have been asked to formulate an aqueous gel containing metronidazole designed for the treatment of vaginosis. Which of the following statements are true?**
   a. The pH of the gel should be circa 7.0.
   b. Benzoic acid would be a suitable preservative for the formulation.
   c. The gel must be isotonic.
   d. The gel must be coloured.

# chapter 8
# Respiratory dosage forms

## Overview

**In this chapter the following points will be discussed/described:**

- an overview/description of respiratory dosage forms and the rationale for their use
- formulation strategies for aerosols and related products that are specifically designed for administration to the respiratory tract
- the advantages and disadvantages of respiratory dosage forms
- considerations for the manufacture of respiratory dosage forms.

## Introduction

Drug delivery to the respiratory tract is principally performed for the treatment of local disorders, e.g. asthma and cystic fibrosis. However, due to the excellent blood perfusion of this organ, large surface area of the alveoli and the thin barrier for absorption, this route of administration may be employed to achieve drug absorption into the systemic circulation without direct passage to the liver. These features have been commercially exploited for the systemic delivery of insulin, ergotamine and other drugs for which oral absorption is inappropriate. There are several methods by which therapeutic agents may be successfully delivered to the lungs, including the use of aerosols, dry-powder inhalers (DPIs) and nebulisers. Uniquely, the performance of the respiratory drug delivery system is dependent on both the formulation and the nature/properties of the delivery system (i.e. the inhalation system).

## Advantages and disadvantages of respiratory drug delivery

### Advantages
- Respiratory drug delivery ensures that the required dose of drug is delivered to the

## KeyPoints

- Administration of dosage forms to the respiratory tract is performed for the treatment of localised disorders, e.g. asthma (β-agonists, corticosteroids), infections (gentamicin, ciprofloxacin). However, due to the extensive blood supply to the alveoli, this route of administration may be used to gain systemic absorption (without first-pass metabolism).
- The size of the inhaled solution/particles is a key determinant of the resulting clinical activity.
- The main types of dosage form that are administered by the respiratory route are solutions (for nebulisation) and particles (the latter either as a drug powder or as an aerosol). The control of particle size of solutions is performed by the nebuliser whereas the control of the particle size for dry powders/aerosols is defined by the size of the particle used in the formulation and the possible aggregation in either the solid or the suspended state.

site of action (where local effect is required). In so doing the incidence of side-effects is minimised.

- Due to the large surface area of the lungs (i.e. the alveoli), the excellent blood supply and the thin nature of the barrier between the lung and the systemic circulation, the lungs act as a portal for systemic drug absorption. Therefore, respiratory drug delivery systems offer a viable alternative to parenteral medications.
- There is a rapid onset of action following respiratory drug delivery. This is particularly beneficial for the treatment of asthma and, more recently, for the lowering of postprandial blood glucose levels.
- The delivery system used in respiratory dosage forms is, in most instances, portable and is therefore convenient for the patient to carry.

## Disadvantages

- In some dosage forms (e.g. conventional aerosols) coordination is required between activating the inhaler and inspiration. Failure to do so results in the deposition of the drug in the upper airways. There are respiratory delivery systems that do not require this coordination; however, many of these are bulky and therefore conspicuous in use.
- Deposition of drug to the lower airways may be impeded in the presence of high volumes of mucus (e.g. due to an infection).
- The physical stability of pharmaceutical aerosols may be problematic.

## Physiology of the respiratory tract

The key physiological aspects of the respiratory tract that are pertinent to respiratory drug delivery are as follows:
- The physiological role of the respiratory tract is in the transfer of oxygen into the blood from inspired air and the removal of carbon dioxide from the blood into the expired air.
- The respiratory tract (Figure 8.1) may be subdivided into several regions:
  - The upper respiratory tract, comprising:
    - nose
    - throat
    - pharynx
    - larynx.
  - The lower respiratory tract, comprising:
    - trachea, dividing into:
    - bronchi, dividing into:

**Figure 8.1** Diagrammatic representation of the respiratory tract.

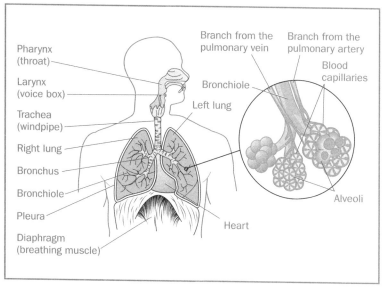

- bronchioles, dividing into:
- alveoli.

- The role of the upper respiratory tract and the trachea, bronchi and bronchioles is in the conductance of air to and from the alveoli. The role of the alveoli is in the transfer of respiratory gases.
- The diameters of the conducting airways sequentially decrease towards the alveoli. This has a profound effect on the deposition of particles/droplets within the respiratory tract.
- The bronchioles divide into a series of alveolar sacs (Figure 8.1), each of which contains circa $2-6 \times 10^8$ alveoli. An extensive capillary network is associated with these alveolar sacs and, furthermore, the diffusion barrier between the blood supply and the alveoli is relatively thin, enabling rapid drug diffusion and absorption into the bloodstream. The combined surface area of the alveoli is large (70–80 m²), which, in addition to the points raised previously, makes the respiratory tract particularly suitable for the systemic absorption of therapeutic agents.
- The epithelial cells in the conducting airways are ciliated and are responsible for the transport of foreign particles upwards towards the upper respiratory tract for subsequent elimination. Furthermore, goblet cells produce mucus which acts to trap foreign particles (prior to removal by the ciliated epithelia).

## Factors affecting the deposition of particles/droplets within the respiratory tract

All respiratory dosage forms deliver the drug to the respiratory tract as either particles or droplets. It is therefore essential that these particles/droplets are deposited to the required site of pharmacological action, e.g. the bronchioles for asthma treatment or the alveoli for systemic absorption. To achieve the optimum pharmacological effect, the pharmaceutical scientist must fully understand and embrace the factors that affect the deposition of the drug within the respiratory tract as these will influence the formulation of respiratory dosage forms. These factors are as follows: (1) the size of the inspired particles/droplets; and (2) the effect of humidity on particle size.

### The size of the inspired particles/droplets

The diameters of the various airways leading to the alveoli sequentially decrease and therefore the location of the deposited particle/droplet is dependent on particle size. The distribution of diameters of particles in an aerosol is non-normal and, as a result, these are normally plotted as log-normal distributions. Due to this heterogeneous distribution of particles, conventional parametric statistics may not be used to summarise the central tendency and variance of the distribution. The size of particles designed for pulmonary administration is defined using a parameter that accounts for this limitation and relates the diameter of the particles to the particular application, namely, the *aerodynamic diameter*. This is the diameter of a spherical particle of unit density (1 g/cm$^3$) that possesses the same gravitational settling velocity as the particle under examination and, in the case of spherical particles, may be expressed as follows:

$$d_s = \sqrt{\left(\frac{\rho}{\rho_o}\right)} \times d$$

where $d_s$ is the aerodynamic diameter; $\rho$ is the density of the particle under examination; $\rho_o$ is the density of the spherical particle (i.e. 1 g/cm$^3$); and $d$ is the diameter of the particle under examination.

Typically the aerodynamic diameter is proportional to the velocity of settling of the particle within the respiratory tract.

In addition, the aerodynamic diameter may be expressed in terms of the *mass median aerodynamic diameter* (MMAD), i.e. the aerodynamic diameter that divides the distribution equally in two parts in terms of the weight of particles.

The variance (and hence the dispersity) of the log-normal distribution of particles in therapeutic aerosols may be expressed in terms of the *geometric standard deviation* ($s_g$). The log-normal distribution will be normally distributed and the geometric standard deviation is therefore the standard deviation of the distribution following the log-normal transformation.

Following inhalation, the deposition of the particles within the respiratory tract is dependent on the size of the inhaled particles. Typically particles with a mass median aerodynamic diameter greater than 10 µm will be trapped in the trachea, whereas particles exhibiting MMADs of circa 5 µm and 2 µm will be deposited within the bronchioles and alveoli. Particles that are trapped in the higher regions of the respiratory tract will be removed from the lung by muciliary clearance (entrapment in the mucus and movement towards the mouth by the action of the cilia). Particles with an MMAD diameter of <1 µm are inhaled to the lowest sections of the lung but are then exhaled, thereby providing no pharmacological effect.

## The effect of humidity on particle size

Following entry within the respiratory tract, the particle will be exposed to a highly humid environment (circa 99%) and this results in the deposition of a layer of moisture on the surface of the particle. The effect of this deposited layer on particle size is dependent on the hydrophilicity/lipophilicity of the particle:

- *Lipophilic inhaled particles.* The adsorbed layer of moisture is negligible and therefore does not affect the MMAD of the inhaled particle.
- *Hydrophilic inhaled particles.* The adsorbed layer of moisture results in the dissolution of the hydrophilic particle. Due to the differences in the vapour pressures of water (moisture) and the solution of therapeutic agent that has formed on the surface of the hydrophilic particle (the latter being lower), further moisture is adsorbed on to the surface of the particle until the vapour pressures of the two systems are similar. The continuing adsorption of moisture on to the surface of the particle results in further dissolution of drug with the attendant increase in effective particle size and possible deposition within the higher regions of the respiratory tract.

## Mechanisms of particle deposition within the respiratory tract

The previous section has outlined the effect of particle size on the deposition of particles within the respiratory tract (and hence on the clinical efficacy of aerosol products). Particle size additionally affects the mechanisms by which the particles are deposited within the respiratory tract. There are four main mechanisms that operate in aerosol products: (1) inertial impaction; (2) gravitational sedimentation; (3) Brownian diffusion; and (4) electrostatic precipitation.

## Inertial impaction

Inertia may be defined as the property of a particle that enables it either to remain at rest or to remain in uniform motion until exposed to an external force. Following inhalation, aerosol particles must (frequently) change direction to ensure deposition at the required site within the respiratory tract. Initially the particles will move horizontally towards the back of the throat; however, the flow of air will change direction both into the trachea and then at the other main tributaries along the respiratory tract until the alveoli are reached. The inertial effect resists the change in direction of flow and, as a result, particles with sufficient momentum (the product of the velocity and the mass) will attempt to maintain their initial path of flow. This resistance results in impaction of the particle with the respiratory tract at sites prior to the bronchioles and alveoli.

The probability of impaction ($Pr_{\text{impaction}}$) may be mathematically defined as:

$$Pr_{\text{impaction}} = \frac{U_t U \sin\theta}{rg}$$

where: $U_t$ refers to the terminal settling velocity (i.e. the velocity of motion whenever the force of the falling particle downwards is equal to the drag force acting in the opposite direction; this results in a net zero force and a constant particle velocity); $U$ refers to the velocity of the air stream following inhalation; $\theta$ refers to the angle of the airflow change; $g$ refers to gravity; and $r$ is the radius of the airway.

Accordingly, the probability of the occurrence of inertial impaction increases as the angle and velocity of airflow increase, whereas the probability decreases in airways of larger radius. Inertial impaction is relevant for particles of larger MMAD (>5 µm) within the larger airways, e.g. nose, mouth, pharynx and larynx. As the airflow velocity decreases within the lower sections of the respiratory tract (due to the effects of branching of the airways providing resistance to the flow of air), the contribution of inertial impaction on the deposition of particles within the deeper regions of the respiratory tract is lower.

## Gravitational sedimentation

As the term implies, gravitational sedimentation refers to the downward movement of particles under the action of gravity. Gravitational sedimentation is an important mechanism for the deposition of particles of small MMAD (typically 1.0–5.0 µm) within the bronchioles and alveoli. Furthermore, the probability of impaction by gravitational sedimentation within these sites may be enhanced by either a steady rate of breathing or holding the breath. This is particularly relevant for the deposition of particles with an MMAD of circa 1.0 µm.

## Brownian diffusion

Brownian motion refers to the random movement of particles within a fluid (liquid or air), which, within the respiratory tract, enables small particles (<0.5 µm) to move towards and be deposited on the walls of the various sections of the respiratory tract. This is not a significant contributor for the deposition of particles from therapeutic aerosols.

## Electrostatic precipitation

This is a phenomenon in which the charge on the surface of a particle may affect the resultant deposition, i.e. a charged particle interacts with a site within the respiratory tract that possesses an opposite charge. Typically, this mechanism is not important for particles of MMAD that are greater than circa 4 µm. Furthermore, the charge on particles may induce an interaction with the plastic surfaces of containers/spacers/inhalers and may therefore compromise the effective delivery of the particles to the required site within the respiratory tract.

> **Tip**
>
> The region of deposition of the inhaled particles/droplets within the respiratory tract directly affects the clinical performance of the dosage form. Therefore, it is important that the formulation scientist designs the dosage form to ensure that, following inhalation, the particle/droplet size is of the correct MMAD to reach the target site within the respiratory tract.

## Formulation of respiratory dosage forms

There are three main types of delivery system for respiratory dosage forms: (1) metered-dose inhalers (MDIs); (2) dry-powder inhalers; and (3) nebulisers. The formulation and design of these systems are described in this section.

## Metered-dose inhalers

MDIs are a commonly used system – in use for over 50 years – for the delivery of therapeutic agents to the respiratory tract. The advantages and disadvantages associated with the use of MDIs are detailed below.

### Advantages
- portable (which is particularly useful for the treatment of acute respiratory conditions, e.g. breathlessness in asthma patients)
- low-cost dosage form
- hermetically sealed and therefore oxidation of the therapeutic agent and microbial contamination are minimised
- effective treatment of respiratory disorders.

### Disadvantages
- MDIs most frequently contain *dispersed* drug and therefore problems may arise concerning the physical stability of the formulation.
- Frequently there is ineffective use of drug. Typically the majority of drug that is emitted from the aerosol fails to reach the proposed site of pharmacological action. This point is further addressed later in this chapter.
- Clinical efficacy is often dependent on the ability of the patient to use the MDI correctly: efficacy is lowered by the inability of the patient to synchronise inhalation and actuation of the dosage form.

The main components of MDIs are the container, the metered valve and the formulation (within the container), as illustrated in Figure 8.2.

### Formulation and operation of an MDI
The formulation component of MDIs comprises two parts: the propellant and the therapeutic agent (which may be present as a solution or, most commonly, as a suspension). Following actuation, a mixture of the propellant and the therapeutic agent is released into the oral cavity of the patient in the form of droplets (circa 40 µm MMAD). The volume of product (typically 25–100 µm) dispensed from the MDI during the actuation process is controlled by a *metering valve*. In this, the depression of the stem of the canister (containing the formulation) facilitates the release of a volume of product into the mouthpiece of the device. Release of the downward pressure on the stem of the canister allows the free volume within the metering valve to be replenished. Removal of liquid propellant following actuation results in a momentary change in the vapour pressure within the canister. To correct this, liquid propellant evaporates to occupy the free volume within

**Figure 8.2**   Illustrated cross-section through a metered dose inhaler showing the location of the formulation and the design of the container.

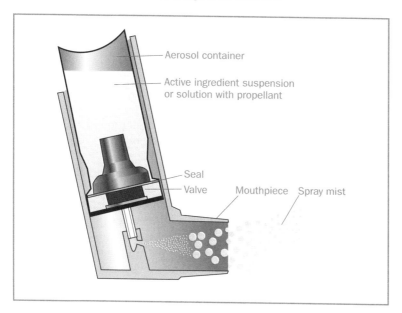

Aerosol container

Active ingredient suspension or solution with propellant

Seal

Valve

Mouthpiece

Spray mist

the canister, thereby maintaining a constant vapour pressure and hence operating pressure.

When formulated correctly the propellant will rapidly evaporate following exposure to atmospheric conditions and the droplets are reduced to solid particles of therapeutic agent, which may then be deposited at the required site in the respiratory tract. The reader will be aware of the particle size requirements for penetration into the lower regions of the respiratory tract and therefore to ensure that the therapeutic agent reaches the required site. There are three prerequisites for this process to occur successfully:

1.  The particle size of the therapeutic agent that has been incorporated into the formulation must exhibit the required diameter and polydispersity. Reduction in the particle size to achieve the required diameter is most commonly performed using milling techniques (Chapter 11). It is difficult to achieve the required polydispersity using such techniques.
2.  Following shaking of the MDI (prior to use), the drug particles must be readily resuspended. Aggregation of drug particles will increase their effective size and will lead to deposition at the incorrect region of the respiratory tract and, if the size of the aggregate is sufficiently large, clogging of the nozzle of the MDI.

3.  As highlighted previously, the formulation exits the MDI as a droplet composed of the dispersed therapeutic agent with the liquid propellant. The purpose of the propellant is therefore to provide sufficient pressure to deliver the dose of drug from the MDI to the upper respiratory tract. However, it is essential that evaporation of the propellant occurs rapidly to ensure that the solid drug particle is free to travel to the lower regions of the lung.

Based on these three requirements, the formulation of MDI involves consideration of the properties of both the propellant and the therapeutic agent.

■  *Propellants used in MDI.* The purpose of the propellant is twofold: to provide the driving pressure to force the therapeutic agent from the MDI to the upper respiratory tract and, secondly, to exhibit the required evaporation rate to facilitate particle delivery to the required site within the respiratory tract. Both the pressure within the metal canister of the MDI and the subsequent rate of evaporation of the propellant are functions of the partial vapour pressure of the propellants within the canister. To achieve these required properties mixtures of propellants are used. Currently there are two main classifications of propellants used, as follows: (1) chlorofluorocarbons (CFCs) and (2) hydrofluorocarbons.
    •  *CFCs.* CFCs have been used as propellants for many years; however, due to concerns with their contribution to the depletion of the ozone layer, their use has been prohibited in all aerosols, with the exception of MDIs (under the proviso that no suitable alternatives are available).
    The main examples of these that have been or are being used as propellants in MDI, including their physicochemical properties, are shown in Table 8.1 and Figure 8.3.

**Table 8.1**   The physicochemical properties of propellants 11 (trichloromonofluoromethane), 12 (dichlorodifluoromethane) and 114 (dichlorotetrafluoroethane)

| Propellant | Boiling point | Vapour pressure at 20°C | Freezing point |
|---|---|---|---|
| 12 | −29.8°C | 568 kPa | −158°C |
| 114 | 3.6°C | 183 kPa | −94°C |
| 11 | 23.7°C | 89 kPa | −111°C |

Propellant 12 may be used as the sole propellant in MDI formulations; however, propellants 11 and 114 are generally used in combinations with other propellants. For example, blends of propellants 11 and 12 or propellants 11, 114 and 12 are used in MDI to provide the correct range of vapour pressures (103–484 kPa) at room temperature and hence provide the required expulsion pressure, droplet size distribution and evaporation rate to ensure optimal clinical performance. Furthermore, mixtures of propellants may be required to ensure the surfactants that are used to stabilise the dispersed drug particles are soluble in the propellant. This is an important consideration in the formulation of a stable MDI product.

Calculation of the vapour pressure of propellant mixtures is performed using *Raoult's law*, which states that the vapour pressure of a mixture of propellants ($p_{total}$) is simply the sum of their partial vapour pressures (e.g. $p_1$ and $p_2$ for propellants 1 and 2, respectively).

$$p_{total} = p_1 + p_2$$

The partial vapour pressure of each component is the product of the mole fraction of that component ($\chi$) and the partial vapour pressure of the propellant ($p_i^0$):

$$p_1 = \chi_1 \times p_1^0$$

This equation can be used to calculate the total vapour pressure of a mixture of propellant 11 and propellant 12 (60:40 ratio), as follows:

- *Determination of the mole fraction of each propellant in the mixture*
  Propellant 11 (molecular weight 137.50 g/mol)

$$n_{11} = \frac{60}{137.5} = 0.44$$

**Figure 8.3**   Structural formulae of (a) dichlorodifluoromethane (propellant 12), (b) dichlorotetrafluoroethane (propellant 114) and (c) trichloromonofluoromethane (propellant 11).

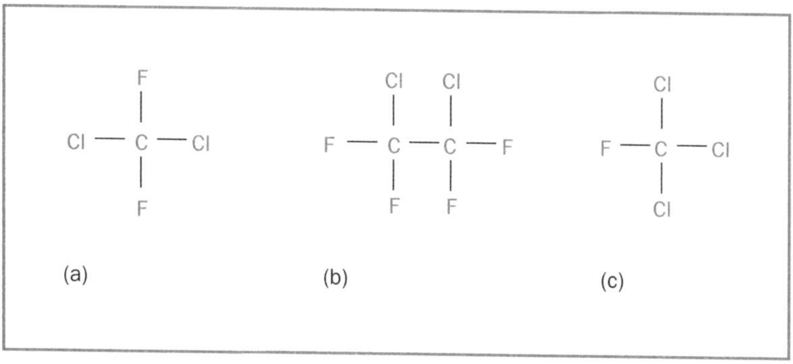

Propellant 12 (molecular weight 121.00 g/mol)

$$n_{12} = \frac{40}{121} = 0.33$$

Total number of moles of propellant

$$n_{total} = n_{11} + n_{12} = 0.44 + 0.33 = 0.77$$

Mole fraction of propellant 11

$$\chi_{11} = \left(\frac{n_{11}}{n_{11} + n_{12}}\right) = \frac{0.44}{0.44 + 0.33} = 0.57$$

Mole fraction of propellant 12

$$\chi_{12} = \left(\frac{n_{12}}{n_{11} + n_{12}}\right) = \frac{0.33}{0.44 + 0.33} = 0.43$$

- *Determination of the partial vapour pressures of each component in the mixture*
  Partial vapour pressure of propellant 11

$$p_{11} = \chi_{11} \times p_{11}^0 = 0.57 \times 89 = 50.86 \text{ kPa}$$

  Partial vapour pressure of propellant 12

$$p_{12} = \chi_{12} \times p_{12}^0 = 0.43 \times 568 = 243.43 \text{ kPa}$$

- *Calculation of the total vapour pressure within the canister*

$$p_{total} = p_{11} + p_{12} = 50.86 + 243.43 = 294.29 \text{ kPa}$$

As the reader will observe, the partial pressure of the propellant mixture may be readily manipulated by altering the ratio of the individual propellants. The partial vapour pressure of the mixture is an important consideration in the formulation of aerosols. The vapour pressure is responsible for the expulsion of the defined volume of product from the aerosol. If the pressure is too high, impaction of the expelled droplet/particle on to the surfaces of the upper respiratory tract will occur, thereby diminishing the clinical performance of the MDI. Reduction of the partial vapour pressure of the system may be achieved by increasing the mole fraction of a less volatile propellant (e.g. propellant 11) into the formulation. However, as the mole fraction of a less volatile propellant is increased, the rate of evaporation of the propellant following actuation decreases and, hence, this may reduce the percentage of the actuated dose of the therapeutic agent that reaches the lower airways.

- *Hydrofluorocarbons (HFCs).* HFCs were developed to address the deleterious effects of CFC propellants on the ozone layer. Unfortunately, although the effects on the environment are lower than for CFCs, these propellants are also damaging to the ozone layer. In spite of this HFCs represent a significant evolution in the development of propellants that are ozone-friendly. Two examples of these that are employed in MDIs are heptafluoropropane and tetrafluoroethane (Figure 8.4).

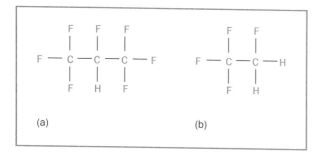

**Figure 8.4** Structural formulae of selected hydrofluorocarbons used as propellants in metered dose inhalers: (a) heptafluoropropane; (b) tetrafluoroethane.

Both heptafluoropropane and tetrafluoroethane may be employed as propellants for MDIs. One potential problem with their use is their extremely hydrophobic properties. As a result, the solubility of the commonly used surfactants (sorbitan trioleate, oleic acid, sorbitan sesquioleate) may not be sufficiently high to stabilise the formulation.

The physicochemical properties of the HFC propellants are similar to CFC, reflecting their applicability as propellants for MDIs. These are summarised in Table 8.2. As before, modification of the vapour pressure of the propellant system may be achieved by blending the HFC propellants (according to Raoult's law).

**Table 8.2** Physicochemical properties of selected hydrofluorocarbon propellants

| Propellant | Boiling point | Vapour Pressure | Freezing point |
|---|---|---|---|
| Tetrafluororethane | −26.2°C | 569 kPa (at 20°C) | −108°C |
| Heptafluoropropane | −16.5°C | 398 kPa (at 20°C) | −131°C |

There are other formulation considerations for MDI. These are summarised below.

## The physicochemical properties of the therapeutic agent

There are two key physicochemical properties that affect the stability and clinical performance of MDI – the solubility of the therapeutic agent in the propellant system and the MMAD (and polydispersity) of the dispersed particles. If the drug shows limited solubility in the propellant that is susceptible to changes in storage conditions, this may result in fluctuations in the solubility of the drug in the propellant that may, in turn, result in recrystallisation of the drug. This will result in the generation of larger particles. It is additionally important that the MMAD (and polydispersity) of the therapeutic agent that is to be incorporated into the propellant system is within the inspirable size range (remembering that the MMAD of particles that will be deposited in the lower airways is circa 1–5 µm). The drug particle size should be manipulated to the required size fraction by milling techniques.

## Maintenance of the physical stability of the MDI

Most commonly the therapeutic agent is present in the MDI as dispersed particles within the liquid propellant and therefore there are similar concerns regarding the physical stability of the formulation. Irreversible aggregation of the particles (caking) will result in problems regarding both the successful deposition of drug within the respiratory tract and the successful discharge of the particles through the orifice of the MDI. To obviate these problems surface-active agents of low hydrophile–lipophile balance are incorporated into the liquid propellant which will adsorb to the dispersed particles and, in so doing, will stabilise the drug suspension (see Chapter 3 for a full description of the mechanism of stabilisation). Examples of surface-active agents that are used for this purpose (generally in concentrations ranging from 0.1% to 3.0% w/w) include:

- sorbitan trioleate
- sorbitan sesquioleate
- oleic acid.

Unlike suspensions designed for oral application, enhancement of suspension stability by modification of the viscosity of the liquid propellant phase is not performed for MDI.

## Enhancement of drug solubility in the propellant

In some aerosol products, e.g. those for topical administration, the pharmaceutical scientist may wish to dissolve the therapeutic agent in the propellant mixture. For this purpose ethanol is frequently added as a co-solvent to facilitate this demand. This approach is infrequently used for MDI as the presence of ethanol will result in the generation of droplets after actuation that require a longer period to evaporate (due to the vapour pressure of ethanol being substantially lower than that of the propellant mixture). This may affect the clinical performance of the MDI.

## Spacer devices for MDIs

Previously in this section, the importance of the evaporation rate of the propellant after actuation to the subsequent deposition of the therapeutic agent within the respiratory tract was detailed. Furthermore, the reliance of conventional MDI on the ability of the patient to coordinate actuation and inhalation was discussed. One method by which these two problems may be minimised is the use of spacer devices (Figure 8.5).

**Tip**

Formulations that are delivered to the respiratory tract using metered-dose inhalers are generally composed of the drug (dispersed or dissolved), a surfactants(s), the propellant(s) and, occasionally, a co-solvent (to enhance drug solubility within the propellant).

**Figure 8.5**   Diagrammatic presentation of a spacer device, showing the clinical use of the spacer with the MDI.

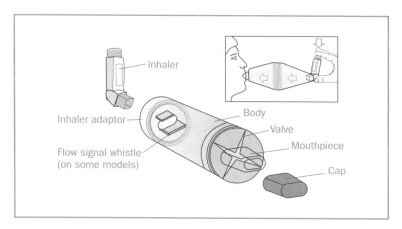

Following actuation the propellant droplets (containing suspended particles of therapeutic agent) will reside within the spacer device, thereby facilitating evaporation of the propellant and allowing the patient to inspire the dose of drug without the need for coordination between actuation of the MDI and inhalation. One disadvantage of spacer systems is their size (rendering them conspicuous in use and relatively non-portable).

## Manufacture of MDIs

There are two methods by which MDIs are manufactured:
(1) pressure filling; and (2) cold filling.

### Pressure filling

In this method the therapeutic agent (and excipients) is mixed with a portion of a propellant which possesses a relatively high boiling point and low vapour pressure, e.g. trichloromonofluoromethane (23.7°C and 89 kPa, respectively) at or below 20°C. This drug/propellant slurry is then filled into the canister and the canister is then crimped with the attached valve. The second more volatile propellant is then filled into the canister through the valve under pressure until the correct propellant blend is attained. Although this technique is straightforward, it is dependent on the use of a less volatile propellant whose boiling point is greater than that of the temperature in the mixing area.

### Cold filling

In cold filling the drug (and excipients) and propellants are mixed under cold conditions (circa −30°C) and filled into the canister (which is subsequently crimped with the attached valve). Propellant is then added into the canister through the valve at this low temperature until the correct mass has been added.

## Dry-powder inhalers

DPIs are respiratory dosage forms in which a powder (containing the therapeutic agent) is inhaled into the respiratory tract. The flow of the powder is activated whenever the patient inhales and therefore no propellant is required. It has been reported that MDIs are more efficient respiratory delivery systems than their dry-powder comparators.

There are a number of designs of these systems:

- *Inhalers in which the drug (and excipients) are present within a hard gelatin capsule.* In these the hard gelatin capsule (containing the therapeutic agent) is located inside the device. Prior to use the gelatin capsule is pierced in two locations to enable airflow through the capsule. As the patient inhales, a rotor is activated, resulting in turbulent airflow (>35 l/min), which carries the powder to the patient.
- *Inhalers in which the drug (and excipients) are present within a blister pack.* In these systems, a multidose unit comprising four or eight individual doses is located within a circular device containing a mouthpiece (Figure 8.6). After the device has been assembled the lid is elevated and the individual disc containing the therapeutic agent (and excipients) is pierced. Inhalation by the patient results in the passage of air through the device, which delivers the bioactive powder to the patient.
- *Inhalers in which the drug (and excipients) are present within the inhaler.* In these systems the inhaler is preloaded with

**Figure 8.6** Diagrammatic representation of a diskhaler device, illustrating the individual components and assembly.

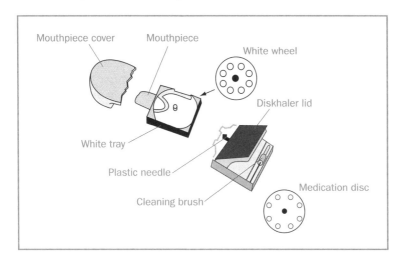

multiple doses of drug. As before, the packaging containing each dose is broken and the therapeutic agent (and excipients) are carried to the patient during inhalation. These systems may be replenished with drug by incorporating a replacement reservoir. In an alternative strategy the drug (*without excipient*) is located within a reservoir (not as individual units). The device is activated (e.g. by twisting the base), resulting in the passage of drug into a chamber. As before, when the patient inhales, air is drawn through the device and the powdered drug will be delivered to the patient.

## Formulation considerations for DPIs

The formulation of DPIs is relatively straightforward, particularly when compared to MDIs. The major formulation concerns for DPIs are as follows:

> **Tip**
>
> DPIs do not use a liquid propellant. The powder enters the respiratory tract as the patient inhales air through the device.

- *Therapeutic agent.* As before, the particle size (MMAD) of the therapeutic agent should be less than 5 µm.
- *Excipients.* Excipients are used in the majority of DPIs for two main reasons:
  - *To facilitate production.* As the dose of therapeutic agent in the DPI is generally low, the addition of an inert excipient may be used to enhance the quality of the filling process.
  - *To improve the flow properties of the therapeutic agent during inhalation.* To ensure optimum clinical efficacy,

it is essential that following inhalation the flow properties of the therapeutic agent are sufficient to enable the dose of drug to be efficiently removed from the device. If the drug particles are aggregated or if there is an interaction between the drug and the plastic interior of the device, this will result in both a variable and inaccurate dose of drug being delivered to the patient. Therefore, to address this problem the therapeutic agent is frequently mixed with an inert excipient, most commonly lactose (MMAD 30–60 µm). The larger particle size of the excipient ensures that it does not enter into the lower airways. Within the device the drug particles are adsorbed on to the larger lactose particles and, in so doing, their flow properties are improved. It is essential, however, that dissociation of the drug/carrier interaction occurs prior to impaction of the lactose carrier within the oral cavity/pharynx.

The interaction between the therapeutic agent and the carrier is dependent on several factors, including:

- the surface areas of the particles
- the surface energies of the particles
- the morphologies of the particles.

These factors are dependent (to a large degree) on the techniques used to process the powders, in particular, the milling conditions (Chapter 11).

## Solutions for nebulisation

Nebulisation involves the application of energy (either a high-velocity gas or by the use of ultrasonic systems) to a solution of a therapeutic agent and results in the formation of droplets of solution, which are then inspired by the patient through a facemask. The energy source is provided by a nebuliser (most commonly a jet or ultrasonic nebuliser). The use of nebulisers is generally reserved for the treatment of acute conditions (e.g. acute asthma, respiratory infection) or in those patients who have difficulties using other respiratory dosage forms.

### Formulation of solutions for nebulisation

The general formulation considerations for solutions for nebulisation are similar to those described in Chapter 1 and therefore in this section only the particular formulation requirements for solutions designed for administration to the respiratory tract will be discussed.

### Vehicle

The vehicle for solutions designed for nebulisation is water for injections (nebuliser solutions are sterile). Due to the possibility of acidic solutions (pH <5) causing bronchoconstriction, the pH of nebuliser solutions is greater than 5. This may be modified via the use of buffers (e.g. citrate, phosphate).

### Co-solvents

Co-solvents may be used in nebuliser solutions to increase the solubility of the drug. However, as in other applications, care should be taken when selecting the concentration of these to ensure that there is no toxicity to the respiratory epithelia. Examples of co-solvents that may be used include:

- propylene glycol
- glycerol
- ethanol.

### Osmolality-modifying agents

The use of hypo- and hyperosmotic solutions within the respiratory tract has been linked to bronchoconstriction and therefore solutions designed for nebulisation should be formulated to be iso-osmotic. As described previously (Chapter 5), modification of the osmolality of nebuliser solutions may be performed by adding the appropriate ionic concentration of, for example, sodium chloride, potassium chloride, mannitol.

### Miscellaneous agents

Other components that may be present in nebuliser solutions include:

- antioxidants (water-soluble examples; however, it should be noted that sulphites may cause bronchospasm)
- preservatives: some multidose nebuliser solutions contain preservatives; however, the vast majority of formulations do not contain these excipients. Generally nebuliser solutions are packaged as single-dose vials/ampoules and, as such, do not require the addition of preservatives.

A diagrammatic representation of formulation considerations for respiratory dosage forms is shown in Figure 8.7.

## Quality control of respiratory products

The three main categories of respiratory products have been defined in this chapter as (a) metered dose inhalers, (b) dry powder inhalers and (c) solutions for nebulisation. The details of the

**Figure 8.7** Formulation considerations for respiratory dosage forms.

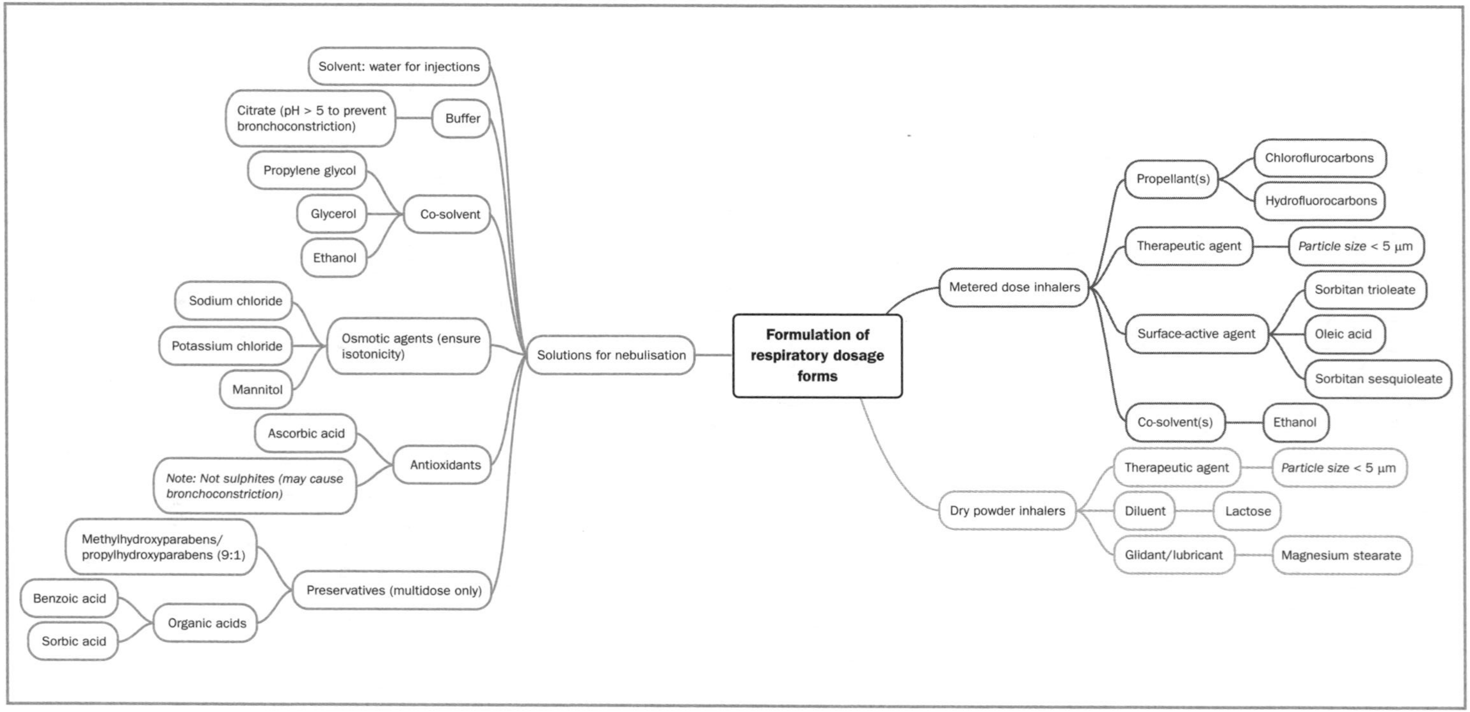

quality control of these products will focus primarily on the first two categories; the quality control of solutions for nebulisation being similar to that for other solutions (as described in previous chapters). Thus, the following analyses are typically applied to respiratory products following manufacture and over the designed period of the shelf-life:

1. **Concentration of therapeutic agent**: Following manufacture the concentration of therapeutic agent must lie within 95–105% of the nominal concentration. Over the shelf-life of the product the concentration of drug must not fall below 90% of the nominal amount.

2. **Appearance**: The appearances of these products are defined in a product specification, adherence to this being examined both following manufacture and storage.

3. **Uniformity of dose**: Assessment of the uniformity of dose of respiratory products is performed by administering a dose from the product, which is transported by the application of a vacuum prior to deposition of the drug on to a collection unit (e.g. filter disc) and quantification using an appropriate analytical method. Collection of a series of doses is performed at different stages of the contents of the device, i.e. at the start (three doses), middle (four doses) and towards the end of the contents (three doses). Typically, the product will pass the test if in nine out of ten results (doses) the drug content lies between 75% and 125% of the average value and all doses range between 65% and 135% of the average content. If two or three values lie outside the 75–125% range then the test is repeated for two more inhalers. The test is passed if not more than three of the thirty doses (3 inhalers ×10 doses each) lie outside the range 75–125% and no value lies outside the range 65–135%.

4. **Number of deliveries per metered dose inhaler**: This test examines the number of deliveries from the inhaler and compares this to the number specified on the label. The total number of deliveries (actuations) must not be less than the number defined in the product specifications.

5. **Fine particle dose and particle size distribution**: Given the importance of particle size in the performance of respiratory products, pharmacopoeial tests are available to characterise the particle size properties generated by such products. There are several methods that are used, namely the (a) glass impinger, (b) the metal impinger, (c) the multi-stage liquid impinger and (d) the Andersen sizing sampler. All three methods examine the deposition of particles/droplets within different chambers and attempt to generate information that relates to the possible clinical performance of these systems.

Movement of the dose through the device is performed using a vacuum. In these methods, the mass fractions of the dose collected at different stages of these devices are collected and quantified, this information being related to the fine particle dose (the mass of active ingredient that is less than 5μm). The reader should consult the appropriate pharmacopoeias for a more detailed description of the design and operation of the various methods described above.

The reader should note that other tests, similar to those described in earlier chapters, are applied to solutions for nebulisation. Furthermore, the list of methods is indicative of the quality control methods that may be employed. The reader should consult the appropriate pharmacopoeias for a more detailed description of the above methods and others that have not been explicitly covered in this chapter.

## Multiple choice questions

1. **Concerning respiratory drug delivery, which of the following statements are true?**
   a. Delivery of therapeutic agents to the respiratory tract is performed primarily for the treatment of local conditions, e.g. asthma, infection.
   b. The treatment of asthma involves the deposition of the therapeutic agent at the alveoli.
   c. Systemic absorption of certain therapeutic agents occurs following respiratory delivery.
   d. Respiratory drug delivery is associated with a slow onset of drug action.

2. **Regarding the physiology of the respiratory tract, which of the following statements are true?**
   a. The diameters of the conducting airways decrease towards the alveoli.
   b. Drug absorption from the respiratory tract occurs at the alveoli.
   c. The surface area of the alveoli is >100 m².
   d. The primary mechanism of elimination of particles >5 μm in diameter is expiration.

3. **Which of the following factors influence the deposition of particles within the respiratory tract?**
   a. The size of the inspired droplets: increasing particle/droplet size results in greater deposition within the lower airways.
   b. Humidity.
   c. The vapour pressure of the propellant.
   d. The patient's inhaler technique.

4.  **Concerning the formulation and use of metered-dose inhalers, which of the following statements are true?**
a.  Oxidation of drugs is minimised.
b.  Metered-dose inhalers are usually formulated to ensure that the drug is soluble within the propellant system.
c.  Metered-dose inhalers require the inclusion of a preservative.
d.  Metered-dose inhalers may be easily formulated to contain water.

5.  **Concerning propellants for metered-dose inhalers, which of the following statements are true?**
a.  The vapour pressure remains constant throughout the lifetime of use of the inhaler.
b.  The use of chlorofluorocarbons has increased in recent years.
c.  The volume of propellant released upon actuation is controlled by a metering valve.
d.  All metered-dose inhalers are filled under atmospheric conditions.

6.  **The following excipients are included in metered-dose inhalers for the following reasons:**
a.  Ethanol – to increase the vapour pressure of the propellant.
b.  Tween 80 – to enhance the solubility of the therapeutic agent within the propellant.
c.  Oleic acid – to stabilise the suspended drug in the propellant.
d.  Antioxidants – to inhibit oxidative degradation of the therapeutic agent.

7.  **Concerning the formulation of dry-powder inhalers, which of the following statements are true?**
a.  Liquid propellants are not required.
b.  The mass median aerodynamic diameter (MMAD) of the therapeutic agent should be less than 5 µm.
c.  Lactose is commonly used to improve the flow properties of the powdered drug.
d.  Following inspiration, both lactose and the powdered drug reach the site of action.

8.  **The advantages of metered-dose inhalers include:**
a.  Administration of high doses of therapeutic agent.
b.  Convenience for the patient.
c.  Requirement for patient's ability to coordinate actuation and inhalation.
d.  Greater efficiency than nebulisers.

9. **Concerning nebulisers, which of the following statements are true?**
   a. Nebulisers require either a compressed-gas source or an ultrasonic device.
   b. Nebulisers are portable and convenient to use.
   c. They require patient coordination.
   d. They are suitable for the delivery of all drugs.

10. **Regarding the formulation of solutions for nebulisation, which of the following statements are true?**
    a. The pH of nebuliser solutions is always greater than 5.
    b. Propylene glycol may be employed as a co-solvent in nebuliser solutions.
    c. Nebuliser solutions should be hypertonic.
    d. In general, preservatives are not required in nebuliser solutions.

11. **A metered-dose inhaler has been formulated containing a propellant mixture, a hydrophilic drug (salbutamol sulphate) and a surface-active agent (Tween 80). The boiling point of the propellant mixture is 6°C. Blockage of the exit orifice of the inhaler was observed following actuation of the device. Which of the following explanations would account for this observation?**
    a. The particle size of the dispersed drug within the metered-dose inhaler was too large.
    b. The drug precipitated out of solution, the large crystals blocking the exit orifice.
    c. The concentration but not the type of surfactant was inappropriate.
    d. The concentration and the type of surfactant were both inappropriate.

12. **Which of the following statements are true regarding the efficacy of metered-dose inhalers containing CFC propellants?**
    a. The efficacy of performance increases as the partial vapour pressure of the propellant increases (up to a defined partial vapour pressure).
    b. The efficacy of performance increases as the concentration of surfactant increases (up to a defined concentration).
    c. The efficacy of performance increases as the solubility of the drug in the propellant increases (up to a defined concentration).
    d. The efficacy of performance increases as the diameter of the exit orifice increases.

13. **Which of the following are advantages of a pressurised metered-dose inhaler?**
   a. Easy delivery of high doses
   b. Convenience
   c. No need for patient coordination
   d. Greater efficiency than a nebuliser.

14. **Which of the following statements are true concerning the physical properties of an aerosol?**
   a. Particles that have the same shape have the same aerodynamic diameter.
   b. Particle density is important when determining aerodynamic diameter.
   c. Particles between 1 μm and 10 μm are important for drug delivery to the airways.
   d. The probability of inertial impaction is not related to particle velocity.

15. **Which of the following statements are true concerning a metered-dose inhaler that has been formulated containing a water-soluble β2 agonist, sorbitan trioleate and a HFC based propellant system?**
   a. Controlled flocculation of suspended drug particles will enhance both the physical stability and clinical performance.
   b. Particle sedimentation does not occur upon storage.
   c. Physical stability and clinical performance will be improved by the addition of Tween 80.
   d. Controlled flocculation may be engineered by the addition of electrolytes.

# chapter 9
# Solid-dosage forms
# 1: tablets

## Overview

**In this chapter the following points will be discussed/described:**

- an overview/description of the various types of solid-dosage forms and the rationale for their use
- formulation strategies for solid-dosage forms
- an overview/description of the methods of manufacture of solid-dosage forms, including post-manufacture processes (e.g. coating)
- the advantages and disadvantages of the different methods that are used to manufacture solid-dosage forms
- a detailed description of the excipients used in the formulation and manufacture of solid-dosage forms.

## Introduction

Solid-dosage forms broadly encompass two types of formulation, namely tablets and capsules (capsules are described in Chapter 10). It has been estimated that solid-dosage forms constitute circa 90% of all dosage forms used to provide systemic administration of therapeutic agents. This highlights the importance of these dosage forms in the treatment and management of disease states. The widespread use of tablets has been achieved as a result of their convenience and also the diversity of tablet types. Examples of tablet types available include: (1) conventional compressed tablets; (2) multiple compressed tablets; (3) enteric-coated tablets; (4) sugar-coated tablets; (5) film-coated tablets; (6) chewable tablets; (7) effervescent tablets; (8) buccal and sublingual tablets; and (9) vaginal tablets.

## KeyPoints

- Solid-dosage forms encompass the largest category of dosage forms that are clinically used.
- There are several types of tablet solid-dosage form that are designed to optimise the absorption rate of the drug, increase the ease of administration by the patient, control the rate and site of drug absorption and mask the taste of a therapeutic agent.
- The formulation of tablets involves the use of several components, each of which is present to facilitate the manufacture or to control the biological performance of the dosage form.
- There are four main methods by which tablets may be manufactured: (1) wet granulation; (2) dry granulation (slugging); (3) direct compression; and (4) roller compaction. There are advantages and disadvantages associated with the use of these methods.

### Conventional compressed tablets

These tablets are designed to provide rapid disintegration and hence rapid drug release, and represent a significant proportion of tablets that are clinically used. The manufacture of these tablets involves the compression of granules or powders (both containing drug) into the required geometry. Following ingestion, the tablets will disintegrate within the gastrointestinal tract (stomach), allowing the drug to dissolve in the gastric fluid and, ultimately, be absorbed systemically.

### Multiple compressed tablets

These are tablets that are composed of at least two layers. Typically there are two designs of multiple compressed tablets: (1) multiple-layered; and (2) compression coated. In the former design the first layer is formed by a relatively light compression of the drug containing powder mix/granules. The next layer is then formed by compression of the powder/granule mix (containing drug) on top of the lightly compressed first layer. Additional layers are formed in a similar fashion. In the second approach the initial layer is prepared by light compression (as described above), removed and located in a second tablet press. The granules/powders of the second coat are fed into the press and allowed to form a constant mass around the surface (and edges) of the pressed tablet prior to compression to form the finished product. It is, of course, possible to prepare tablets containing more than two layers although, in so doing, the complexity of the manufacturing process is dramatically increased.

There are several applications of the use of multiple compressed tablets, including:

- the separation of drugs into separate layers that may be incompatible when formulated as a (single-layer) conventional tablet
- the delivery of therapeutic agents at different rates or to different sites within the gastrointestinal tract from a single tablet. The dissolution of the layers of the tablet or, indeed, the dissolution/diffusion of the drug from the layer may be controlled by the inclusion of polymeric excipients
- the production of tablets that are coated. This is important in cases where the drug has a bitter taste or where the drug is irritant to the stomach (e.g. non-steroidal anti-inflammatory drugs) or is chemically unstable under acidic conditions.

### Enteric-coated tablets

These are tablets that are coated with a polymer that does not dissolve under acidic conditions (i.e. the stomach) but does dissolve under the more alkaline conditions of the small intestine (i.e. pH > 4). Enteric polymers are primarily employed as coatings of conventional tablet dosage forms and, by inhibiting the dissolution of the therapeutic agent within the stomach, offer protection against possible drug degradation (e.g. erythromycin) or irritation of the gastric mucosa (e.g. non-steroidal anti-inflammatory drugs). Following dissolution of the coating, the tablet will disintegrate and the drug will dissolve in the gastric fluids (thereby facilitating absorption).

Examples of polymers that are used for this purpose include: (1) cellulose actetate phthalate/cellulose acetate butyrate; (2) hydroxypropylmethyl cellulose succinate; and (3) methacrylic acid co-polymers (Eudragit).

## Cellulose acetate phthalate/cellulose acetate butyrate
These are cellulose derivatives in which a proportion of the hydroxyl groups have been esterified with either phthalic acid or butyric acid. Dissolution of the polymer occurs in solutions where the pH > 6.

## Hydroxypropylmethyl cellulose succinate
This is a cellulose derivative in which a proportion of the hydroxyl groups on hydroxypropylmethyl cellulose have been esterified with either acetic acid or succinic acid. Dissolution of this polymer occurs within the intestine, as it is insoluble in the stomach.

## Methacrylic acid co-polymers (Eudragit)
These are a range of co-polymers based on methacrylic acid in which the pendant carboxylic acid groups have been esterified to form ethyl methacrylate, dimethylaminomethacrylate, butylmethacrylate and methyl methacrylate. The presence of the wide range of functional groups enables the production of co-polymers that exhibit a range of solubilities. One example of the Eudragit family that is employed as an enteric coating is poly(methacrylic acid-co-methylmethacrylate) (Figure 9.1).

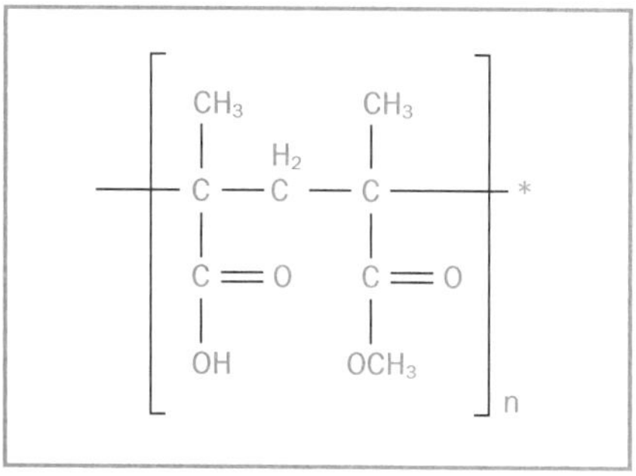

**Figure 9.1**  Structural formula of poly(methacrylic acid-co-methylmethacrylate).

Two commercial co-polymers are available which resist dissolution in the stomach but dissolve at higher pH, namely Eudragit L 100 (poly(methacrylic acid-co-methylmethacrylate), 1:1 molar ratio of monomers), which is soluble in the intestinal fluid from pH 5.5, and Eudragit S 100 (poly(methacrylic acid-co-methylmethacrylate), 1:2 molar ratio of monomers), which is soluble in the intestinal fluid from pH 7.

## Sugar-coated tablets

These are conventional tablets that have been coated with a concentrated sugar solution to improve the appearance of the formulation and/or to mask the bitter taste of the therapeutic agent. The use of sugar coatings has dramatically decreased due to the advent of film-coated tablets (as a result of the improved mechanical properties of the latter coating).

## Film-coated tablets

These are conventional tablets that have been coated with a polymer or a mixture of polymers (and, when required, a plasticiser to render the coating flexible). Film coatings show improved mechanical properties when compared to sugar coatings and, furthermore, film coatings may be deposited over embossed markings on the tablet surface. Film coatings are generally less elegant than sugar coatings. Examples of polymers that are used to film-coat tablets (and which dissolve in the stomach to enable tablet disintegration and drug dissolution) include:

- hydroxypropylmethylcelluose
- hydroxypropylcellulose
- Eudragit E 100 (a co-polymer of butylmethacrylate, 2-dimethyl-aminoethylmethacrylate and methylmethacrylate, 1:2:1)

In addition to improving the appearance of conventional tablets, film coatings are employed to control the rate and duration of drug release or to target drug release to certain regions of the gastrointestinal tract, e.g. the colon. If the film coating is insoluble, the tablet will retain its shape during transit along the gastrointestinal tract. Drug release occurs by diffusion through the insoluble coating and subsequent partitioning into the gastrointestinal fluids. Examples of polymers that may be used for this purpose include: (1) ethylcellulose; (2) Eudragit RS and RL.

### Ethylcellulose

Ethylcellulose is an ethyl ether of cellulose in which there is complete substitution of the hydroxyl groups with ethoxy groups (Figure 9.2). It is insoluble in aqueous solutions at all pH values. When used as a coating, a plasticiser (e.g. tributyl citrate, triethyl citrate) is required to lower the glass transition temperature of the polymer and hence to enhance the flexibility of the coating.

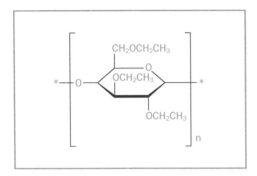

**Figure 9.2**   Structural formula of ethylcellulose.

## Eudragit RS and RL

Eudragit RS and RL are methacrylate co-polymers that are insoluble, with the RL grade being more permeable than the RS grade. Chemically the RL grade is composed of poly(ethyl acrylate-co-methyl methacrylate-co-triethylammonioethyl methacrylate chloride), the ratios of the three monomers being 1:2:0.2, respectively. Similarly, the RS grade is a co-polymer of the above three monomers, with a different ratio of monomers (1:2:0.1, respectively).

The use of film coatings to target drug release within the gastrointestinal tract requires the use of polymers that dissolve within certain pH ranges. For example, enteric coatings offer drug targeting to regions of the gastrointestinal tract in which the pH is greater than 5.5 (e.g. Eudragit L 100). Targeting drug release to the colon involves the use of polymer coatings that dissolve at higher pH values (>7), e.g. Eudragit S 100.

## Chewable tablets

As indicated by the name, these tablets are chewed within the buccal cavity prior to swallowing. The main applications for this dosage form are:

- for administration to children and adults who have difficulty in swallowing conventional tablets
- antacid formulations in which the size of the tablet is normally large and the neutralisation efficacy of the tablet is related to particle size within the stomach.

Conversely, chewable tablets are not conventionally used if the drug has issues regarding taste acceptability.

## Effervescent tablets

Effervescent tablets are added to aqueous solutions where they will rapidly disintegrate and produce either a drug suspension or an aqueous solution. The disintegration of the tablet is due to a chemical interaction that occurs between two components: (1) an organic acid (e.g. citric acid); and (2) sodium bicarbonate in the presence of water. The evolution of carbon dioxide from this reaction results in tablet disintegration. The patient then consumes the solution/suspension. The main advantage of the use of effervescent tablets is the production of a dosage form from which the therapeutic agent is more rapidly absorbed than from alternative solid-dosage forms (e.g. conventional tablets). Conversely, the main disadvantages of this type of dosage form are the possible (un)availability of water and the need to package these tablets in moisture-impermeable packaging (typically aluminium foil), to inhibit the interaction between the acid and sodium bicarbonate due to the presence of environmental moisture.

## Buccal and sublingual tablets

Buccal and sublingual tablets are dosage forms that are held within the oral cavity and slowly dissolve; the drug being absorbed across the buccal mucosa to produce a systemic effect. The type of tablet dictates the location within the oral cavity. Accordingly buccal tablets are positioned between the cheek and the gingiva whereas sublingual tablets are positioned underneath the tongue. These tablets are employed to achieve either rapid absorption into the systemic circulation (e.g. glyceryl trinitrate sublingual tablets) or, alternatively, to enable systemic drug absorption in situations where oral drug delivery is inappropriate, e.g. nausea. Drug absorption across the buccal mucosa avoids first-pass metabolism. Typically buccal and sublingual tablets should be formulated to dissolve slowly in vivo (and not disintegrate), to be retained at the site of application and should not contain components that stimulate the production of saliva.

## Vaginal tablets

These are ovoid-shaped tablets that are inserted into the vagina (using a special inserter). Following insertion, retention and

### Tips

- The different types of solid-dosage forms are designed to provide flexible platforms for the delivery of therapeutic agents to the gastrointestinal tract.
- Coating tablets with polymers offers opportunities for the release of the therapeutic agent at particular regions within the gastrointestinal tract, e.g. the small intestine, colon.
- Effervescent tablets are used to enhance the absorption of poorly soluble drugs.
- The use of multilayered tablets has decreased in recent years.

slow dissolution of the tablet occur, releasing the therapeutic agent to provide the local pharmacological effect (e.g. for the treatment of bacterial or fungal infection). Vaginal tablets may also be used to provide systemic absorption of therapeutic agents. In a similar fashion to buccal/sublingual tablets, it is important that dissolution, and not disintegration, of the tablet occurs in vivo, as disintegration will reduce tablet retention within the vagina.

## Advantages and disadvantages of tablets as dosage forms

As stated in the introduction, tablets are the most popular dosage form used today and therefore there are several advantages associated with their use. However, it is also important to highlight the disadvantages associated with their use.

### Advantages
- Tablets are convenient to use and are an elegant dosage form.
- A wide range of tablet types is available, offering a range of drug release rates and durations of clinical effect. Tablets may be formulated to offer rapid drug release or controlled drug release, the latter reducing the number of daily doses required (and in so doing increasing patient compliance).
- Tablets may be formulated to release the therapeutic agent at a particular site within the gastrointestinal tract to reduce side-effects, promote absorption at that site and provide a local effect (e.g. ulcerative colitis). This may not be easily achieved by other dosage forms that are administered orally.
- Tablets may be formulated to contain more than one therapeutic agent (even if there is a physical or chemical incompatibility between each active agent). Moreover, the release of each therapeutic agent may be effectively controlled by the tablet formulation and design.
- With the exception of proteins, all classes of therapeutic agents may be administered orally in the form of tablets.
- It is easier to mask the taste of bitter drugs using tablets than for other dosage forms, e.g. liquids.
- Tablets are generally an inexpensive dosage form.
- Tablets may be easily manufactured to show product identification, e.g. exhibiting the required markings on the surface.
- The chemical, physical and microbiological stability of tablet dosage forms is superior to other dosage forms.

### Disadvantages

- The manufacture of tablets requires a series of unit operations and therefore there is an increased level of product loss at each stage in the manufacturing process.
- The absorption of therapeutic agents from tablets is dependent on physiological factors, e.g. gastric emptying rate, and shows interpatient variation.
- The compression properties of certain therapeutic agents are poor and may present problems in their subsequent formulation and manufacture as tablets.
- The administration of tablets to certain groups, e.g. children and the elderly, may be problematic due to difficulties in swallowing. These problems may be overcome by using effervescent tablet dosage forms.

## Manufacture of tablets

There are four main methods by which tablets are manufactured: (1) wet granulation; (2) dry granulation; (3) direct compression; and (4) roller compaction (chilsonisation).

The choice of manufacturing process employed is dependent on several factors, including the compression properties of the therapeutic agent, the particle size of the therapeutic agent and excipients and the chemical stability of the therapeutic agent during the manufacturing process. Generically, the manufacture of tablets may be considered to consist of a series of steps:

- mixing of the therapeutic agents with the excipients
- granulation of the mixed powders (note: this is not performed in direct compression)
- mixing of the powders or granules with other excipients (most notably lubricants)
- compression into tablets.

The details of each of these steps will vary depending on the manufacturing method used.

### Excipients used in the manufacture of tablets

The choice of excipients required for tablet manufacture is dependent on the process employed. Typically the following excipients are used in the manufacture of conventional tablets: (1) diluents/fillers; (2) binders; (3) disintegrants; (4) lubricants; (5) glidants; and (6) miscellaneous.

### Diluents/fillers

Diluents are employed in the formulation of tablets (by all methods) to increase the mass of the tablets that contain a low concentration of therapeutic agent and thereby render the manufacturing process more reliable and reproducible. Diluents must exhibit good compression properties and be inexpensive. Examples of diluents for tablets are: (1) anhydrous lactose; (2) lactose monohydrate; (3) spray-dried lactose; (4) starch; (5) dibasic calcium phosphate; (6) microcrystalline cellulose (MCC); and (7) mannitol.

### Anhydrous lactose

Anhydrous lactose is either pure anhydrous β lactose (*O*-β-D-galactopyranosyl-(1-4)-β-D-glucopyranose) or a mixture of anhydrous β lactose (70–80%) and 20–30% anhydrous α-lactose (*O*-β-D-galactopyranosyl-(1-4)-α-D-glucopyranose) (Figure 9.3).

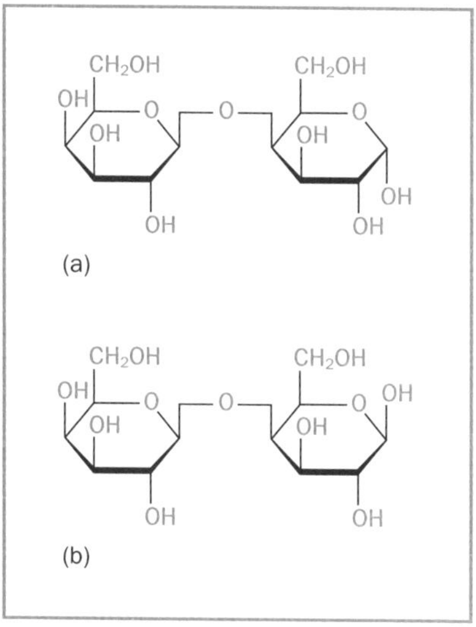

**Figure 9.3** Structural formulae of (a) anhydrous α lactose and (b) anhydrous β lactose.

Anhydrous lactose is available in a range of particle sizes and is commonly used as a diluent in wet granulation and dry granulation processes. It is a crystalline material.

### Lactose monohydrate

Lactose monohydrate is composed of the monohydrate of α-lactose and, whilst crystalline in nature, there may be varying proportions of the amorphous form. Lactose monohydrate is available in a wide range of grades that offer differing physical properties, e.g. particle size distribution/polydispersity and density (bulk and tapped).

### Spray-dried lactose

Spray-dried lactose is a mixture of crystalline α-lactose monohydrate (80–90%) and 10–20% amorphous lactose. It is prepared by spray drying a suspension of α-lactose monohydrate.

The suspension is composed of circa 80–90% suspended α-lactose monohydrate, the remainder being in solution (which subsequently forms the amorphous portion of the spray-dried material). The specific use of spray-dried lactose for the manufacture of tablets by direct compression is described at a later stage in this chapter.

### Starch

Starch is a polysaccharide composed of amylose and amylopectin, which is used as a diluent (and also as a binder and disintegrant) in tablet formulations. A pregelatinised grade is also available in which the granules of starch have been physically and chemically modified to produce a free-flowing powder. An example of a commercially available pregelatinised starch is Starch 1500LM (which has a low moisture content).

### Dibasic calcium phosphate

This is a commonly used diluent in tablet formulations and is available as different hydrate forms and in a range of particle sizes. It is a basic excipient and may therefore chemically interact with acidic components in the presence of moisture. It has excellent flow and compression properties.

### Microcrystalline cellulose (MCC)

MCC is a crystalline powder that is prepared by the controlled acid hydrolysis of cellulose. Several grades of MCC are commercially available that differ in their physicochemical properties, e.g. density, flow properties and particle size distribution. In addition to its use as a diluent, MCC may also be used as a binder in wet granulation (see below) and a disintegrating agent. Two commonly used grades of MCC are Avicel PH 101 (powder) and PH 102 (granules).

### Mannitol

Mannitol is a polyol that is commonly employed as a diluent in tablets, particularly chewable tablets, due to the inherent sweetness and negative heat of solution. It exhibits excellent flow properties (Figure 9.4).

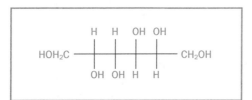

**Figure 9.4** Structural formula of mannitol.

## Binders

Binders are predominantly (but not exclusively) polymeric components that are employed in the production of tablets by the wet granulation method of manufacture. In this role, binders are either added as a solution or as a solid into the powder mix (following which the granulating fluid, typically water, is added). This process is described in the next section. The properties of commonly used binders are summarised in Table 9.1.

**Table 9.1**    Examples and properties of commonly used binders used in the manufacture of tablets by wet granulation

| Binder | Concentration | General comments |
|---|---|---|
| Hydroxypropylmethylcellulose | 2–5% w/w | The required concentration is dependent on molecular weight (grade) used |
| Polyvinylpyrrolidone | 0.5–5% w/w | The required concentration is dependent on molecular weight (grade) used |
| Hydroxypropylcellulose | 2–6% w/w | The required concentration is dependent on molecular weight (grade) used |
| Sucrose | 50–67% w/w* | *Added as a syrup Produces hard tablets whose properties are affected by moisture |
| Microcrystalline cellulose | 20–90% w/w | Often used as a binder and diluent |
| Acacia | 1–5% w/w | Produces hard tablets |

## Disintegrants

Disintegrants are employed in tablet formulations to facilitate the breakdown of the tablet into granules upon entry into the stomach. If the formulated tablet is hydrophobic and/or it has been manufactured using a high compression force, the rate of water uptake into, and hence disintegration of, the tablet will be unacceptably low. In these situations disintegrants are an essential formulation component, enabling tablet disintegration to occur within the specifications defined in the various pharmacopoeias (typically disintegration of conventional tablets must occur within 15 minutes). There are several mechanisms by which disintegrants elicit their effect:

■ Disintegrants may *increase* the *porosity* and *wettability* of the compressed tablet matrix. In so doing gastrointestinal fluids may readily penetrate the tablet matrix and thereby enable tablet breakdown to occur. In this approach it is necessary for the disintegrant to be homogeneously dispersed throughout the tablet matrix (requiring concentrations between 5% and 20% w/w). Examples of disintegrants that operate in this fashion include:

- starch
- MCC
- sodium starch glycolate (Figure 9.5).

**Figure 9.5**  Structural formula of sodium starch glycolate. Sodium starch glycolate is the sodium salt of the carboxymethyl ether of starch and is employed at circa 5% w/w as a disintegrant within tablet formulations.

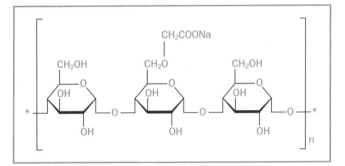

Sodium starch glycolate is the sodium salt of the carboxymethyl ether of starch and is employed at circa 5% w/w as a disintegrant within tablet formulations.

■ Disintegrants may operate by *swelling* in the presence of aqueous fluids, thereby expediting tablet disintegration due to the increase in the internal pressure within the tablet matrix. Disintegrants in this category are typically hydrophilic polymers. To operate effectively it is important that the incorporated disintegrant does not cause a significant increase in viscosity, thereby decreasing the diffusion of gastrointestinal fluids into the tablet matrix. Examples of excipients that facilitate tablet disintegration by swelling in the presence of aqueous fluids include:

- sodium starch glycolate
- croscarmellose sodium (a cross-linked sodium carboxymethylcellulose) (usual working range 0.5–5.0% w/w)
- crospovidone (a cross-linked polyvinylpyrolidone) (usual working range 2–5% w/w)
- pregelatinised starch (usual working concentration of 5% w/w).

■ Tablet disintegration may also be mediated by the production of gas whenever the tablet contacts aqueous fluids. This is the mechanism of disintegration of effervescent tablets.

## Lubricants

During compression lubricants act at the interface between the face of the die and the surface of the tablet to reduce the friction at this interface during ejection of the tablet from the tablet press. Inadequate lubrication of this interface results in the production of tablets with a pitted surface and is due to the inability of the tablet surface to detach from the surface of the tablet die. The appearance of the tablet is an important pharmacopoeial (and consumer) requirement and, therefore, inadequate lubrication will lead to rejection of the tablet batch.

There are two main categories of lubricant: (1) insoluble; and (2) soluble.

### Insoluble lubricants

Insoluble lubricants are added to the final mixing stage prior to tablet compression. The concentration of lubricant used is an important consideration in the subsequent disintegration and dissolution of the drug from the tablet: high lubricant concentrations result in reduced rates of disintegration and dissolution and possible failure of the quality control tests for these parameters. Conversely, insufficient concentrations of disintegrant, as stated above, may lead to tableting defects. In addition to concentration, the time of mixing of the lubricant with the granules/powders and the particle size of the lubricant will affect the performance of the lubricant. Overmixing (in terms of both shearing stress and time) may adversely affect tablet disintegration and drug dissolution. In particular, mixing of the disintegrant and the insoluble lubricant together should be avoided as this leads to the formation of a film of lubricant on the surface of the disintegrant, which subsequently reduces the wettability of, and water uptake by, the disintegrant, leading to compromised disintegration. Furthermore, it has been shown that the efficacy of the lubricant is enhanced if the surface area is increased (i.e. the particle size of the lubricant is decreased). Examples of insoluble lubricants that are commonly used are as follows:

- *Magnesium stearate.* Magnesium stearate is not a pure compound, being a mixture of magnesium salts of a range of fatty acids, predominantly (but not exclusively) stearic and palmitic acid. The lubricant properties of magnesium stearate are particularly susceptible to variations in the physical properties, e.g. specific surface area/particle size, crystalline structure, moisture content and the composition of fatty acids. Information regarding the chemical properties and physical composition of the batch of magnesium stearate used must therefore be available to the formulator prior to use as a tablet lubricant. Bearing these factors in mind, the concentration range of magnesium stearate employed is 0.25–0.50% w/w.

- *Stearic acid*. Similar to magnesium stearate, stearic acid is a mixture of stearic (>40% w/w) and palmitic acids, the sum of the two acid components being greater than 90% w/w. As above, there is a wide range of grades of stearic acid that differ in their physical and chemical properties, e.g. melting range, acid, iodine and saponification values. The concentration range of stearic acid employed as a lubricant is 1.0–3.0% w/w.
- *Glyceryl behenate*. Glyceryl behenate is a mixture of glyceryl fatty acid esters, in which the concentration of monoesters (glyceryl monodocosanoic acid) should be between 12% and 18% w/w. The remaining concentration is composed of glyceryl didocosanoic acid and glyceryl tridocosanoic acid. It is employed as a tablet lubricant within the concentration range 1.0–3.0% w/w.
- *Glyceryl palmitostearate*. Glyceryl palmitostearate is a mixture of mono-, di- and triglycerides of fatty acids containing between 16 and 18 carbon atoms. It is used as a lubricant in tablet manufacture at concentrations between 1.0% and 3.0% w/w.

### Soluble lubricants

Soluble lubricants are principally employed to overcome the possible deleterious effects of their insoluble counterparts on the time required for tablet disintegration and hence drug dissolution. The use of soluble lubricants does overcome these issues; however, for the most part, the efficacy of insoluble lubricants is superior to soluble lubricants. Examples of soluble lubricants include:

- *Polyethylene glycol (PEG)*. Typically, PEGs, e.g. PEG 4000, 6000 and 8000, that exhibit large average molecular weights are used as lubricants in the manufacture of tablets. The efficacy as a lubricant is less than that of magnesium stearate.
- *Polyoxyethylene stearates*. The polyoxyethylene stearates belong to a group of polyethyoxylated derivatives of stearic acid that differ in the number of oxyethylene repeating groups along the polymer chain and whether present as a mono- or distearate. All members of this group are non-ionic surfactants that may be employed as tablet lubricants within the concentration range 1.0–2.0% w/w.
- *Lauryl sulphate salts*. Salts of lauryl sulphate (e.g. sodium, magnesium) are anionic surfactants that may be employed as lubricants within the concentration range 1.0–2.0% w/w. In addition to their action as lubricants, the dissolution of

poorly soluble drugs from tablets will be enhanced due to the surface-active activity of these agents.

## Glidants

Glidants act to enhance the flow properties of the powders within the hopper and into the tablet die in the tablet press. The reduced friction between the powders/granules and the surfaces of the hopper and dies has been suggested to be due to the ability of the particles of the glidants to locate within the spaces between the particles/granules. To achieve this effect it is therefore necessary for the glidant particles to be firstly small and, secondly, to be arranged at the surface of the particles/granules. Glidants are typically hydrophobic and therefore care should be taken to ensure that the concentration of glidants used in the formulation does not (in a similar fashion to lubricants) adversely affect tablet disintegration and drug dissolution. Examples of glidants used in tablet manufacture include: (1) talc; and (2) colloidal silicon dioxide.

### Talc

Chemically talc is hydrated magnesium silicate (typically $Mg_6(SiO_2)_4(OH_4)$) and is a crystalline material of small particle size. It is typically used as a glidant in tablet formulations at concentrations between 5.0% and 30.0% w/w. Whilst talc is non-toxic whenever ingested orally, inhalation of large masses of talc may result in the formation of granulomas. This has restricted the use of talc in many pharmaceutical formulations.

### Colloidal silicon dioxide

Colloidal silicon dioxide is commonly used in tablet formulations as a glidant (0.1–0.5% w/w) due to a combination of its hydrophobic properties and low (colloidal) particle size (typically <15 nm).

## Miscellaneous excipients used in the formulation of tablets

In addition to the principal tablet components described above, the successful formulation of tablets may require the incorporation of other classes of excipients. These include: (1) adsorbents; (2) sweetening agents/flavours; (3) colours; and (4) surface-active agents.

### Adsorbents

Adsorbents are used whenever it is required to include a liquid or semisolid component, e.g. a drug or a flavour, within the tablet formulation. As the production of tablets requires solid components, the liquid/semisolid constituent is adsorbed on to a solid component which, in many cases, may be one of the other components in the tablet formulation (e.g. diluent) during mixing.

If this approach is not possible, an adsorbent is specifically included in the formulation. Examples of these include: (1) magnesium oxide/carbonate; and (2) kaolin/bentonite.

### Magnesium oxide/carbonate

Magnesium oxide is a solid adsorbent that is commercially available in two forms termed light (less dense, 5 grams, occupying circa 40–50 ml) and heavy (more dense, occupying 10–20 ml). Magnesium carbonate consists of several forms (hydrate, the basic form and anhydrous). Both salts are employed as adsorbents in tablets within the concentration range 0.5–1.0% w/w. In addition, magnesium oxide may also be used as a glidant whereas magnesium carbonate is used as a diluent. Both salts are basic and therefore may interact with acidic drugs.

### Kaolin/bentonite

These are natural materials that are composed of hydrated aluminium silicate. Unlike kaolin, bentonite is a colloidal material and therefore the particle size of this excipient is lower than for kaolin. These excipients are employed as adsorbents within the concentration range 1.0–2.0% w/w.

### Sweetening agents/flavours

Sweetening agents and flavours (in accordance with other dosage forms) are employed to control the taste and hence the acceptability of tablets. These agents are of particular importance if the conventional tablet contains a bitter drug or, more importantly, if the tablet is a chewable tablet. The excipients that are used to increase the sweetness and improve the flavour of tablets have been described in previous chapters.

### Colours

Coloured tablets are generally formulated either to improve the appearance or to identify the finished product uniquely. In some formulations the drug is coloured and, when manufactured, may result in the tablet exhibiting a speckled, and hence, inelegant appearance. To obviate this problem, an appropriate colouring agent is included in the tablet formulation. It is important that the colour is distributed equally throughout the tablet and this is normally achieved by adding a water-soluble colour to the granulation liquid in the wet granulation method of manufacturing tablets.

### Surface-active agents

Surface-active agents may be incorporated into tablets to improve the wetting properties of hydrophobic tablets and hence increase the rate of tablet disintegration. In addition, surface-active agents may increase the aqueous solubility of poorly soluble drugs in the gastrointestinal tract and, as a result, the rate of dissolution of the active agent will increase. One of the most popular choices of surface-active agent for this purpose is sodium lauryl sulphate. However, it should be noted that the surface-active agent should not interact with the therapeutic agent as this may affect the dissolution rate of the drug.

**Tips**

- Tablets will generally contain several excipients that are required to ensure that the tablets may be successfully manufactured and will provide the required biological performance.
- Certain excipients are common to the vast majority of tablets, e.g. lactose, stearates, microcrystalline cellulose.

## Methods used for the manufacture of tablets

Tablets are commonly manufactured by one of the following manufacturing processes:

- wet granulation
- dry granulation (slugging or roller compaction)
- direct compression.

The above methods may be subdivided into two distinct classes depending on the requirement for a liquid in the manufacturing process. Consequently, in wet granulation a liquid is required in the manufacture of the granules (which are then compressed into tablets), whereas in the other methods no liquids are required. In general, the choice of method for the manufacture of tablets is dependent on a number of factors, including:

- the physical and chemical stability of the therapeutic agent during the manufacturing process
- the availability of the necessary processing equipment
- the cost of the manufacturing process
- the excipients used to formulate the product.

The key aspects of each manufacturing method are summarised in the following section.

### Manufacture of tablets by wet granulation

A stepwise summary of the manufacturing steps used in the manufacture of tablets by the wet granulation method is detailed below.

#### Stage 1: mixing of the therapeutic agent with the powdered excipients (excluding the lubricant)

This step involves the introduction of the powdered excipients and drug (excluding the lubricant) into a powder mixer. The mixing speed and time must be sufficient to ensure that a homogeneous mixture is produced, i.e. in which the concentration of each component in each region of the mixer/sample is identical. The efficiency of mixing is enhanced by the use of powders that have similar average particle size/distribution, although this is often not the case in many mixing operations. Generally the shear rate required to mix pharmaceutical powders is low. Examples of mixers that are used within the industry include:

- *Planetary bowl mixer.* The planetary bowl mixer (Figure 9.6) is a vertical mixer in which the powders are mixed in a bowl using a rotating mixing attachment. The mixing arm is connected to the mixer by means of a shaft whose movement is offset. In addition, the mixing shaft rotates around the mixing bowl to ensure that the contents of the bowl are thoroughly mixed. The movement of the mixing shaft and mixing arm is termed planetary due to the similarity with the movement of the planets around the sun.

**Figure 9.6** The planetary bowl mixer showing the motor, the shaft, the mixing geometry and the bowl.

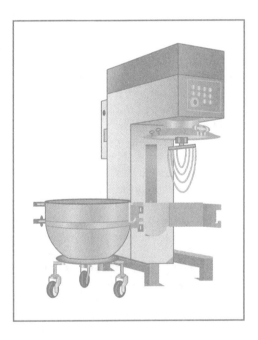

- *Rotating drum mixers.* In rotating drum mixers, the drum is attached to a drive shaft, which is then attached to a motor. The drive shaft rotates and, in so doing, the drum – and hence the powdered contents – is rotated. As shown in Figure 9.7 there are a number of available mixing geometries.
- *High-speed mixers.* High-speed mixers are commonly employed in the pharmaceutical industry due to their

**Figure 9.7**  Examples of mixing geometries for the laboratory drum mixer: (a) double cone mixer; (b) Y-cone (or V-cone) mixer; (c) cube mixer.

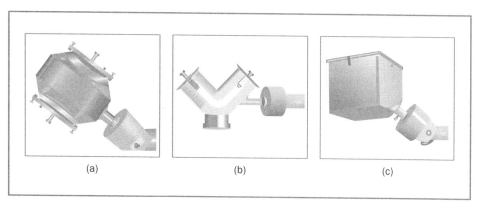

(a)    (b)    (c)

excellent powder-mixing properties and their combined use as a mixer and wet granulator. The powders are contained within a stainless-steel mixing bowl, on the bottom of which is a horizontal three-bladed mixing head (impeller) that is capable of high-speed rotation. Additional mixing of the powder bed is performed through the (horizontal or vertical) introduction of a second impeller into the powder bed. The combined actions of the two impellers offer excellent mixing properties. Following mixing the powder mixture is discharged through a screen.

■ An example of a high-speed (Diosna) mixer and a typical mixing head are shown in Figure 9.8(a, b).

■ *Ribbon/trough mixers.* In ribbon/trough mixers, mixing is performed via the rotation of blades that are housed within a central drive shaft. Efficient mixing may be achieved in short mixing times using, if required, high shear rates. Ribbon and trough mixers differ in the nature of the mixing geometries (as shown in Figure 9.9).

### Stage 2: wet granulation of the powder mix

Granulation is a unit operation in which mixed powders are aggregated into and retained as larger particles (circa 0.2–4.0 mm in diameter). Ideally the chemical composition of the granules should be homogeneous. There are several advantages for the use of granules in tablet manufacture, as follows:

■ *Prevention of segregation of powder components during the tableting process or during storage.* Following mixing, the individual particulate components of powders will tend to segregate due to differences in the particle size and density of each component; the degree of segregation increases as the time of storage increases. If the segregated powder is then

**Figure 9.8** (a) A high speed laboratory Diosna mixer/granulator showing the mixing bowl and the attached motor assembly. (b) A typical mixing head for a high-speed mixer/granulator, showing the main horizontal three blade impeller and the side-entry horizontal two blade impeller. (c) Diagrammatic representation of a Collette-Gral mixer/granulator showing the vertical arrangements of the main three-bladed impeller and the secondary two-bladed impeller. In addition this diagram shows the spray inlet for the granulation fluid when used as a mixer/granulator.

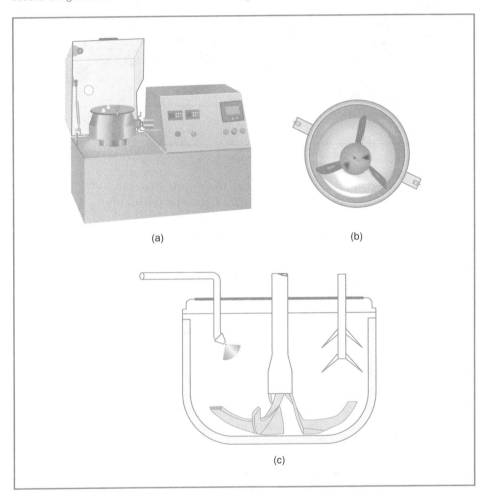

(a)

(b)

(c)

compressed into tablets, the intrabatch chemical composition of the dosage form will vary, resulting in issues regarding the quality of the product. This problem is obviated if the powders are granulated, as there is no component segregation within the granule.

- *Enhancement of the flow properties.* Due to differences in particle size/shape and cohesive/electrostatic properties, the

**Figure 9.9**  Commercially available (a) ribbon and (b) trough mixers, illustrating the mechanism of mixing associated with the different mixing geometries used in each mixer. Caption (a) refers to ribbon mixers whereas caption (b) illustrates a typical geometry that is used in trough mixers.

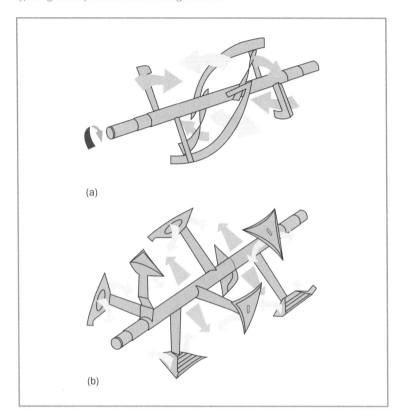

(a)

(b)

flow properties of mixed powders from the tablet hopper into the tablet die may be inappropriate and irregular, once again leading to issues regarding the intrabatch variability of tablet dosage forms. The flow properties of granules are both improved and more reproducible than those of mixed powder systems and are relatively unaffected by the adsorption of low amounts of moisture (during storage). Conversely, the adsorption of moisture on to the surface of powders may result in caking, which may ultimately adversely affect powder flow.

■ *Enhancement of the compaction properties.* In comparison to powder mixes, the compaction properties of granules are improved due to the presence of the binder component on the surface of the granules, leading to greater intergranule adhesive interactions.

- *Lower incidence of dust production.* One problem associated with the processing of powders into tablets is possible dust production. This problem is obviated with the use of granules.

### Description of wet granulation

Granulation is a unit operation in which mixed powders are simultaneously mixed with a suitable fluid, e.g. water, isopropanol or ethanol (or mixtures thereof). To achieve cohesion between the powders, it is necessary to include a binding agent within the formulation, either in the solid state within the powder mix or dissolved in the binding fluid. Powder mixing, in conjunction with the cohesive properties of the binder, enables the formation of granules. A description of commonly employed strategies for wet granulation is outlined below.

### Oscillating granulator

In this strategy, the mixing of the formulation components is performed using a relatively low shear mixer, e.g. a planetary bowl mixer. The granulating fluid (which may contain the appropriate concentration of binder) is then sprayed on to the powder mass whilst maintaining mixing. The wetted powder mass is then passed into an oscillating granulator, which forces the powder mass through a metal screen under the action of an oscillatory stress. This process is illustrated in Figure 9.10.

**Figure 9.10**  Diagrammatic representation of the (a) oscillatory action and (b) orientation of sieves in an oscillatory granulator.

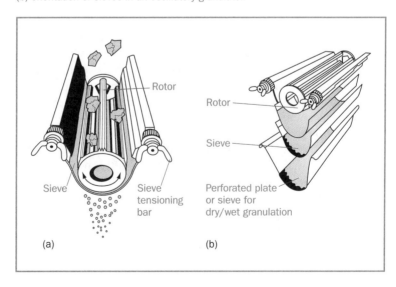

Providing the correct mass of granulation fluid has been employed, granules are produced following pass through the oscillatory granulator. The size of the granules produced using this method is dependent on the size of the metal screen used within the granulator. It is worth noting that the powdered binder may either be added to the powder mix or dissolved within the granulation fluid and added to the powder mix. In the former scenario the granulation fluid is composed only of the chosen solvent. Due to the complete dissolution within the granulation fluid prior to addition to the powder mix, the cohesive properties of granules produced by the latter method are greater.

### High-speed mixers/granulators

More recently, it has become common practice to perform the powder-mixing and wet granulation steps within a single operation. Typically, high-speed mixers/granulators are employed, e.g.

- the Diosna mixer/granulator (Figure 9.8a, b)
- Collette–Gral mixer/granulator (Figure 9.8c).

The design and operation of these have been described previously in this chapter.

### Fluidised-bed granulation

In this technique, the mixed powders are fluidised within a chamber by the action of a positive (temperature-controlled) stream of air. Whilst in this dispersed state, the particles are sprayed with the granulation fluid (usually containing dissolved binder), causing the particles to interact with (adhere to) each other. The fluidised-bed granulator contains filters to prevent the escape of fluidised particles into the environment, thereby enabling these to be recycled into the fluidised granules. Following completion of the spraying step, fluidisation of the granules is maintained to enable evaporation of the solvent used in the granulation fluid. A schematic representation of fluid-bed granulation is presented in Figure 9.11.

### Extrusion

Extrusion is a technique in which a premixed powder mass (containing the required formulation components and to which the required mass of granulation fluid has been added) is placed into the barrel of the extruder via a hopper. In the barrel, the wet mass is moved horizontally from the end adjacent to the hopper by the turning motion of either a single-screw or a twin-screw system. This action also serves to enhance the mixing of the granulation fluid within the wet mass. Following passage through the barrel of the extruder, the wet mass is passed through

**Figure 9.11**  Diagrammatic representation of the operation of fluid-bed granulators. In figure (a) the powders are suspended by a vertical flow of air from the bottom of the granulator. The granulation fluid is sprayed on to the powders from the top of the granulator. In figure (b) the powders are similarly suspended by a vertical upward flow of air but the airflow is tangential enabling flow in a circular fashion. The granulation fluid is sprayed into this tangential airflow. (Adapted from http://www.glatt.com/e/01_technologien/01_04_09.htm.)

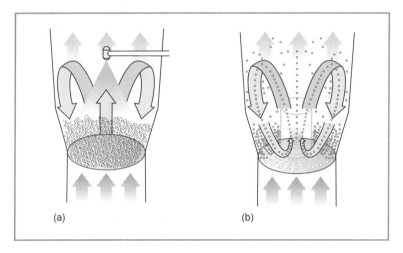

(a)                                    (b)

a perforated plate (termed the die) into lengths. To obtain granules successfully, it is important that:

- there is good dispersion of the granulation fluid through the wet mass
- the mechanical properties of the wet mass (and hence extrudate) should facilitate deformation
- the formulation should be sufficiently cohesive to form granules but not adhesive (to prevent aggregation of formed granules).

When formulated and processed correctly, the extruded strands should break to produce granules of uniform particle size.

### Stage 3: processing granules into tablets

Following the production of (drug-containing) granules, further unit operations are required for the successful production of tablets. These are as follows:

- *Drying of the granules.* Following production, granules are then dried to facilitate further processing. There are several types of drier that are used within the pharmaceutical industry for this purpose, including:
  - *The shelf or tray drier.* The shelf or tray drier is similar in design to a conventional oven. In this the wet granules are uniformly located on a series of horizontally arranged,

shallow trays. Air enters the drier where it is warmed by heaters prior to being passed across the surface of the granules. The moisture is then passed from the drier through an outlet. An example of an industrial tray drier is shown in Figure 9.12.

**Figure 9.12** An industrial tray drier.

The drying efficiency of this type of system may be enhanced by the use of a vacuum. In this design there is no inlet or outlet for air and, following the introduction of the granules into the shelves, the door is closed and a vacuum created using a vacuum pump. The removed moisture is condensed into a liquid using a condenser and collected prior to disposal. The use of the vacuum design lowers both the time and temperature of the drying process.

- *The fluidised-bed drier.* The design and operation of the fluidised-bed drier for the production of granules have been addressed in the previous section. When used as a drier, the granules are suspended in the fluidised warmed air, enabling relatively rapid drying. There are several advantages of the use of fluidised-bed driers for the drying of granules that have rendered this technique popular within the pharmaceutical industry, including:

- Excellent heat transfer to and mass (moisture) transfer from the granules, due primarily to the suspension in, and drying of, individual granules in the warmed airflow. This enables granules to be rapidly and reproducibly dried.
- Accurate control of the drying conditions may be achieved using a fluidised-bed drier, e.g. airflow, temperature.

Conversely, limitations of this technique include:

- Attrition of granules whilst suspended in the warmed airflow.
- Loss of particles from the drier. Particles are normally prevented from escaping by means of a bag filter.
- Generation of static electricity within the fluid-bed drier (due to the rapid movement of particles), which may lead to explosions (particularly if a volatile solvent has been used as the granulation fluid).

▪ *Milling of the granules (reduction in the granule size).* Following drying, the next stage in the production of tablets by wet granulation usually involves milling of the granules to the required particle size. This step is employed for two main reasons:

- to control the particle size (and particle size distribution), which, in turn, improves both the flow of the granules into the tablet die and the fill of the granules within the die
- the choice of the granule size is determined by the size of the die (and hence the final tablet size). In general the granule size decreases as the tablet size decreases.

The theoretical aspects of milling are described in Chapter 11.

Reduction in the size of granules may be performed using a range of particle size reduction methods, including:

▪ *Oscillating granulator.* In this method the dried granules are forced through a screen of defined mesh size, as described previously. The mesh size of the screen is usually smaller than that used for granulation.

▪ *Quadro Comil.* In the Quadro Comil (Figure 9.13) the dried granules are placed inside a conical chamber (containing a screen of defined mesh size). The granules are then passed through the screen in a centrifugal manner by the action of a rotating impeller, after which the granules are collected.

▪ *Mixing of the granules with lubricant.* The penultimate stage in the manufacture of tablets using wet granulation involves a final mixing of the dried granules with the lubricant. This step is normally conducted using the same equipment that was

**Figure 9.13** The laboratory Quadro Comil: (a) encased drive unit, conical chamber and powder collection zone; (b) conical chamber, screen and rotating impeller. (Courtesy of Dr Brendan Muldoon, Actavis.)

employed to mix the powders at the start of the process. It is important to note that both the time and shear rate of mixing in this step will affect the final performance of the tablet, in terms of both the quality of the tablet produced and the subsequent drug release.

- *Compression of the formulation into tablets.* This unit operation will be described at a later section in this chapter.

## Stages of wet granulation

During wet granulation particles interact with each other to form aggregates (granules); the extent of this interaction is a major contributor to the performance of the finished dosage form. There are two primary stages involved in particle–particle interactions during wet granulation, as follows:

- *Particle–particle interactions facilitated by the formation of **liquid** bridges.* As may be expected, the presence and, importantly, volume of the granulation medium contribute greatly to the granulation process. The interactions between particles are dependent on the mass of free water available and these are characterised into four categories (as depicted in Figure 9.14):
  - *Pendular* state. The pendular state (Figure 9.14a) is established at lower moisture levels and relates to the formation of liquid bridges (connections) between adjacent particles. The attractive forces between the particles are due to both interfacial and hydrostatic effects.
  - *Funicular* state. As the mass of water increases, the mass of the liquid bridges – and, accordingly, the adhesive forces – increases between the particles (Figure 9.14b).

**Figure 9.14**  Diagrammatic representation of particle–particle interactions during wet granulation: (a) pendular, (b) funicular, and (c) capillary states; (d) overwetted (undesirable) state.

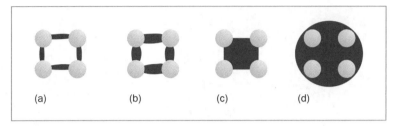

(a)                    (b)                    (c)                    (d)

- *Capillary* state. Further increases in the mass of available water result in the displacement of air from the free volume between the particles (Figure 9.14c). The adhesive interaction between the particles is now principally due to a capillary effect at the liquid/air interface (i.e. at the surface of the growing granule).
- The final state (Figure 9.14d) is the *overwetted* state and this is pharmaceutically unacceptable due to the associated processing (mixing and drying) problems. Following drying, individual granules will not be produced and, in addition, due to the excessive mass of binder associated with these particles, the disintegration and dissolution properties of the dosage form may be unduly extended. It should be noted that, if an insufficient mass of granulating fluid is used in wet granulation (i.e. underwetted), the granules will disintegrate during the subsequent processing steps and, if processed into tablets, the tablets will exhibit poor mechanical strength.

■ *Particle–particle interactions facilitated by the formation of* **solid** *bridges*. Solid bridges, consisting predominantly of the polymeric binder, are formed following drying. As drying progresses a polymeric network (resembling the pendular structure) is formed, which largely contributes to the mechanical properties of the resultant granule. Overdrying of the granule results in destruction of these bridges (due to the compromised effects on their mechanical properties). This is the main mechanism of action of polymeric binders. Under certain conditions, crystallisation of the binder on to the surface of the dispersed particles may occur on drying (e.g. sugar), leading to a hardening effect. If the drug is soluble in the granulation fluid, crystallisation may additionally occur, which, in turn, may influence the physicochemical properties of the formulation, e.g. by affecting the hardness of the formed tablet or by affecting drug dissolution.

## Advantages and disadvantages of wet granulation

Wet granulation is a popular technique within the pharmaceutical industry for the manufacture of tablets. There are several advantages and disadvantages associated with this technique.

### Advantages

- Reduced segregation of formulation components during storage and/or processing, leading to reduced intra- and interbatch variability.
- Useful technique for the manufacture of tablets containing low concentrations of therapeutic agent.
- Employs conventional excipients and therefore is not dependent on the inclusion of special grades of excipient (cf. the requirement for spray-dried excipients in the direct compression method of tablet manufacture).
- Most manufacturing plants are built around wet granulation tablet manufacture.
- Tablets produced by wet granulation are amenable to post-processing unit operations, e.g. tablet-coating techniques.

### Disadvantages

- Often several processing steps are required.
- Solvents are required in the process: this leads to a number of concerns, e.g.:
  - Drug degradation may occur in the presence of the solvent. This is particularly relevant if water is used as the granulation medium due to the susceptibility of some drugs to hydrolysis. To overcome this concern, a hydroalcoholic (water/alcohol) or an alcohol (ethanol or isopropanol) granulation medium should be used.
  - The drug may be soluble in the granulation fluid. During the drying process the drug will then precipitate/crystallise, resulting in possible changes in the polymorphic form. If the drug and some excipients are soluble in the granulation medium, subsequent drying will result in deposition of these components on the surface of the insoluble particles and, in so doing, this may enhance the hardness of the granule (and hence both the tablet and biological properties).
  - Heat is required to remove the solvent. This may result in the degradation of thermally labile therapeutic agents. In addition, drying is a costly operation and, furthermore, if alcohols are used as the granulation medium, there are issues regarding solvent recovery and flammability.

# Tips

## Manufacture of tablets by dry granulation

Similar to wet granulation, the manufacture of tablets by dry granulation involves the production and subsequent compression of granules into the final dosage form. In dry granulation no solvent is required; the aggregation of particles into granules is facilitated by the application of high stresses to the mixed powders. There are two methods by which dry granules are formed: (1) slugging; and (2) roller compaction.

### Slugging

In this technique the powders are mixed (as described previously) and then compressed into a primordial *oversized* tablet using a tableting press that is capable of applying a high stress (to ensure that aggregation of the particles and then aggregation of granules occur during compaction). Following this the tablet is milled to produce granules of the required size (using conventional milling equipment, as described previously).

### Roller compaction

In roller compaction the formulation ingredients are mixed (as described previously) and are then compressed using a roller compactor. In this the powders are fed from a hopper on to a moving belt and then transported to, and compressed by, the passage between the narrow gap between two (oppositely) rotating rollers to produce a sheet/film of compressed material. The compressed sheet is then milled to produce granules of the required size (again using conventional milling equipment, as described previously).

## Excipients used in dry granulation

The types of excipients used to prepare tablets (and the rationale for their use) using dry granulation methods are similar to those described for wet granulation. Typically, the following excipients are required:

- *Diluent/filler*, e.g.:
  - anhydrous lactose or lactose monohydrate
  - starch
  - dibasic calcium phosphate
  - microcrystalline cellulose (MCC)

- *Disintegrants*, e.g.:
  - starch
  - MCC
  - sodium starch glycolate
  - croscarmellose sodium
  - crospovidone.
- *Lubricants*, e.g.:
  - stearates (magnesium stearate, stearic acid)
  - glyceryl fatty acid esters (glyceryl behenate, glyceryl palmitostearate)
  - PEG
  - polyoxyethylene stearates
  - sodium lauryl sulphate.
- *Glidants*, e.g.:
  - talc
  - colloidal silicon dioxide.
- Miscellaneous excipients (colours, sweetening agents, etc.).

### Stages of dry granulation

In dry granulation, the aggregation of particles is facilitated by the application of high compression stresses and therefore the stages/ mechanisms of granule formation are different from those described for wet granulation. In dry granulation particle–particle interactions occur due to: (1) electrostatic forces; (2) van der Waals interactions; and (3) melting of components within the powder mix.

#### Electrostatic forces

These are thought to play a role in the initial cohesive interaction between particles; however, they do not contribute greatly to the granule strength

#### Van der Waals interactions

Van der Waals interactions play a major role in the interactions between particles in the dry state. As the reader will be aware (see Chapter 2), van der Waals forces increase in magnitude as the distance between particles decreases. Under normal circumstances, particles are unable to interact due to the inability of the particles to locate at close distances to one another. However, this problem is obviated in dry granulation by the application of high stresses to the powder mix.

#### Melting of components within the powder mix

During dry granulation processing the powders will be exposed to shear stresses that may result in the partial melting of excipients (of low melting point). Upon cooling, solidification of the excipients will occur, resulting in increased interactions between adjacent particles. This mechanism does not strongly contribute to particle–particle interactions in dry granulation.

Following dry granulation, the granules are processed using a series of unit operations (identical to those used in wet granulation). These are as follows:

- milling of the granules (reduction in the granule size)
- mixing of the granules with lubricant
- compression of the formulation into tablets.

This unit operation will be described later in this chapter.

## Advantages and disadvantages of dry granulation

The popularity of dry granulation for the manufacture of tablets has decreased in recent years, having been superseded by direct compression (see following section). However, both slugging and roller compaction are still employed in tablet manufacture. As with wet granulation, there are several advantages and disadvantages associated with these techniques.

### Advantages

- Both roller compaction and slugging require conventional (i.e. non-specialist) grades of excipients.
- These methods are not generally associated with alterations in drug morphology during processing.
- No heat or solvents are required.

### Disadvantages

- Specialist equipment is required for granulation by roller compaction.
- Segregation of components may occur postmixing.
- There may be issues regarding powder flow.
- The final tablets produced by dry granulation tend to be softer than those produced by wet granulation, rendering them more difficult to process using post-tableting techniques, e.g. film coating.
- Slugging and roller compaction lead to the generation of considerable dust. Therefore, containment measures are required. Furthermore, there may be a reduction in the yield of tablets.

## Manufacture of tablets by direct compression

The manufacture of tablets using wet granulation or dry granulation methods is, in light of the requirement for a series of unit operations, both time-consuming and potentially costly. A potentially more attractive option for the manufacture of tablets involves powder mixing and subsequent compression of the powder mix, thereby obviating the need for granulation (and related unit operations). This process is termed direct compression. The mechanisms of particle–particle interactions in tablets produced by direct compression are similar to those operative in tablets produced by dry granulation and roller compaction.

## Stages in the manufacture of tablets by direct compression

A summary of the various steps used in the manufacture of tablets by direct compression is detailed below.

### Premilling of formulation components

In the preparation of tablets by direct compression, both the particle size and particle size distribution of the therapeutic agent and the excipients are important determinants of the compression properties of the powder mix. Whilst the excipients may be purchased to a particular particle size specification, frequently the particle size properties of the drug may require modification by milling, using, for example, the Quadro Comil or a high-energy mill, the Fitzmill. The operation of the Quadro Comil has been previously outlined. The operation of the Fitzmill is illustrated in Figure 9.15.

In this process the powder is fed into a chamber where it is exposed to a rotating blade system. The particle size is reduced by the cutting action of the blades or the impaction of the particles by the blades and the screen on the periphery of the chamber. The milled particles are then discharged through the screen. The particle size/particle size distribution may be controlled by altering the rotor (and hence blade) speed, the morphology of the blades and the mesh size of the screen.

### Mixing of the therapeutic agent with the powdered excipients (including the lubricant)

This step involves introducing all the powdered excipients and drug (including the lubricant) into a powder mixer. The types of mixers used for this process are identical to those that have been previously described in this chapter and include:

- planetary bowl mixer
- rotating drum mixers, e.g. the Y-cone, cube or double-cone mixers
- high-speed mixers, e.g. the Diosna mixer.

**Figure 9.15** The operation of the Fitzmill. (Adapted from http://www.fitzmill.com/pharmaceutical/size_reduction/theory_applications/theory_applications_pD37F.html.)

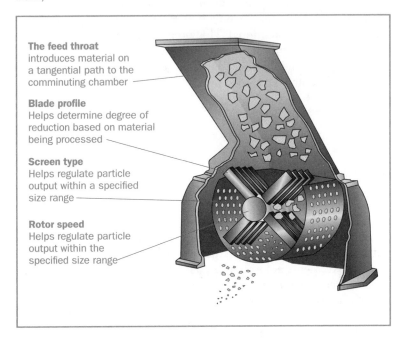

The feed throat
introduces material on
a tangential path to the
comminuting chamber

Blade profile
Helps determine degree of
reduction based on material
being processed

Screen type
Helps regulate particle
output within a specified
size range

Rotor speed
Helps regulate particle
output within the
specified size range

### *Compression of the mixed powders into tablets*
This unit operation will be described at a later section in this chapter.

## Excipients used in the manufacture of tablets by direct compression
In a similar fashion to the other methods of tablet manufacture, the production of tablets necessitates the inclusion of several types of excipients. The rationale for the inclusion of these excipients is identical to that explained previously. However, and most importantly, to achieve the correct powder flow and compression properties, certain grades of excipients are specifically employed for the manufacture of tablets by direct compression. These grades have typically been prepared by specific methods (e.g. spray drying) to achieve the correct physicochemical properties (e.g. particle size/distribution and flow properties). Examples of excipients that are employed for the formulation and manufacture of tablets by direct compression include: (1) diluent; (2) compression aid; (3) distintegrant; and (4) lubricants and glidants.

### Diluent
Examples of commonly used diluents include:

- spray-dried lactose (e.g. Lactopress Spray-Dried, Lactopress Spray-Dried 250, Pharmatose DCL 11, Pharmatose DCL 14)
- dicalcium phosphate (e.g. Encompress grades)
- mannitol (granular or spray-dried grades, e.g. Pearlitol)
- sorbitol
- MCC (e.g. Avicel PH 102).

### Compression aid
Examples of commonly used compression aids include MCC (e.g. Avicel PH 102).

### Disintegrant
Examples of commonly used disintegrants include:

- pregelatinised starch (e.g. Starch 1500)
- sodium starch glycolate (e.g. Explotab, Primojel)
- croscarmellose sodium (e.g. Ac-Di-Sol, Explocel)
- crospovidone (e.g. Polyplasdone XL, Polyplasdone XL-10, Kollidon CL, Kollidon CL-M).

### Lubricants and glidants
The types of lubricants and glidants that are employed for the manufacture of tablets by direct compression are similar to those used in other tablet manufacture methods and include:

- lubricants (e.g. magnesium stearate, stearic acid, sodium stearyl fumarate)
- glidants (e.g. talc, colloidal silicon dioxide).

## Advantages and disadvantages of tablet manufacture by direct compression
The advantages of the direct compression method may at this stage be obvious; however, it is important to note that this method does suffer from several disadvantages. As for other methods, this section details the main advantages and disadvantages of this manufacturing method.

### Advantages
- There are fewer processing steps (unit operations) and therefore the method is potentially more cost-effective than other methods.

## Tips

- Direct compression is of interest to the pharmaceutical industry for the manufacture of tablets due to the low number of processing steps and the absence of liquid.
- Tablets produced by direct compression are often softer than their counterparts that have been produced by wet granulation and therefore they may be difficult to film-coat.
- The cost of excipients for the manufacture of tablets by direct compression is greater than the cost of those used in other methods of tablet manufacture.

- Direct compression does not require the use of water or other solvents. This therefore negates potential problems regarding the stability of therapeutic agents in the presence of the solvents. In addition, heating (a costly unit operation) is not required in direct compression.
- Lubrication is performed in the same vessel as powder mixing, thereby reducing both transfer losses and contamination of equipment.

### Disadvantages
- Specialist (and more expensive) excipients are required. The excipients that are used for direct compression processing are typically specially processed (e.g. spray-dried) for this application to achieve the correct physicochemical properties.
- The quality of the final dosage form is dependent on the powders being easily mixed and remaining homogeneously mixed. Segregation of the mixed components is minimised by ensuring that the excipients and the dosage forms exhibit similar morphologies, densities and particle size/distribution properties.
- There may be issues regarding powder flow into the tableting machine.
- The final tablets produced by direct compression tend to be softer than those produced by wet granulation, rendering them more difficult to process using post-tableting techniques, e.g. film coating.
- If the loading of the therapeutic agent in the final formulation is high (> circa 10% w/w), the compression properties of the powder mix are significantly affected by the compression properties of the therapeutic agent. In some cases the physical properties of the chosen drug may be inappropriate for compression using this technique.
- Direct compression is not used if a colourant is required in the formulation due to the mottled appearance of the resulting dosage form.

## The compression process in tableting
The final stage of tableting involves the compression of the granules/powders between the punch and die of the tablet press. There are several identifiable stages within the compression process, which are individually described below.

### Stage 1: the filling of the die with the granules/powders
The formulated powders or granules are fed (usually by gravitational flow) into the die section, i.e. in the space between the lower and upper punches, from the hopper on the tablet press.

The volume of space occupied by the granules/powders is defined by the position of the lower punch and the die plate (also where the feed shoe of the hopper resides). Accordingly, the mass of solid material within this space (and hence the size of the tablet) is altered by increasing/decreasing the position of the lower punch.

Stage 1 is illustrated in Figure 9.16.

**Figure 9.16**  Diagrammatic representation of the initial stage in tablet formation.

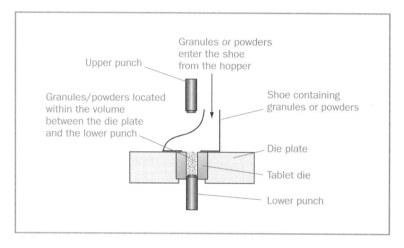

## Stage 2: compression of the powder/granule bed

This stage commences with the retraction of the shoe to facilitate access of the upper punch to the powder/granule bed. The upper punch then descends to the powder/granule bed, at which point a stress is applied, thereby compressing the powder/granule bed into a tablet. In general the lower punch remains static during this stage; however, the manufacturing cycle may be designed to enable the lower punch to move vertically upwards during the compression phase. In addition to the choice and concentration of formulation components, changing the stress applied to the powder/granule bed by the upper punch alters the tablet hardness.

Stage 2 is illustrated in Figure 9.17.

## Stage 3: tablet ejection

Following compression, the upper punch is elevated to its original position. At this stage the lower punch moves upwards until it is flush with the die plate. The shoe of the hopper is then moved across the die plate where it pushes the tablet from the lower press. Simultaneously the lower punch descends to its starting position and granules/powders fill the volume between the shoe

**Figure 9.17**   Diagrammatic representation of the compression stage in tablet formation.

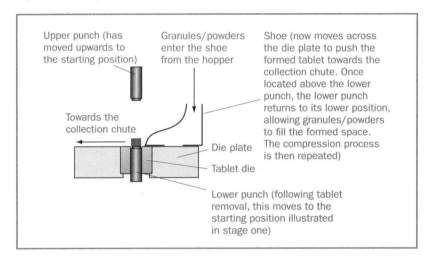

and the lower plate. At this point the manufacturing sequence returns to stage 1 once more.

This final stage is illustrated in Figure 9.18.

**Figure 9.18**   Diagrammatic representation of the final stage in tablet formation.

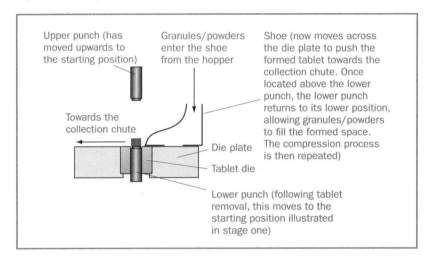

## Types of tablet presses

There are two basic designs of tablet presses used in the pharmaceutical industry: (1) single-punch presses and (2) rotary presses.

### Single-punch presses

The operation of the single-punch press has been described in the previous section. This tablet press is composed of one set of punches and a die (Figure 9.19).

**Figure 9.19** Examples of punches and dies used in a single-punch tablet press. The tablets produced using these punches will be (a) circular and (b) oval, respectively. (Courtesy of Dr Brendan Muldoon, Actavis.)

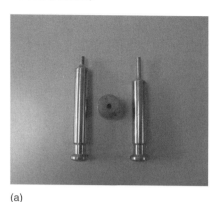

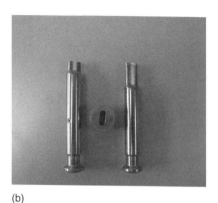

(a)                                                    (b)

Typically the output from a single-punch tablet press is approximately 200 tablets per minute. The single-punch tablet press is frequently employed for pilot-scale manufacture of tablets and, in addition, it is used for the preparation of primordial tablets in dry granulation (slugging).

### Rotary tablet presses

Rotary tablet presses are employed for the large-scale manufacture of tablets, often producing up to 10 000 tablets per minute. As with the single-punch tablet press, the powder/granule bed is compressed between two punches; however, the mechanism by which this process is performed is different. Rotary tablet presses are composed of a series of upper and lower punches (up to 60 per machine) that are housed within a circular die table that rotates in a circular motion. Both punches (upper and lower) are lowered and elevated by the action of an upper and lower roller. The powders/granules are fed from the hopper on to the upper surface of the die table. These are then transported by a feed frame into the die, where they are subsequently compressed by the simultaneous movement of the upper and lower punches. As before, the tablets are removed from the rotating die table into a chute, from which they are collected. A diagrammatic representation of the rotary tablet press is shown in Figure 9.20.

## Compression of powders and granules

Ultimately the preparation of tablets requires the compression of, and subsequent cohesion between, the powders/granules. It is therefore of interest to consider both the various processes that occur within the powder/granule bed during compression and, additionally, the nature of the interactions between the

**Figure 9.20** Diagrammatic representation of the operation of the rotary tablet press.

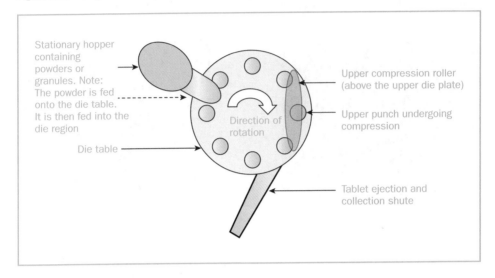

Stationary hopper containing powders or granules. Note: The powder is fed onto the die table. It is then fed into the die region

Die table

Direction of rotation

Upper compression roller (above the upper die plate)

Upper punch undergoing compression

Tablet ejection and collection shute

compressed particles. As may be expected, the fate of a powder bed under compression will differ from a granule bed under compression and, accordingly, these two scenarios will be discussed separately. Following the application of a stress, solid particles will undergo deformation, the nature of which is dependent on the magnitude of the applied stress and the time of loading. A generic relationship between applied stress and the strain (proportional elongation) is illustrated in Figure 9.21.

**Figure 9.21** The effect of stress on the resultant strain of solid particles.

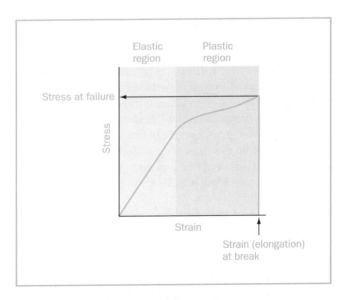

Elastic region

Plastic region

Stress at failure

Stress

Strain

Strain (elongation) at break

Certain regions may be identified in this figure: (1) elastic region; (2) plastic region; and (3) fragmentation (the point at which the material fractures).

### Elastic region
This is the initial *linear* region of the relationship between stress and strain during which the particles undergo elastic deformation. The observed deformation is due to compression of the molecular structure (either crystalline or amorphous) of the particles. The release of the stress within this region will result in the return of the structure of the solid to the equilibrium position. If the compression properties of powders/granules are predominantly elastic, this will result in delamination and subsequent tablet failure. For similar reasons the stresses employed for the manufacture of tablets are most commonly greater than those required for elastic deformation.

### Plastic region
This is the second *non-linear* region of the relationship between stress and strain where the particles undergo plastic deformation. Irreversible deformation occurs within this range of applied stresses and is due to the movement of molecules in the direction of the stress. The successful compression of particles requires the inclusion of components that undergo plastic deformation, as there is limited structural recovery of such materials within the plastic region, resulting in good cohesion between compressed particles. Whilst plastic deformation is often employed for the successful compression of particles into tablet, the quality of the compressed tablet may be affected by the inclusion of excipients that do not exhibit this type of deformation.

### Fragmentation
Fragmentation of particles occurs following the application of a defined stress (termed the ultimate tensile stress) that results in the destruction of intraparticle bonds. In so doing the particle is fractured into a number of smaller particles. Continued application of the applied stress will result in further fragmentation. The surface area of the powder is thus increased, thereby increasing the possible sites for particle–particle interactions. Tablets formed by fragmentation of powders are relatively insensitive to the effects of other excipients; however, the strength of the tablet formed under these increased stresses is often dependent on the nature of fragmentation (e.g. particle size range).

To consider the effects of compression on tablet formation, it is convenient to discuss the fate of powders and granules independently.

**Compression of a *powder* bed during tableting**

This may be defined by a number of stages.

- *Stage 1: rearrangement of the powder bed upon the application of a stress.* Following the application of the initial stress, the particles in the powder bed will undergo rearrangement to minimise the free space between particles. The extent of this rearrangement is dictated by both the size of the particles and frictional forces that operate between the particles.

- *Stage 2: deformation of the powders under the applied stress.* During this stage the powders will undergo deformation (elastic, plastic or fragmentation) as a result of exposure to the applied stress. The physicochemical properties of the powders will affect the nature of the predominant deformation type.

  It is important to note that the two main mechanisms of tablet formation in powder systems are plastic deformation and fragmentation. As illustrated in Figure 9.21, the nature of powder deformation is dependent on the magnitude of the applied stress and, in addition, the physicochemical properties of the materials under compression.

- *Stage 3: bonding of the compressed powders.* Following the application of the required stress, interparticle bonding occurs, resulting in the production of a tablet. There are two predominant bonding mechanisms in tablets prepared by direct compression: (1) adsorption and (2) diffusion.

**Compression of a *granule* bed during tableting**

As before, the events that surround the compression of granules into tablets may be described as a series of stages.

- *Stage 1: rearrangement of granule structure.* Following exposure to stress by the movement of the upper punch, the distribution of the granules is rearranged within the space between the two punches. The extent of this movement is small due to the coarse and regular nature of the granules.

- *Stage 2: granule deformation and bond formation.* As the loading is increased, the granules will undergo initially elastic and then plastic deformation. The granule structure is retained (i.e. the granules are reduced to individual particles), but the stress acts to change both granule shape and porosity due to movement of the individual particles within the granule.

- *Stage 3: formation of intergranule bonds.* As before, the application of the required stress results in the formation of intergranule bonds and hence tablets. Similar to compressed powders, there are two main mechanisms responsible for intergranule bonding, as follows:

- *Bonding by adsorption.* As the reader will be aware, granules may be formed by wet granulation (involving the presence of a binder) or dry granulation methods. The interaction of granules produced by dry granulation methods (roller compaction or slugging) involves adsorption bonds that are formed between close contact of the granule surfaces; van der Waals forces are responsible for bond formation. This type of interaction is referred to as substrate–substrate interactions. In granules formed by wet granulation, theoretically all the granules have been coated with the polymeric binder. Therefore, in this scenario, adsorption bonds may be formed in a similar fashion. However, a range of surface interactions are possible:
  - binder-coated granules with binder-coated granules
  - binder-coated granules with uncoated (substrate) granules
  - uncoated (substrate) granules with uncoated (substrate) granules.
- *Bonding by diffusion.* In the previous section it was noted that bonding of particles by diffusion requires molecular mobility at the surface of the particle. For uncoated granules this interaction is mostly facilitated by the melting (and solidification) of certain components under the effect of applied stress. However, the interaction of binder-coated particles involves the interaction (by diffusion) of the binder molecules on the surfaces of interacting granules. As before, this increased molecular mobility is a result of the effects of the applied stress on the viscoelastic properties of the binder.

## Tablet defects

Typically there are three main types of tablet defect observed: (1) pitting; (2) capping; and (3) lamination. *Pitting* refers to the production of pit marks on the surface of the tablet and is accredited to insufficient lubricant at the tablet/punch interface or punches with a roughened surface. This defect may be corrected by increasing the concentration of lubricant in the formulation or by altering the mixing conditions (time and rate of mixing), whereas punches should be polished regularly to prevent such adhesions. *Capping* and *lamination* (illustrated in Figure 9.22) refer to mechanical splitting of the tablet. In the former scenario, the top (cap) of the tablet is fractured whereas in lamination the fracture may occur within the main body of the tablet.

**Tip**

To ensure the production of tablets with optimal mechanical properties, it is important that tablet compression is facilitated by both elastic and plastic deformation.

**Figure 9.22**   Diagram illustrating capping and lamination in tablets.

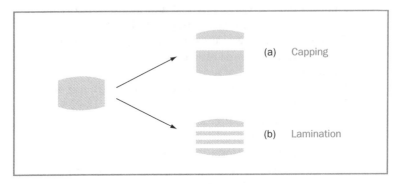

These two tablet defects occur during the ejection stage of the manufacturing process and are accredited to stress-induced fracture. Alteration in the formulation/manufacture to increase the strength of the tablets may resolve these problems. For example, the type and concentration of the binder may be altered, whereas capping and lamination of tablets may occur if the geometry of the punches is intricate. Whilst increasing the compression pressure is known to increase the mechanical strength of tablets, it should be noted that the application of excessive stresses often results in a reduction in the tablet strength, which may result in capping and lamination.

A diagrammatic representation of formulation considerations for tablets is shown in Figure 9.23.

## Post-processing effects on tablet properties

Tablets are known to undergo post-processing alterations in their physicochemical properties, particularly under defined conditions of humidity and temperature. Storage may affect the mechanical properties of the tablet due to the following processes:

■ Adsorption of water on to the tablet surface and within the pores of the tablet may lead to dissolution and then crystallisation of tablet components, which in turn act as solid bridges and hence increase the powder–powder or granule–granule interactions.

■ Adsorption of water on to the tablet surface and within the pores of the tablet may induce crystallisation of amorphous excipients or change the polymorphic state of components within the tablet.

Conversely, in the presence of moisture, tablet strength may be decreased due to:

**Figure 9.23** Formulation considerations for tablets.

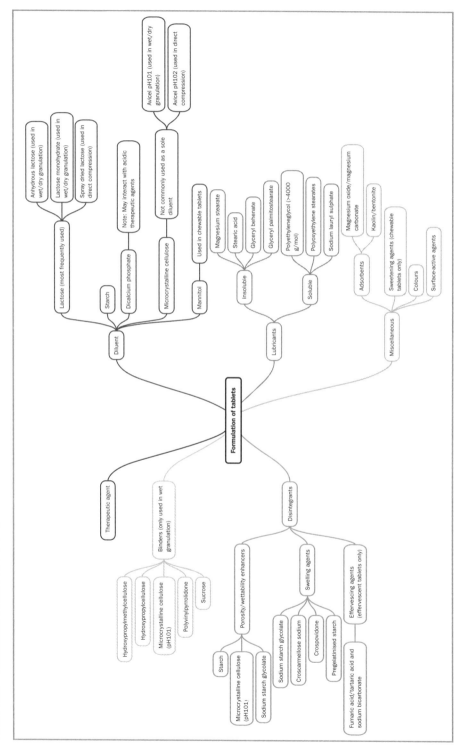

- disruption of the granule–granule or powder–powder bonds due to dissolution of formulation components
- stress relaxation of viscoelastic components. Polymers are viscoelastic and, as such, will undergo stress relaxation as a function of time following removal of the compression stress. This relaxation may be visualised as movement/rearrangement of the polymeric chains, which may, in turn, compromise the mechanical properties of the tablet.

## Coating of tablets

At the start of this chapter a description of the various types of commercially available tablets was provided. The reader will have observed that tablets are often formulated with an external coating. A summary of the rationale for tablet coatings is provided below:

- to protect the drug from degradation in the stomach (an enteric coating)
- to prevent drug-induced irritation at a specific site within the gastrointestinal tract, e.g. the stomach for non-steroidal anti-inflammatory drugs
- to provide controlled release of the drug throughout the gastrointestinal tract
- to target drug release to a specific site in the gastrointestinal tract, e.g. the delivery of drug to the colon for the treatment of inflammatory conditions
- to mask the taste of drugs
- to improve the appearance of the tablet.

Examples of the materials that are commonly used as tablet coatings have been described in an earlier section within this chapter.

### General description of tablet coating

The main steps involved in the coating of tablets are as follows:

- The tablets (or granules) are placed within the coating apparatus and agitated.
- The coating solution is sprayed on to the surface of the tablets.
- Warm air is passed over the tablets to facilitate removal of the solvent from the adsorbed layer of coating solution on the surface of the tablets.
- When the solvent has evaporated, the tablets will be coated with the solid component of the original coating solution.

### Coating formation

Coating solutions are available in two main formulation types: (1) solutions; and (2) emulsions.

### Coating solutions

Coating solutions contain the coating material (polymers or sugar), the coating solvent and other excipients that are required to improve the formulation or performance of the tablet coating, e.g. colourants/opacifiers, plasticisers (to render the film flexible). The choice of the solvent/solvent blend is generally made according to the physicochemical properties of the coating material (i.e. the compatibility of the material with the solvent); however, other considerations include the volatility and the flammability of the solvent. The concentration of the coating material within the solution is also a consideration. Increasing the concentration of coating material within the solvent will reduce the processing time; however, by increasing the concentration of material, the viscosity of the solution may be unacceptably high to achieve the correct spray properties during coating.

### Coating emulsions

More recently emulsions have been developed as tablet-coating systems. In these the polymer is dissolved in a volatile organic phase (with plasticiser and colourants/opacifiers, as required) and this is emulsified within an external aqueous phase. The use of coating emulsions has been favoured in recent years due to environmental concerns associated with conventional organic coating solutions.

   The initial stage in the coating process involves the deposition and subsequent spreading of the atomised coating solution/emulsion on the surface of the tablet (or granule). To achieve a uniform surface distribution of the coating solution/emulsion on the tablet, consideration of the wetting properties of the solution/emulsion on the surface of the tablet is required. Furthermore, the surface rugosity (roughness) and porosity contribute to the tenacity of the interaction initially between the coating solution/emulsion and the tablet but also influence the bond strength between the formed coating and the tablet surface. Following spreading, evaporation of the solvent initially enables coalescence of the organic droplets, and hence initial film formation on the surface of the tablet. As drying continues, the saturation solubility of the coating material in the solvent is exceeded and the solid coating is formed on the surface of the tablet. It should be noted that contact, spreading, droplet coalescence and solvent evaporation occur almost instantaneously.

## Tablet coating in practice

There are several designs of systems that are used in industrial practice to coat tablets (or granules). Examples of these systems are described below.

## Pan coater

The pan coating system is generically composed of a metal pan (drum) into which the tablets are placed and that may be rotated at a range of speeds. The coating solution is sprayed on to the surface of the tablets within the pan whilst the drum is rotated. Simultaneously warm air is passed over the surface of the tablets to facilitate the evaporation of the solvent in which the coating material has been dissolved. A schematic diagram of the operation of a basic pan coater is shown in Figure 9.24.

**Figure 9.24**   Diagram of the operation of the standard pan coater.

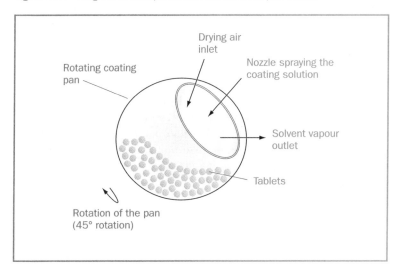

Control of the coating process is obtained by modifying the following parameters:

- rotation rate of the drum/pan
- airflow rate
- temperature of the air
- concentration of sugar/polymer within the coating solution/ emulsion.

More recently, pan coaters have been developed in which the pan is perforated (e.g. the Accela-Cota and Hi-Coater systems). In these systems the warmed air is passed into the drum and through the tablet bed before being exhausted (with the solvent from the coating solution) via the perforated drum. In the Driacoater system, the drum is composed of a series of perforated fins (typically 8 per drum) from which the warmed air is provided. As the drum rotates, the tablets in the tablet bed are mixed by and collected on the fins before being suspended in the warmed air.

The tablets are then dropped into the tablet bed and the process is repeated. The warmed air is then exited from the rear of the pan.

## Air suspension coaters

Air suspension coaters are highly efficient coating systems in which the coating solution is sprayed on to tablets (or granules) that have been suspended in a positive (warmed) airflow. This ability simultaneously to suspend and coat tablets leads to high coating efficiency. Typically the tablets are initially suspended in the centre of the chamber and then move to the periphery of the chamber before falling to the bottom, at which stage the process is continuously repeated. The coating solution is fed into the fluidisation chamber (usually at the bottom of the chamber) as an atomised spray that has been generated either by passage of the coating solution through a nozzle under high pressure or by passage of the coating solution through a nozzle at low pressure, at which point the solution comes into contact with two high-pressure air streams. A diagrammatic representation of the fluidised-bed coating process is shown in Figure 9.25.

**Figure 9.25**   Diagrammatic representation of fluidised-bed coating.

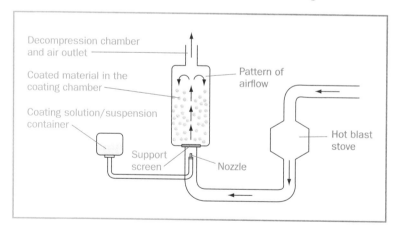

## Process variables in fluidised air coating

There are several process factors that control both the efficiency of the coating process and the quality of the formed coat. These are: (1) evaporation rate of the solvent; (2) fluidised air volume; (3) specific humidity; and (4) coating spray rate and duration.

## Evaporation rate of the solvent

The rate of evaporation of the solvent directly affects both the quality (in particular the mechanical properties) of the tablet coating and the time required to form the tablet coating. Whilst it is important to process the coating in the minimum time, increasing the evaporation rate of the solvent decreases the time available for polymer–polymer interactions to occur. Therefore, if solvent evaporation rate is too rapid, the mechanical properties of the films will be compromised due to the adverse effects on polymer interactions. Both the solvent vapour pressure and the process temperature affect the rate of evaporation of a solvent. Therefore, a low process temperature is normally employed for coating solutions/solvents containing a solvent of high vapour pressure, e.g. dichloromethane. Small fluctuations in processing temperature will have greater effects on the quality of the tablet coatings prepared using organic solvents than when processed using an aqueous coating solution.

### Fluidised air volume
The fluidised air volume will affect both the velocity of the droplets of coating solutions/emulsions and their fluidised pattern within the coating chamber.

### Specific humidity
It is important to control the specific humidity within the warmed air and hence in the coating chamber to ensure that the quality of the tablet coating is optimised. If the relative humidity in the coating chamber is high, evaporative cooling by the solvent may occur. This will, in turn, lower the temperature of the air to below the dew point, resulting in the condensation of water on to the tablet surface. This will interfere with the coating process, resulting in poor adhesion of hydrophobic coatings to the tablet surface and visual imperfections in the formed coating. Therefore, control (but not elimination) of the relative humidity within the coating process is required. The presence of humidity within the coating chamber may be useful in dispelling static electricity that may occur after the coating process has been completed.

### Coating spray rate and duration
The coating spray rate is controlled within the coating process and is selected according to the solubility of the coating solvent in the air volume and the viscosity of the atomised droplets. In addition, it is important to ensure droplet integrity (i.e. minimise droplet aggregation) during the process. It should be noted that excessive spray rates will produce coatings that exhibit poor adhesion to the tablet surface.

Typically the coating process will involve several passes through the coating apparatus. Therefore one method by which the thickness of the coating on the tablet may be modified is to increase the time spent within the coating chamber.

Alternatively, the concentration of coating material may be increased within the coating solution. The viscosity of the solution must be considered to ensure that the increased viscosity does not compromise the atomisation process, and specifically the droplet size.

## Problems associated with tablet coatings

There are several problems associated with tablet coatings, including: (1) poor adhesion of the coating to the tablet; (2) tablet abrasion; (3) filling tablet markings; (4) rough surface; (5) formation of cracks in the coating; and (6) variations in the colour of the coating.

### Poor adhesion of the coating to the tablet

This phenomenon may be due to:

- high relative humidity within the coating chamber when coating tablets using an organic solvent system
- high coating spray rate
- concentration of polymer in the coating solution/emulsion is too low
- temperature of air is too low, resulting in a slow rate of solvent evaporation (particularly valid for coating systems that employ solvents of low vapour pressure, e.g. water)
- air fluidisation rate or pan rotation rate is too slow
- the tablet substrate has minimal curvature. Typically curved surfaces are easier to coat than flat surfaces.

### Tablet abrasion

The coating process involves exposing the tablets to shearing stresses that are generated as a result of collisions with other tablets and also with the walls of the coating chamber. This may result in damage to the tablet surface. This problem may occur due to: (1) inappropriate tablet hardness; (2) irregular tablet shape; (3) tablet bed is too heavy during coating; and (4) the speed of rotation of the pan or the air fluidisation rate is excessive.

#### Inappropriate tablet hardness

Tablet hardness may be improved by increasing the compaction pressure or binder concentration (in wet granulation). Generally the hardness of tablets produced by wet granulation is greater than by other methods and therefore the tablets produced by this method are generally suitable for coating

#### Irregular tablet shape

Irregular tablet shapes are more prone to abrasion than regular tablet shapes. This problem may therefore be overcome by changing the tablet shape.

### Tablet bed is too heavy during coating

This leads to increased tablet–tablet contact. To correct this problem the loading of tablets within the coating chamber is reduced.

### The speed of rotation of the pan or the air fluidisation rate is excessive

By reducing the speed of pan rotation or air fluidisation rate, tablet–tablet contact is decreased.

## Filling tablet markings

Manufacturers may wish to identify their product with a particular mark/name (performed by using a tablet punch that has been embossed with the specified mark). If the coating conditions are unsuitable, the coating will excessively deposit within the mark/name and, in so doing, the marking will be partially obscured. This may occur due to:

■ the use of deep markings
■ use of an excessive volume of coating solution
■ air temperature is too low
■ pan rotation speed/fluidisation flow rate is too low.

## Rough surface

One of the major problems of tablet coating is the production of tablets that exhibit a rough surface. This phenomenon is often associated with drying of the coating droplets prior to reaching the surface of the tablet. To correct this problem the spray rate may be increased and the inlet air temperature decreased.

## Formation of cracks in the coating

The formation of cracks in tablet coatings is principally due to the use of an inappropriate coating formulation. The reader is aware that plasticisers are employed to lower the glass transition temperature of polymer coatings. This in turn renders the film more flexible and less brittle. Therefore cracking in polymer coatings may indicate that either the plasticiser concentration should be increased or, alternatively, a different plasticiser that is more compatible with the polymer chosen for the coating should be considered. In certain situations cracking of polymer coats may occur due to the use of a polymer that has a low stress resistance and is therefore prone to stress failure. To rectify this situation either the molecular weight of the polymer under investigation should be increased or, alternatively, a different polymer should be used that has a greater resistance to the applied stress (i.e. an increased ultimate tensile strength).

## Variations in the colour of the coating

Tablets that have been coated with a polymer containing a colourant should show uniform colour. Variations in the colour of a tablet coating may be due to:

- improper mixing of the colour within the coating formulation
- uneven coating process, resulting in regional differences in the thickness of the applied coating
- migration of coloured components within the tablet core into the coating. This may be resolved by the use of a coloured coating that will mask the effects of the migration or by the use of a coating in which the components within the table core are insoluble.

## Quality control of tablets

Akin to other dosage forms, pharmaceutical tablets have specifications regarding the properties of the product following manufacture (finished product specification) and over the period of storage. Typically, following manufacture and over the designed period of the shelf-life, the following tests are applied to pharmaceutical tablets:

1. **Uniformity of content**: Following manufacture the individual mass of the therapeutic agent in 10 tablets is determined. To pass the uniformity of content test, the individual content of each individual tablet must lie within the range 85–115% of the average content. If the mass of therapeutic agent in more than one tablet lies outside this range or if the mass of therapeutic agent in one tablet lies outside an extended range of 75–125% of the average content then the batch has failed the uniformity of content test.

   However, if the mass of therapeutic agent of one tablet lies outside the 85–115% range but within the 75–125% of the average content, the drug contents of 20 further tablets is determined. The batch will pass uniformity of content then if not more than one of the individual contents of the 30 tablets is outside the range of 85–115% and none is outside 75–125% of the average content range.

   In terms of the average concentration of drug within the tablets, the concentration of therapeutic agent must lie within 95–105% of the nominal concentration following manufacture. Over the shelf-life of the product the concentration of drug must not fall below 90% of the nominal amount.

2. **Uniformity of tablet mass**: In this test, the masses of twenty individual tablets are determined and the average mass determined. An upper boundary is then set regarding the

number of tablets that deviates from a stated percentage deviation. For example, for tablets (uncoated and film-coated):

(a) if the average mass is ≤80 mg, a 10% deviation is allowed (i.e. not more than two tablets may exceed a 10% deviation

(b) if the tablet mass is >80 mg and ≤250 mg, the deviation is 7.5%

(c) if the tablet mass is 250 mg or greater, the percentage deviation is 5%.

Upon storage, the average mass of tablets must lie within a defined range.

3. **Dissolution analysis**: Dissolution analysis quantifies the release of therapeutic agent into a liquid (dissolution) medium under in vitro conditions. In these tests the tablets are placed in vessels of known dimensions (one tablet per vessel), exposed to a defined volume of fluid and stirred using a range of stirring geometries. There are a number of types of dissolution apparatuses, notably the basket apparatus (in which the tablet is housed within a rotating mesh basket), the paddle apparatus, in which stirring of the dissolution medium is performed using a flat paddle, and a flow-through cell dissolution apparatus. The stirring kinetics, dimensions of the apparatuses and temperature of the dissolution medium (36.5–37.5°C) are accurately controlled to ensure reproducibility. The mass of drug dissolved from each of six tablets at a defined time is determined and compared to the product specifications. The dissolution fluid is selected according to the type of tablet. For example, dissolution testing of enteric coated tablets would involve exposure to 0.1 N HCl for circa 2 hours, followed by exposure to phosphate buffered saline (pH 6.8). In some cases the dissolution of the drug from the tablet in a single dissolution medium may be suitable.

4. **Appearance**: The appearance of the tablets is examined and compared to the description on the product specification.

5. **Disintegration analysis**: The disintegration of tablets is measured using the relevant pharmacopoeial disintegration test. The disintegration test apparatus allows the disintegration time of six replicate tablets to be determined. In this test each tablet is placed in a tube, which has a plastic mesh base that allows the movement of fluid into the interior of the tube. The tube assembly, which is housed within a fluid reservoir (bath), is raised and lowered at a regular rate and at a defined temperature. The disintegration of the tablets after a defined time period is determined. The composition of the fluid within the fluid reservoir is chosen to reflect the type of tablet under investigation. For example:

(a) *Uncoated tablets*: Purified water (termed water R) used and the time of the test is 15 minutes. The tablets comply with the test if all six tablets have disintegrated.

(b) *Coated tablets*: Purified water is used and the time of the test is 60 minutes. If any of the tablets have not disintegrated under these conditions, the test is repeated using six fresh tablets and replacing purified water with 0.1 N HCl. The tablets comply with the test if all six tablets have disintegrated. If the coating on the tablets is thin (film-coated tablets), the exposure time is lowered to 30 minutes.

(c) *Chewable tablets*: These are not required to conform to the disintegration test.

(d) *Effervescent tablets*: The disintegration of effervescent tablets is normally performed by analysing tablets individually. Hence, in this test, an effervescent tablet is placed in a beaker containing purified water at between 15°C and 25°C. To pass the test, the tablet must have disintegrated following the complete evolution of the gas. This is then repeated for five replicate tablets.

(e) *Soluble tablets*: The disintegration of soluble tablets is performed at 15–25°C using purified water as the liquid medium within the disintegration apparatus. To pass the test, the tablets must have disintegrated within 3 minutes.

(f) *Enteric tablets*: Analysis of the disintegration of enteric tablets is performed in two stages. Initially disintegration (six tablets) is performed using the disintegration apparatus with 0.1 N HCl chosen as the liquid medium and an operation time of 2 h. The tablets are then visually examined and no tablet should show signs of disintegration. The medium is then exchanged for phosphate buffer solution (pH 6.8) and the test continued for a further 1 h. All tablets must have disintegrated after this period to comply with the test.

6. **Friability test**: This is a non-pharmacopoeial test that evaluates the resistance of the dosage form to applied stresses. The test involves initially weighing a fixed number of tablets. These are then placed in a rotating drum which has a side shelf. The drum rotates along the $x$ axis and in so doing, the tablets are collected on the shelf, moved to the top of the rotation cycle before being dropped from the top of the drum to the floor. This process continues for a fixed number of revolutions following which, the weights of the tablets are measured. The percentage weight loss is then compared to a product specification.

The reader should note that the list of methods is indicative of the quality control methods that may be employed. The reader should consult the appropriate pharmacopoeias for a more detailed description of the above methods and others that have not been explicitly covered in this chapter.

## Multiple choice questions

1. **Concerning roller compaction as a method for tablet manufacture, which of the following statements are true?**
   a. Roller compactions requires the application of heat.
   b. Roller compaction is suitable for the manufacture of tablets containing high-potency (i.e. low-dose) drugs.
   c. Roller compaction produces soft tablets.
   d. Roller compaction requires only conventional excipients.

2. **Concerning direct compression as a method for tablet manufacture, which of the following statements are true?**
   a. It is a simple process, involving a lower number of unit operations in comparison to wet granulation.
   b. The morphology of particles is not important in direct compression.
   c. Direct compression employs spray-dried excipients only.
   d. Coloured tablets may be easily produced by direct compression.

3. **Concerning the pan coating of tablets, which of the following statements are true?**
   a. The coating solution may be aqueous or organic in nature.
   b. Hydrophilic polymers may be used to coat tablets using this technique.
   c. The mass of tablets in the pan affects the quality of the coating.
   d. The speed of rotation of the pan does not affect the quality of the coating.

4. **Concerning wet granulation, which of the following statements are true?**
   a. Alcohol may be employed as a granulating fluid.
   b. Only spray-dried excipients may be used.
   c. Wet granulation may be unsuitable for thermolabile drugs or drugs that are prone to hydrolysis.
   d. Tablets produced by this method are prone to capping.

5. **Concerning pharmaceutical tablets, which of the following statements are true?**
   a. Pharmaceutical tablets may include a disintegrant to improve tablet strength.

b. Pharmaceutical tablets may include magnesium stearate to prevent tablet adhesion to punches.

c. Pharmaceutical tablets include diluents such as colloidal silicone dioxide.

d. Pharmaceutical tablets may involve a granulation stage in order to prevent segregation of the powder mix.

6. **The dissolution rate of drugs from tablets may be increased by which of the following?**

a. Increasing the compression stress during tableting.

b. Increasing the concentration of lubricant.

c. Increasing the concentration of disintegrant.

d. Film-coating tablets with ethylcellulose.

7. **Concerning film coatings, which of the following statements are true?**

a. Film coatings may enhance the dissolution rate of therapeutic agents in the gastrointestinal tract.

b. Enteric film coatings should dissolve in the stomach.

c. Aqueous film coatings, e.g. hydroxypropylcellulose, may be employed to target drug release at the colon.

d. Film coatings may be employed to mask the taste of unpalatable therapeutic agents.

8. **Concerning pan coating of tablets, which of the following statements are true?**

a. Only organic polymer solutions may be employed to coat tablets using the pan coating method.

b. Pan coating is a more effective method than fluidised bed coating.

c. The quality of the tablet coating is improved by increasing the temperature of the inlet air.

d. The quality of the tablet coating is affected by the rotation speed of the pan.

9. **Which of the following excipients are employed in tablets for the stated reasons?**

a. Lactose – as a diluent.

b. Sucrose – as a binder.

c. Hydroxypropylmethylcellulose – as an enteric coating.

d. Polyethylene glycol – as a lubricant.

10. **Regarding the various types of tablets, which of the following statements are true?**

a. Enteric-coated tablets release the therapeutic agent within the small intestine but not in the stomach.

   **b.** Effervescent tablets are used to enhance the dissolution rate and hence absorption rate of poorly soluble drugs.

   **c.** Sugar-coated tablets are employed to reduce drug degradation within the stomach.

   **d.** Film-coated tablets may be used to target drug release to the colon.

**11.** **You have been asked to manufacture the tablet formulation shown in Table 9.2.**

**Table 9.2**  Tablet components to be formulated

| Component | % w/w |
|---|---|
| Oxytetracycline (as the dihydrochloride salt) | 20% |
| Polyvinylpyrrolidone | 15% |
| Avicel PH 102 | 30% |
| Magnesium stearate | 5% |
| Fumed silicone dioxide | 0.5% |
| Lactose | ad 100% |

   **Regarding the manufacture of this formulation, which of the following statements are true?**

   **a.** The preferable manufacturing method is wet granulation.

   **b.** The preferable manufacturing method is dry granulation.

   **c.** The preferable manufacturing method is direct compression.

   **d.** The preferable manufacturing method is roller compaction.

**12.** **Regarding the formulation described in question 11, which of the following statements are true?**

   **a.** Avicel PH 102 is included to facilitate downstream coating of the tablet.

   **b.** Magnesium stearate is included as a lubricant and at the required concentration it will retard drug release.

   **c.** Fumed silicon dioxide is included as a complexation agent to enhance the solubility of oxytetracycline.

   **d.** Magnesium stearate is included as a lubricant at the optimal concentration.

**13.** **A manufacturing method is required to manufacture a tablet formulation in which the drug undergoes rapid degradation by hydrolysis and is additionally thermolabile. Which of the following considerations are valid?**

   **a.** Wet granulation should not be used due to the issues regarding hydrolysis.

b. Wet granulation should be used if the contact time with water is short.
c. Wet granulation may be used if an alcoholic solvent is used.
d. Wet granulation is not the method of choice.

14. **During manufacture of a tablet (containing 10% w/w drug), extensive capping is observed. Which of the following would account for this manufacturing issue?**
a. The high concentration of drug.
b. Insufficient concentration of lubricant in the formulation.
c. Use of flat-faced punches.
d. Excessive compression force.

15. **The physical properties of a coating on a tablet normally depend on which of the following?**
a. Coating thickness.
b. The rate of drying.
c. The type of coating equipment.
d. The polymer concentration of the coating solution.

# chapter 10
# Solid-dosage forms 2: capsules

## Overview

**In this chapter the following points will be discussed/described:**

- an overview/description of the various types of capsules, their manufacture and the rationale for their use
- formulation strategies for hard and soft capsules
- an overview/description of the methods of manufacture of hard and soft capsules
- the advantages and disadvantages of the use of capsules
- a detailed description of the excipients used in the formulation and manufacture of capsules.

## Introduction

Capsules are solid-dosage forms that are most commonly composed of gelatin and are designed to contain a drug-containing formulation. Two types of capsule are available – hard and soft gelatin capsules. These differ in both their mechanical properties and in capsule design. Hard gelatin capsules are less flexible and are composed of two pieces, termed the capsule and the body, whereas soft gelatin capsules are more flexible and are composed of a one-piece capsule shell. A wide range of formulation types may be included within the interior of the capsule. For example, powders, tablets, semisolids and non-aqueous liquids/gels may be filled into *hard* capsules, with powders being the most common formulation option. Soft gelatin capsules are usually filled with non-aqueous liquids containing the therapeutic agent either dispersed or dissolved within this carrier. Capsules offer the pharmaceutical scientist an alternative method for the formulation of solid-dosage forms.

## KeyPoints

- Capsules are solid-dosage forms that are available in two types – hard (two-piece) or soft (one-piece). The major component of the capsule shell is gelatin, although other polymers have been investigated as capsules (e.g. starch, hydroxypropylmethylcellulose).
- Capsules may be filled with a range of formulation types, including powders, tablets, semisolids (hard capsules) and liquids (soft capsules).
- Akin to tablets, the formulation that is included within the capsule contains several components, each of which is present to facilitate the manufacture or to control the biological performance of the dosage form.

## Advantages and disadvantages of capsule formulations

### Advantages
The formulation of capsules may be preferred for several reasons:
- The use of capsules avoids many unit operations that are associated with the manufacture of tablets, e.g. compression, granulation, drying.
- Capsules (generally soft gelatin capsules) may be formulated to increase the oral bioavailability of poorly soluble therapeutic agents. This is particularly the case when formulated as a liquid-filled hard gelatin or soft gelatin capsule.
- Capsules are a convenient method by which liquids may be orally administered to patients as a unit dosage form.
- Capsules are difficult to counterfeit.
- The stability of therapeutic agents may be improved in a capsule formulation.
- Capsules are a convenient means of formulating substances of abuse, e.g. temazepam.

### Disadvantages
The disadvantages of capsule formulations include:
- the requirement for specialised manufacturing equipment
- potential stability problems associated with capsules containing liquid fills
- problems regarding the homogeneity of fill weight and content may be associated with capsule formulations.

## Materials and manufacture of capsules

Capsules are primarily (but not exclusively) manufactured using gelatin; however, the suitability of other materials as replacements, e.g. hydroxypropylmethylcellulose and starch, has been investigated. Gelatin is a mixture of proteins that is extracted from animal collagen (derived from animal skins, sinews and/or bovine bones) by either partial acid or partial alkaline hydrolysis. From these processes two types of gelatin are obtained, termed type A and type B. Type A is obtained by an acid treatment of pig skin (HCl, $H_2SO_4$, $H_2SO_3$ or $H_3PO_4$, pH 1–3 for approximately 1 day) whereas type B is obtained using an alkaline treatment of demineralised bones (immersion in a calcium hydroxide slurry for 1–3 months), following which gelatin is extracted using a series of hot-water washes (of successively increasing temperatures). The gelatin solutions are then cooled to form a gel; subsequent evaporation of water results in the production of dried gelatin. The isoelectric points of type A and type B gelatin differ (between

7 and 9 and between 4.7 and 5.3, respectively), resulting in differing solubilities as a function of pH. Typically the molecular weight range for gelatin is 15 000–250 000. Whilst the two types of gelatin are often used independently, mixtures of the two types are commercially available. The manufacture of the two types of gelatin is summarised in Figure 10.1.

**Figure 10.1**   Schematic for the production of gelatin.

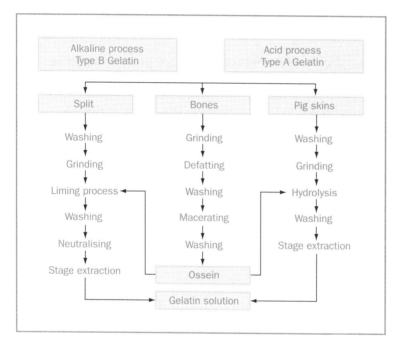

The grade of gelatin is defined by the *bloom strength*, which is defined as the weight (in grams) required to depress a plunger (of defined diameter, 12.7 mm) to a defined depth (4 mm) within an aged gelatin gel (6.66% w/w in water).

The use of gelatin as a capsule material is principally due to the excellent physicochemical and biological properties of this material, including:

- non-toxic material, being used widely as a component of foods. More recently, the production of gelatin using bovine sources has received considerable attention due to the possible transmission of bovine spongiform encephalopathy (BSE). Therefore, if gelatin has been manufactured from bovine sources, it is important that this has been produced from countries in which the incidence of BSE is low – termed grades 1 (highly unlikely) or 2 (unlikely but not excluded)

- soluble in biological fluids at room temperature (note: gelatin capsules do not dissolve but swell when immersed in an aqueous solution <30°C)
- excellent mechanical properties, most notably exhibiting good film, and hence capsule-forming properties
- excellent rheological properties at elevated temperatures. At 50°C, gelatin acts as a mobile liquid (termed a sol), thereby enabling the production of capsules by dip processing (see later)
- undergoes a sol–gel transition at relatively low temperatures. Therefore, gelatin is readily converted to the rigid (gel) state by allowing warmed solutions of this material to cool.

## Description and manufacture of hard gelatin capsules

Hard gelatin capsules are composed of two halves, termed the cap and the body. During manufacture of the dosage form, the formulation is filled into the body (using a range of different mechanical techniques) and the cap is pushed into place. The two halves of the capsule are joined, the cap overlapping with the body. Due to the tight fit between the two halves, separation of the cap and body does not normally occur under normal storage conditions or in clinical use. However, other capsule designs are available in which the two halves form a mechanical seal via indentations on the outside of the body and/or inside the cap (Figure 10.2).

**Figure 10.2**   Diagram of the Conisnap capsule, illustrating the locking mechanism between the two capsule halves.

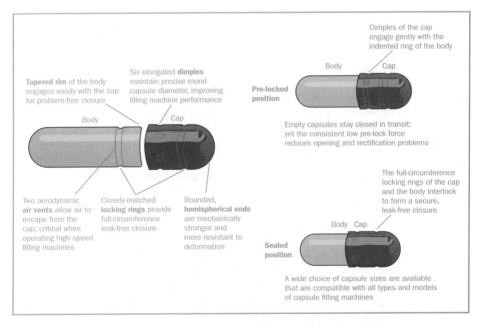

Sealing of hard gelatin capsules may also be performed using two further methods:

(1) *Gelatin band sealing.* In this method a dilute solution of gelatin is applied to the centre of the capsule (between the two halves) which, once dried, produces a hermetic seal.

(2) *Hydroalcoholic solvent seal.* A hydroalcoholic solution (1:1 water/ethanol) is applied to the centre of the capsule (between the two halves). This softens the capsule and, following heating to 45°C, the interface fuses to produce a seal.

The manufacture of hard gelatin capsules is performed using a dip-coating method. The various stages of this are as follows.

### Preparation of the gelatin solution

Initially a concentrated solution of gelatin is prepared (circa 35–40% w/w) in demineralised hot water with stirring. Following dissolution of the polymer a vacuum is then applied to the mixing vessel to remove entrapped air.

### Incorporation of other constituents of the capsule

Other excipients may be included within the heated gelatin solution, e.g.:

■ *Colourants.* Most frequently hard gelatin capsules are coloured to enhance the aesthetic properties and also to act as a means of identifying the product. For this purpose the chosen dye is added at the required concentration into the heated gelatin solution.

■ *Wetting agents/lubricants.* It is permissible to formulate hard gelatin capsules with sodium lauryl sulphate (≤0.15% w/w) as an included component to increase the wetting properties of the capsule shell following contact with an aqueous solution (and hence enhance the dissolution properties of the formulation contained within the capsule). Furthermore, the presence of sodium lauryl sulphate in the heated gelatin solution will also enhance the wetting of the (hydrophilic) gelatin solution on the metal pins during the manufacturing process. This leads to the production of gelatin capsules of uniform thickness.

### Control of the viscosity of the gelatin solution

Following the inclusion of all components within the heated gelatin solution, the viscosity of the solution is then modified (reduced). Control of the viscosity is important as this regulates the thickness of the capsule (generally circa 100 µm). As the viscosity is lowered the capsule thickness will decrease.

### Dip-coating the gelatin solution on to metal pins (moulds)

The machine used to manufacture capsules consists of two sets
of bars, each containing a series of pins (aligned in columnar
formation) that have been lubricated prior to use. The pins (one
set for the production of the cap and one for the body of the
capsule) are dipped into a pan that contains the heated gelatin
solution (maintained at 35–45°C). Following adsorption of the
gelatin solution on to the surface of the pins, the bar containing
the pins is removed and rotated. The reduction in the temperature
of the gelatin and the rotating action cause the gelatin to gel
on the surface of the pins in a uniform manner. The pins are
then advanced through a series of air driers in which air of the
required humidity is passed across the surface of the gelatin film.
Following this, the (hardened) capsules are removed from the pins
and cut to the appropriate size prior to joining the two halves of
the final capsule.

### Properties of the final capsule

The final capsules should exhibit a water content of 13–16% w/w.
This is an important consideration in light of the effects of water
content on the mechanical (and hence in-use) properties of the
capsules. Water acts as a plasticiser for gelatin to ensure that the
mechanical properties of the capsule are sufficiently robust so that
the capsule does not either crack or permanently deform during
manufacture, handling or storage. If the water content is lower
than the above specification, the capsule shell will become brittle
and will crack when exposed to the appropriate stress. Conversely,
if the water content is excessive, the capsule will undergo plastic
flow upon exposure to stress and will lose shape. It is therefore
important that the desired water concentration in the capsule is
achieved within the drying phase of the manufacturing process.
It is additionally preferable that capsules should be stored under
conditions that do not adversely affect this parameter.

A wide range of capsule sizes is available that can
accommodate fill volumes between 0.20 ml and 0.67 ml.

The production of coloured capsules requires the addition of
the appropriate colour and opacifying agent (e.g. titanium dioxide)
in the heated gelatin solution during the capsule-manufacturing
process.

## Formulation considerations for hard gelatin capsules

The fill for hard gelatin capsules may be formulated as either
a powder (or granule) containing the required drug or as a

liquid into which the drug is either dispersed or dissolved. The formulation considerations for both of these strategies are individually described below.

## Solid capsule fills

As the reader will observe, many of the excipients used for the formulation of solid fills for capsules are commonly used in tablet formulations for similar purposes. Therefore, to prevent repetition, only the basic details of the excipients (including their properties and rationale for use) will be described.

### General properties of solid fills

There are several general properties of powders that should be recognised when formulating these as capsule fills, as follows:

- The particle size distributions of the various components of the powder mix (including the therapeutic agent) should be similar both to ensure homogeneous mixing and to minimise segregation.
- It is preferable that the particle size distribution of the powder blend is both monomodal and exhibits low polydispersity (low standard deviation) to ensure predictable and reproducible flow during the filling process. Conversely, multimodal and polydisperse powder mixes will exhibit a tendency to segregate, with resultant problems associated with the homogeneity of the mix and a gradual increase in fill mass as a function of filling time.
- Problems may occur during the filling of particles with an irregular shape (e.g. needle shape).

### Filling of hard gelatin capsules

There are two main methods by which powders are filled into hard gelatin capsules, termed *dependent* and *independent* methods. The design and operation of these are described below.

- *Dependent method (dosing system).* In this method the lower half of the capsule is placed into slots that are located within a revolving turntable. The upper part of the capsule is also housed in a similar turntable. The turntable containing the lower half of the capsule (which may be rotated at a range of speeds) is rotated under a hopper that contains the powder formulation and, as a result, the powder falls into the capsule. The flow of the powder through the hopper and the homogeneity of the powder mix are maintained by the circular movement of an auger. At the end of the operation the two capsule halves are brought together to form the finished dosage form (Figure 10.3).

The mass of powder that is dispensed into each capsule is dependent on the length of time that the hopper spends above the capsule (which is itself dependent on the speed of rotation of the turntable). At the end of the filling process the filled capsules are removed from the turntable.

**Figure 10.3** Diagrammatic representations of the design and operation of the dependent dosing system for filling hard gelatin capsules with powders.

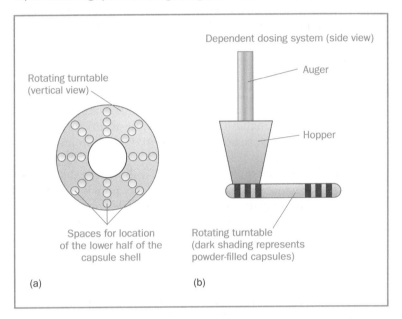

- *Independent method (dosing system).* The independent method of capsule-filling involves the physical transfer of a plug of the mixed powder into the capsule. In this method a tube, which contains a spring-loaded piston, is depressed into a powder bed enabling a volume (plug) of powder to enter the tube. The settings of the spring-loaded piston control the volume of powder that enters the tube. If required, the bonding between the particles within the plug may be enhanced by the application of a compressive pressure. The tube (containing the plug of powder) is then elevated out of the powder bed, rotated and located above the lower half of the capsule and the plug of powder is dispensed into the capsule by the depression of the piston.

The filling of liquids/semisolid formulations into hard gelatin capsules is most frequently performed using a volumetric dosing system. If the rheological properties of the fill are particularly responsive to temperature, filling is performed at temperatures in which the fill is in the liquid state. It should be noted that hard gelatin capsules containing liquid/semisolid fills may be prone to leakage and, therefore, treatment of the capsule shell at the join between the body and the cap is required to prevent this problem.

## Formulation of the fill of hard gelatin capsules

As detailed above, there are three main classes of fill that are employed in hard gelatin capsules, namely powders and liquids/semisolids. The formulation considerations for these categories are considered separately in the following sections.

### Powders

In general, powder formulations for inclusion within hard gelatin capsules should exhibit the following properties:

- *Homogeneity of mix* (described above).
- *Good flow properties*. Filling of the capsules requires the reproducible flow of the powder from the powder bed, through the filling apparatus and into the capsule. As the filling is performed according to volume, it is important that the packing of the particles is also reproducible, as variations in this property will result in differences in the mass of powder filled into each capsule. Acceptable powder flow also requires that clumping of the powders does not occur. Finally, as the filling process may require plug formation (within the dossator), the powder bed should be compressible. Typically the flow properties and the packing properties of the powders are assessed using the following techniques:
  - *Angle of repose (f)*. In this method powder is passed through a funnel until the angle of inclination of the powder is too small to overcome cohesive forces between the particles. The angle of repose is the angle that the powder makes with the horizontal plane. The technique provides an indication of the flow properties of the powder and, specifically, is a measure of cohesion within the powder mass. The tangent of the angle of repose is frequently referred to as a measure of the internal friction of the powder bed.

    If the measured angle exceeds 50°, the flow properties of the powder are poor. Typically an angle of circa 25° is indicative of a powder that would be expected to exhibit suitable flow for manufacturing process. Powder that exhibits high angles of repose will require the addition of a glidant to reduce particle–particle cohesion.
  - *Torque rheometry*. Torque rheometry is a rheological technique in which a stress is applied to the powder

bed (by means of a mixing head) and the subsequent deformation (rate of shear) of the powder bed is determined. Powder beds that should demonstrate high cohesion will require greater shearing stresses to initiate and maintain flow.

This technique is often applied to the characterisation of the wet granulation process and, in particular, to examine the mixing requirements and end-point of the granulation process.

- *Tap density*. Tap density measures the volume occupied by the powder bed, both before $(r_0)$ and following $(r_f)$ the application of a consolidation stress (generally shaking at a defined rate and for a defined period). The ratio of the density of the powder bed before to that after shaking is referred to as a Hausner ratio. Generally a Hausner ratio of circa 1.2 is acceptable, whereas when the Hausner ratio exceeds 1.6 the powder may be problematic to fill into capsules due to the unnecessarily high cohesive interactions between the particles. These cohesive properties may result in erratic filling due to arching and related phenomena.

- *Compatibility between the formulation components and between the formulation components and the capsule.* The excipients that are employed in the formulation of powder fills for hard gelatin capsules are similar to those employed for the formulation and manufacture of tablets. Accordingly the main physicochemical properties of these excipients have been described in Chapter 9. In particular the following excipients are used for the formulation of powder fills:
  - *Diluents*. Diluents are employed to increase the working mass of the powder bed and thereby enhance the reproducibility of the filling process. In addition the diluents may offer additional properties, most notably their flow properties and their ability to undergo compression. Examples of excipients that are employed in the formulation of powder fills include:
    - lactose (monohydrate)
    - maize starch
    - microcrystalline cellulose.
  - *Lubricants/glidants*. Lubricants are used to reduce the interaction of the powders with the metal dossator and/or other metal components of the filling machine, whereas glidants are used to lower the interparticle attraction, thereby reducing clumping and aiding powder flow. The types of lubricants and glidants used in the formulation of powder fill capsules are identical to those employed for tablet manufacture and include:

- magnesium stearate (and other stearates) as a lubricant
- colloidal silicon dioxide as a glidant.
- *Disintegrants.* Disintegrants are employed (as before) to break up the powder mass following release into the stomach. Examples that are used for this purpose include:
  - maize starch
  - microcrystalline cellulose
  - sodium starch glycolate
  - crospovidone
  - croscarmellose.
- *Surface-active agents.* Surface-active agents are employed within powder fill hard gelatin capsules to increase the wetting properties of the powder bed following release within the gastrointestinal tract. Their inclusion is particularly important if the formulation contains significant concentrations of hydrophobic components, e.g. lubricants and glidants. To achieve rapid drug release the powder fill should be readily dispersed within the gastrointestinal contents, a feature that is enhanced by the presence of surface-active agents of high hydrophile–lipophile balance, e.g. sodium lauryl sulphate. In addition, the presence of sodium lauryl sulphate within the hard gelatin capsule shell is allowed and acts similarly to enhance the uptake of fluid into the capsule shell.

### Liquid/semisolid fills for hard gelatin capsules

Liquid/semisolid fills for hard gelatin capsules may be subdivided into various categories:

- *Lipophilic liquids/oils containing dissolved or dispersed therapeutic agent.* Examples of the types of liquids that are commonly used in this category include:
  - vegetable oils (e.g. sunflower, maize, olive)
  - fatty acid esters (e.g. isopropyl myristate).
- *Water-miscible liquids containing dissolved/dispersed therapeutic agent.* Examples of the types of liquids that are commonly used in this category include:
  - polyethylene glycols (PEGs) that are solid at room temperature but will liquefy upon heating (e.g. higher-molecular-weight PEGs)
  - liquid polyoxyethylene–polyoxypropylene block co-polymers (Pluronics).

A major concern in the choice and hence formulation of solvents for liquid fill formulations is the effect of the formulation on the stability of the capsule. As stated previously, the equilibrium moisture content of hard gelatin capsules should be 13–16%

to ensure that the capsule exhibits the optimal mechanical properties. Solvents that are hygroscopic when filled into hard gelatin capsules will enhance moisture uptake into and result in splitting of the capsule shell. The moisture uptake of lipophilic solvents, solid PEGs and Pluronics is low and therefore these solvents are preferred for liquid fill formulations for hard gelatin capsules.

To stabilise the liquid fill formulations for hard gelatin capsules, other excipients will be required, e.g.:

■ *Surface-active agents.* These are included in liquid fills for hard gelatin capsules to:
  • solubilise the therapeutic agent within the solvent
  • stabilise the suspended therapeutic agent
  • enhance the dissolution of poorly water-soluble therapeutic agents within the gastrointestinal tract.
■ *Viscosity-modifying agents.* These are included in liquid fills for hard gelatin capsules to:
  • stabilise the suspended therapeutic agent
  • modify the viscosity of the formulation to optimise filling of the capsule. Typically, viscosity values within the range 0.1–25 Pa/s are considered to be acceptable as liquid fills for hard gelatin capsules. If the viscosity range is lower than this, there will be a loss in the capsule contents due to splashing of the liquid from the capsule during filling. Conversely, liquids of viscosities greater than 25 Pa/s may cause filling problems due to the inability of pumps to fill liquids of this (high) viscosity reproducibly. The coefficient of variation in the volume of liquid fills of high viscosity (>25 Pa/s) will exceed 0.03.
■ *Stabilisers* (e.g. antioxidants, colours).

As the reader will observe, the formulation of the above categories has been addressed in previous chapters. Therefore, for the purpose of this chapter, only the points relevant to the formulation of hard gelatin capsules have been considered.

A diagrammatic representation of the formulations considerations for hard gelatin capsules is shown in figure 10.4.

## Soft gelatin capsules

Soft gelatin capsules are capsules in which the mechanical properties of gelatin have been manipulated by the addition of a plasticiser

### Tips

■ Many of the excipients that are used in the formulation of the powder fill of hard gelatin capsules are also used in the formulation of tablets.
■ Water cannot be used as a liquid fill for hard gelatin capsules. Fill materials for hard gelatin capsules are non-aqueous in nature.

**Figure 10.4** Diagrammatic representation of formulation considerations for hard gelatin capsules.

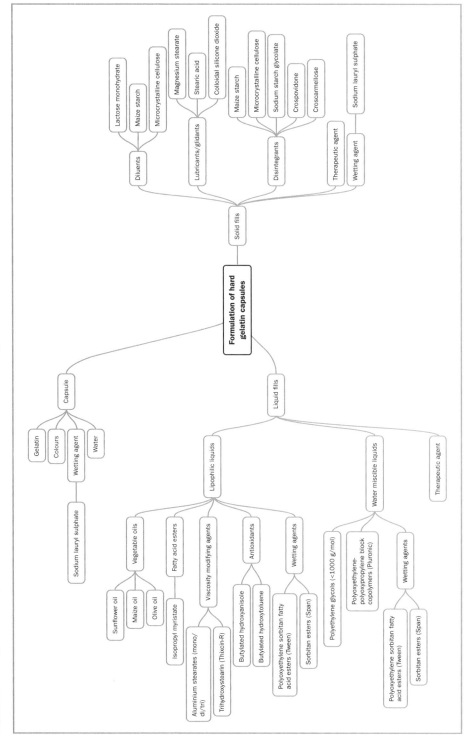

(most notably glycerol or other polyhydric alcohols, e.g. sorbitol), resulting in a more flexible capsule. The popularity of soft gelatin capsules has increased in recent years due to the ability to formulate liquid-based formulations that provide a greater $C_{max}$ than tablet formulations. This has particular applications in the treatment of acute conditions, e.g. pain.

A wide range of shapes of soft gelatin capsule is available, including round, oval, oblong, tubular and various other miscellaneous shapes. The various components of the soft gelatin capsule shell are as follows: (1) gelatin; (2) plasticising agents; (3) water; and (4) miscellaneous excipients.

### Gelatin

Typically type B (alkali-processed) gelatin is used; however, type A (acid-processed) may also be employed for the manufacture of soft gelatin capsules.

### Plasticising agents

As detailed previously, the mechanical properties of the soft gelatin capsule are controlled by the inclusion (and concentration) of plasticisers. For this purpose polyhydric alcohols, principally glycerol or sorbitol or mixtures of these, are used. The concentration of plasticiser is generally 20–30% w/w of the wet mass (refer to the manufacturing method below). This concentration range is an important determinant of the mechanical properties of the soft gelatin capsule. Concentrations in excess of 30% w/w will result in the capsule being too flexible and tacky whereas capsules in which the concentration of plasticiser is below 20% w/w will be too brittle. In both scenarios splitting of the capsule shell may occur either on storage or during use. The hardness of the finished (dried) soft gelatin capsule is specifically defined by the ratio of the plasticiser to gelatin, with ratios of 0.4:1.0 and 0.8:1.0 offering extremes of the typical mechanical properties (hard and soft, respectively).

### Water

Water is required both during the manufacturing process (to facilitate manufacture) and in the finished product to ensure that the capsule is flexible. Initially during manufacture of the soft gelatin capsules the concentration of water is between 30% and 40% w/w of the wet mass; however, following formation of the capsule, the product is exposed to controlled drying, resulting in a final capsule water content of 5–8% w/w. If the capsule is overdried a brittle product will result.

## Miscellaneous excipients

As is the case for hard gelatin capsules, soft gelatin capsules may be coloured or opaque, the chosen colour(s)/opacifiers being added during the manufacturing process. Titanium dioxide is primarily used as an opacifier for capsules. In addition, if required, flavouring agents may be added to the capsule shell.

## Manufacture of soft gelatin capsules

The manufacturing method for soft gelatin capsules was originally patented by RP Scherer in 1933. Whilst variations of this patent have been published, the main principles of the original Scherer method are still employed for the manufacture of soft gelatin capsules. The main features of the method are illustrated in Figure 10.5.

**Figure 10.5**   Illustration of the manufacturing method for soft gelatin capsules.

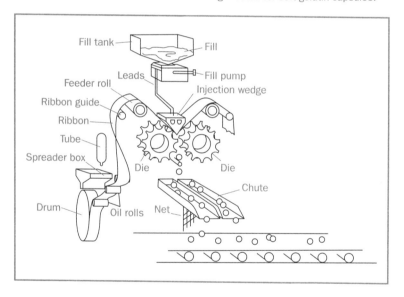

In this method the wet mass formulation is initially prepared containing gelatin, plasticiser(s), water and other excipients, as required. Following this the gelatin solution is fed on to two drums via a spreader box, at which stage ribbons of gelatin are produced. The two sets of ribbons are then fed between two rotary dies (generally lubricated with mineral oil) to form pockets whilst, simultaneously, a metered volume of the capsule fill material is dispensed into the forming pocket. The two halves of capsule (containing the fill material) are sealed by the application of

# Tips

- Liquids are used as the fill materials for soft gelatin capsules; however, water is not employed as a fill material due to the detrimental effects on the physical stability of the capsule.
- Soft gelatin capsules may be formulated to enhance the rate of dissolution of poorly soluble therapeutic agents within the gastrointestinal tract.
- Drugs of abuse are often formulated as soft gelatin capsules.

heat (37–40°C) and pressure, detached from the gelatin ribbon and collected. Following collection, the capsules are washed to remove any lubricant (mineral oil) from the surface of the capsule, dried to remove circa 60–70% w/w water before being allowed to equilibrate under defined conditions of humidity (20–30% relative humidity) and temperature (21–24°C).

Following equilibration, the water content within the capsule shell will be at the required level. It should be noted that other related methods (e.g. the reciprocating die process) are available for the production of soft gelatin capsules; however, a full description of these is beyond the scope of this text.

## Formulation of the fill of soft gelatin capsules

As the reader will have observed from the description of the manufacture of soft gelatin capsules, the fill material for soft gelatin capsules is primarily liquid-based (although powder filling equipment is available). The therapeutic agent may be dissolved or dispersed within the fill material. There are several categories of fill material: (1) lipophilic liquids; (2) self-emulsifying systems; and (3) water-miscible liquids.

### Lipophilic liquids

Lipophilic liquids are commonly used as fill materials for soft gelatin capsules and incorporate both vegetable oils (e.g. soyabean oil) and fatty acid esters. Only a limited number of therapeutic agents are soluble in these materials and therefore, when used as a fill material, the formulation will either require the inclusion of co-solvents and/or surface-active agents (see below) or, alternatively, the fill may be formulated as a suspension (requiring the inclusion of viscosity-modifying agents and/or surfactants).

### Self-emulsifying systems

Self-emulsifying systems are lipophilic liquids that contain a non-ionic emulsifying agent (e.g. the Tween series). Following release into the gastrointestinal tract, the fill material rapidly emulsifies into small droplets (with high surface area) and, in so doing, enhances the dissolution and hence absorption of the therapeutic agent.

### Water-miscible liquids

Liquids in this category include high-molecular-weight alcohols, e.g. PEG 400, PEG 600, non-ionic surface-active agents (e.g. Tweens) and Pluronics (polyoxyethylene–polyoxypropylene block co-polymers). Small concentrations of ethanol and water (generally less than 10%) may be tolerated within soft gelatin capsules. As before, the selected therapeutic agent may be dissolved or, if the solubility is exceeded, dispersed within the solvent, the latter necessitating the addition of surface-active or viscosity-modifying agents to stabilise the formulation.

As required, additional formulation excipients will be required in the above liquid systems: the details of these have been outlined in previous chapters. Examples of the categories of excipient include:

- viscosity-modifying agents
- surface-active agents
- colours
- co-solvents (and other solubilising agents).

A diagrammatic representation of the formulation considerations for hard gelatin capsules is shown in Figure 10.6.

## Quality control of capsules

The quality control procedures for capsules are similar to those for tablets and therefore, when applicable, reference will be made to the information presented in the previous chapter. Typically, following manufacture and over the designed period of the shelf-life, the following tests are applied to pharmaceutical tablets:

1. **Uniformity of content**: Following manufacture the individual mass of the therapeutic agent in 10 capsules is determined. To pass the uniformity of content test, the individual content of not more than one capsule is outside the range 85–115% of the average content and no capsules lie outside 75–125% of the average content.

   If the content of two or three capsules lies outside the 85–115% range but within the 75–125% of the average content, the drug content of twenty further capsules is determined. The batch will pass uniformity of content then if not more than three of the individual contents of the 30 capsules are outside the range 85–115% and none is outside 75–125% of the average content range.

   In terms of the average concentration of drug within the tablets, the concentration of therapeutic agent must lie within 95–105% of the nominal concentration following manufacture. Over the shelf-life of the product the concentration of drug must not fall below 90% of the nominal amount.

2. **Uniformity of capsule mass**: If specified, a similar methodology to that described for tablets may be applied.

**Figure 10.6** Diagrammatic representation of formulation considerations for soft gelatin capsules.

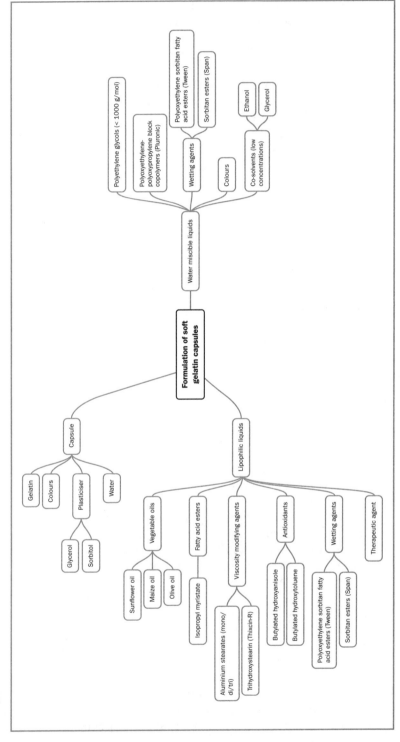

3. **Dissolution analysis**: If specified, dissolution analysis for capsules is performed. The process is identical to that described for tablets where, in most cases, the basket geometry is used. The dissolution medium is chosen to reflect the clinical usage of the preparation.

4. **Appearance**: The appearance of the capsules is examined and compared to the description on the product specification.

5. **Disintegration analysis**: The disintegration of capsules is measured using an identical methodology to that described for tablets. However, if the product specification defines the need for dissolution analysis then there is no requirement for disintegration analysis to be performed. The composition of the fluid within the fluid reservoir is defined in the product specification.

The reader should note that the list of methods is indicative of the quality control methods that may be employed. The reader should consult the appropriate pharmacopoeias for a more detailed description of the above methods and others that have not been explicitly covered in this chapter.

## Multiple choice questions

1. **Regarding capsules, which of the following statements are true?**
   a. Capsules are more straightforward to manufacture than tablets.
   b. Capsules may be formulated to increase the bioavailability of poorly soluble drugs.
   c. Capsules may be coloured to aid identification.
   d. The stability of therapeutic agents when formulated as capsules is always greater than that of tablets.

2. **Concerning hard gelatin capsules, which of the following statements are true?**
   a. Gelatin is a polysaccharide that is derived from animal sources.
   b. There are two types of gelatin, each exhibiting different isoelectric points.
   c. The grade of gelatin is defined by the bloom strength.
   d. Gelatin is freely soluble in water at room temperature.

3. **Regarding hard gelatin capsules, which of the following statements are true?**
   a. Hard gelatin capsules are composed of two halves that are joined following filling.

b. Hard gelatin capsules may be filled with aqueous liquids.
c. Hard gelatin capsules may be filled with non-aqueous liquids.
d. Hard gelatin capsules may be filled with aqueous drug suspensions.

4. **Concerning soft gelatin capsules, which of the following statements are true?**
a. Soft gelatin capsules generally contain a disintegrant.
b. Soft gelatin capsules may be formulated with a powder fill.
c. Soft gelatin capsules often contain emulsions.
d. Soft gelatin capsules may exhibit favourable drug bioavailability due to lipid breakdown.

5. **Concerning the composition and manufacture of hard gelatin capsule shells, which of the following statements are true?**
a. Hard gelatin capsule shells contain water.
b. Hard gelatin capsule shells may contain colouring agents.
c. Hard gelatin capsule shells always contain low concentrations of surfactants.
d. Hard gelatin capsule shells are prepared using a polymerisation process.

6. **Solid fills for hard gelatin capsules frequently contain which of the following excipients for the designated purpose?**
a. Glidants – to improve powder flow.
b. Lactose – as a diluent.
c. Magnesium stearate – as a lubricant.
d. Starch – as a disintegrant.

7. **Ideally, solid fills for hard gelatin capsules should exhibit which of the following properties?**
a. Homogeneity of mix.
b. A low angle of repose.
c. A high Hausner ratio (>1.6).
d. Low shearing stresses during powder flow.

8. **Which of the following are suitable as liquid fills for hard gelatin capsules?**
a. Polyethylene glycols.
b. Arachis oil.
c. Fatty acid esters.
d. Purified water.

9. **Regarding soft gelatin capsules, which of the following statements are true?**

a. The capsule shell contains glycerol as a plasticising agent.
b. The capsule shell contains approximately 20% water.
c. The capsule shell is preferentially composed of type A gelatin.
d. The capsule shell may contain colours or opacifiers.

10. **Which of the following excipients are common components of the fill of soft gelatin capsules?**
a. Alcohol.
b. Lipophilic liquids, e.g. soyabean oil.
c. Water-miscible liquids, e.g. surface-active agents.
d. Lubricants, e.g. magnesium stearate.

11. **A soft-gelatin capsule formulation has been designed composed of a therapeutic agent (5%w/w) dissolved in a vegetable oil. In clinical trials the bioavailability of this formulation was poor. Which of the following excipients may be added to enhance bioavailability?**
a. A second vegetable oil.
b. Isopropyl myristate.
c. A surfactant with a high HLB.
d. A mixed surfactant system.

12. **Which of the following parameters affect the dissolution of poorly-soluble drugs from hard gelatin capsules into the stomach?**
a. Particle size of the drug: larger particles enhancing the drug release rate.
b. Particle size of the drug: smaller particles enhancing the drug release rate.
c. The Hausner ratio of the powder during filling.
d. Shearing stresses during capsule filling.

13. **Which of the following statements are true regarding softgel formulations**
a. The fill material may be an aqueous suspension of drug.
b. The fill material may be a polyol, e.g. poly(ethylene glycol).
c. The capsule shell may be coloured.
d. The fill material should not be coloured.

14. **Which of the following statements are true regarding hard gelatin capsules?**
a. They may be filled with oil in water emulsions.
b. They may be filled with powders.
c. They may be used as a component of inhalation dosage forms.
d. Tablets may be filled into hard gelatin capsules.

15. **Which of the statements are true regarding the manufacture of hard gelatin capsule products?**
    a. The capsule should not be coloured.
    b. Lubricants may be required in the formulation.
    c. Aqueous solutions may be used as the fill material.
    d. Hard gelatin capsules may be coated in a downstream process.

# chapter 11
# Introduction to pharmaceutical engineering

## Overview

**In this chapter the following points will be discussed/described:**

- an overview/description of the unit operations that are used in the manufacture of pharmaceutical products
- theoretical and practical aspects of liquid mixing
- theoretical and practical aspects of filtration
- theoretical and practical aspects of particle size reduction (milling)
- theoretical and practical aspects of drying.

## General introduction

The preparation of dosage forms, such as those described in earlier chapters, is a fundamental examinable component of all undergraduate pharmacy degrees. Within the laboratory the equipment with which the bespoke products are manufactured serves to achieve the same end point as production scale equipment, namely the production of products of defined specifications. However, the manner in which products are manufactured at production (industrial) scale is markedly different from the laboratory counterpart. For example, in the laboratory the manufacture of a liquid dosage form (final volume 1 litre) frequently involves the use of a beaker and a small-scale stirrer, e.g. magnetic or overhead stirrer. However, it is impractical to use such equipment whenever the production batch size is 4000 litres. This chapter will provide a brief overview of the theoretical and practical aspects of the large-scale manufacture of key dosage forms, notably liquids, suspensions, creams and ointments. The manufacture of solid

## KeyPoints

- All pharmaceutical products are manufactured using specialised manufacturing equipment.
- There are several different steps that may be involved in the manufacturing of pharmaceutical products including, mixing, filtration, drying and milling; the choice of the methods being dependent on the nature of the product being manufactured.
- Knowledge of the theoretical aspects of these techniques allows a full appreciation of their practical applications.

dosage forms has been briefly addressed in previous chapters and will not be specifically covered in this chapter. In particular, this chapter will provide an overview of four key unit operations that are used for the manufacture of the aforementioned dosage forms, namely: (1) mixing; (2) filtration; (3) milling; and (4) drying.

## Mixing of pharmaceutical products

### Introduction

Of the four named unit operations, mixing is the only process that is employed in the manufacture of all dosage forms, independent of their type. Consequently the importance of this process to the pharmaceutical industry cannot be overstated. This section will provide an overview of the theoretical concepts and practical applications of this unit operation. It should be stated, however, that mixing is an academically demanding topic, with certain theoretical aspects being outside of the scope of this textbook. Therefore the theory behind certain aspects of mixing has been condensed to accommodate the applied nature of this chapter.

The manufacture of pharmaceutical products involves the mixing of a wide range of materials including:

- mixing of two or more fluids, e.g. mixing two liquid solvents, mixing a drug solution into a liquid vehicle
- mixing of a disperse phase within a vehicle, e.g. in the manufacture of a drug suspension, cream, ointment, paste
- mixing solids, e.g. mixing components prior to tablet production using direct compression, mixing components prior to wet granulation
- gas mixing, e.g. purging liquids with nitrogen gas to minimise drug degradation by oxidation.

The overall aim of any mixing process is to ensure that the resultant mixed product is homogeneous. Accordingly it is important for the reader to gain an understanding of fluid flow. Fluid flow may be characterised in terms of the rheological properties of the system under investigation. Initially a block of liquid is considered as an infinite series of parallel plates, with the bottom plate fixed in opposition (i.e. unable to move). As shown in Figure 11.1, if a horizontal stress (force per unit area) is applied to the top plate, there will be a movement of each of the lower plates, the velocities ($v$) of which are dependent on the distance from the bottom plate ($r$).

In this situation it may be shown that the shearing stress ($\sigma$) is directly proportional to the rate of shear (the velocity gradient), $\left( \dfrac{\mathrm{d}v}{\mathrm{d}r} \right)$, the proportionality constant being referred to as the

**Figure 11.1**  Representation of fluid flow following the application of a horizontal stress.

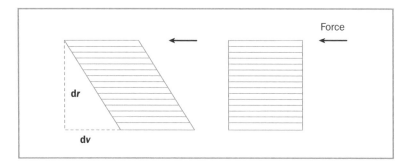

viscosity ($\eta$). This type of flow is referred to as Newtonian flow (Figure 11.2) and is exhibited primarily by simple liquids and dilute suspensions. Mathematically this is defined as:

$$\sigma = \eta\dot{\gamma} \tag{11.1}$$

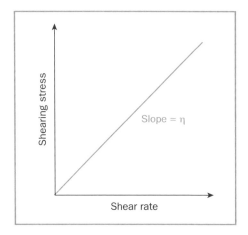

**Figure 11.2**  Flow curve (rheogram) for a Newtonian system.

where $\dot{\gamma}$ refers to the rate of shear, $\eta$ to the viscosity and $\sigma$ to the shearing stress.

Many pharmaceutical systems do not exhibit this linear relation between shearing stress and rate of shear and are referred to as non-Newtonian. There are several sub-groups of non-Newtonian systems but the most relevant to pharmaceutical systems are Bingham flow, pseudoplastic flow and dilatant flow.

Two major features of the Bingham flow rheogram shown in Figure 11.3 may be identified. Firstly, the flow does not pass through the origin but intersects the y axis. This value is referred to as the yield stress and is the stress required to initiate flow.

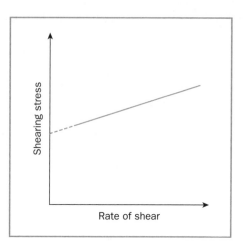

**Figure 11.3**   Representative rheogram for a system exhibiting Bingham flow.

Secondly, once flow has commenced the system behaves akin to a Newtonian system in that shear stress is directly proportional to the rate of shear. Mathematically this is defined as:

$$\sigma = \sigma_y + \eta\dot{\gamma} \tag{11.2}$$

where $\sigma_y$ refers to the yield stress, $\dot{\gamma}$ refers to the rate of shear and $\sigma$ refers to the shearing stress. This flow phenotype is frequently exhibited by flocculated suspensions.

In the pseudoplastic flow system shown in Figure 11.4 flow commences at the origin, indicating that flow commences upon the application of the shearing stress. The system does not exhibit a constant viscosity (the tangent to the curve) but decreases as the rate of shear increases. Consequently these systems are referred to as *shear-thinning*. The viscosity (ratio of the shearing stress to

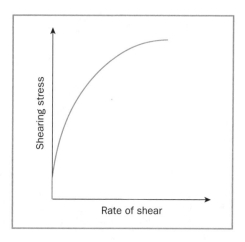

**Figure 11.4**   Representative rheogram for a system exhibiting pseudoplastic flow.

the rate of shear) at each rate of shear is referred to as an apparent viscosity. Mathematically this relationship is defined as:

$$\sigma = K\dot{\gamma}^{n} \tag{11.3}$$

where $K$ is termed the consistency, $n$ is an index of pseudoplasticity, $\dot{\gamma}$ refers to the rate of shear and $\sigma$ refers to the shearing stress. This flow phenotype is frequently exhibited by many pharmaceutical systems, e.g. gels, creams, ointments and pastes.

In the dilatant flow system shown in Figure 11.5 system flow commences at the origin, once more indicating that flow commences upon the application of the shearing stress. Akin to pseudoplastic flow, the system does not exhibit a constant viscosity (the tangent to the curve) but conversely this increases as the rate of shear increases. Consequently these systems are referred to as *shear-thickening*. The viscosity (ratio of the shearing stress to the rate of shear) at each rate of shear is referred to as an apparent viscosity. This flow phenotype is exhibited primarily by concentrated, deflocculated suspensions.

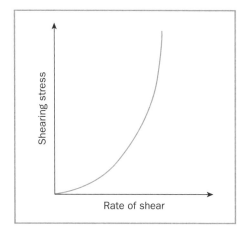

**Figure 11.5** Representative rheogram for a system exhibiting dilatant flow.

## Mechanisms of mixing

In pharmaceutical systems several different systems may be required to be mixed, including:

- liquid–liquid systems in which the two liquids are miscible (e.g. mixing of drug solutions with the vehicle, mixing the vehicle with a co-solvent) and additionally where the two liquids are immiscible, e.g. in the preparation of disperse systems, e.g. creams.
- gas–liquid systems, e.g. whenever a preparation is purged with an inert gas to reduce oxidation.

- solid–liquid mixing, e.g. mixing of suspensions
- solid–solid mixing in the preparation of solid dosage forms.

To optimise the choice of mixing method it is important to understand both the physical properties of the systems to be mixed and, in addition, the nature of how fluids move during mixing. Mixing requires convective flow to ensure that there are no stagnant areas in the mixing vessel and also requires sufficient energy to ensure that mixing is homogeneous. There are two primary mixing mechanisms that operate in the manufacture of pharmaceutical systems, termed laminar and turbulent. The dominant mechanism depends on the properties of the systems being mixed and, in many cases, both types of mixing may be observed within the mixing vessel (at different regions).

### Laminar mixing

Laminar mixing is characterised by the flow of molecules on parallel paths and was shown initially by Reynolds in his study of fluid mechanics. Figure 11.6 illustrates what happens when a fluid is allowed to flow through a horizontal, cylindrical tube and into this, from above, is fed a dye as a fine jet into the main body of the cylinder. At low flow rates the dye is observed to flow as a continuous, horizontal straight line within the main body of the cylinder, with minimal disturbance to the integrity of this flow of dye. As the flow rate of the main liquid through the cylinder is increased, the integrity of the line of dye is disturbed,

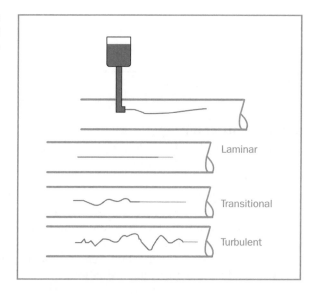

**Figure 11.6**  Diagrammatic illustration of laminar and turbulent flow.

the extent of this disturbance increasing as the flow rate increases until there is uniform mixing of the dye throughout the liquid. By reducing the flow rate of liquid the continuous flow of a thin line of dye can be re-established. The conditions that facilitate the flow of the continuous line of dye are termed laminar flow.

Laminar mixing occurs principally in systems that have a high viscosity (>10 N s m⁻² or >1 poise). The high viscosity of these systems acts to dissipate the inertial energy derived from the mixer and consequently these systems are mixed using impellers that cover a significant proportion of the mixing vessel (this point will be addressed later). Under these conditions laminar mixing operates by deforming and lengthening the fluid elements adjacent to the impeller. Each passage into the shearing region results in a furtherance of the deformation and elongation of the system being mixed until there is homogeneous mixing of the system. An illustration of laminar mixing following the application of a rotational stress is shown in Figure 11.7.

**Figure 11.7**   Illustration of laminar mixing: (a) represents the initial unmixed state; (b) represents the mixing of the solution following the initial rotations of the mixer.

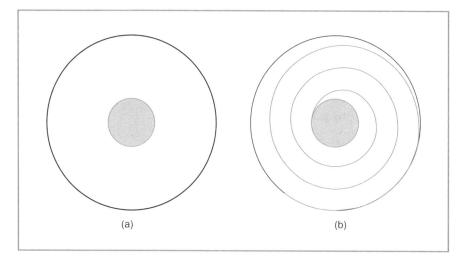

(a)                                        (b)

## Turbulent mixing

In Reynolds' experiments, it was noted that the thin stream of dye within the flow of fluid in the cylinder was disturbed as the flow rate of fluid in the cylinder increased (Figure 11.6). This disturbance was visualised as a series of eddy currents (sections of the fluid that moved in a direction that differed from the main body of fluid flow). As the flow rate increased there was uniform mixing of the dye within the main body of fluid flow; the flow being termed turbulent flow.

Turbulent mixing occurs in low viscosity systems ($<10$ N s m$^{-2}$), in which the inertial energy of the impeller may be readily transferred to and circulate within the bulk of the fluid. Mixing based on turbulence occurs rapidly, is most rapid in the area adjacent to the impeller and may be considered to act as a series of interconnecting eddy currents that increase in scale (size distribution) and intensity (velocities at which the eddies move) with increasing distance from the walls of the mixing vessel.

### The Reynolds number

Use of the Reynolds number (Re) is a straightforward method to characterise the flow of fluids; it is calculated as follows:

$$Re = \frac{\rho ND^2}{\eta} \tag{11.4}$$

where $\rho$ is the density, $\eta$ is the viscosity, $N$ is the speed of agitation and $D$ is the diameter of the impeller.

The Reynolds number is therefore the ratio of the inertial to viscous forces in the system. Low values of Re indicate laminar flow, e.g. $<2000$ for the flow of fluids through a circular pipe, whereas high values of Re ($>4000$) indicate turbulent flow. The observant reader will by now have realised that there is a transition region between laminar flow and turbulent flow in which both mechanisms operate. This is illustrated in Figure 11.6.

### Fluid flow patterns in mixing vessels

There are two main categories of fluid flow patterns that are observed during the mixing of fluids in mixing vessels, namely axial and radial. A diagrammatic representation of these fluid flow types is shown in Figure 11.8.

Axial flow is generally produced by propeller systems. One drawback of such systems is vortex formation in operation which may result in a reduced rate of agitation and the failure of the flow pattern to uniformly mix all regions within the vessel. These problems are commonly addressed by the use of baffles or by the use of an off-centre mounted agitator. Figure 11.9 illustrates the flow patterns for a top-entry, off-centred propeller mixer (without baffles).

Similarly optimised mixing patters may be achieved using a side-entry propeller mixer (Figure 11.10).

Radial flow is typically associated with the use of flat-bladed turbines, an example of which is shown in Figure 11.11.

**Figure 11.8** Diagrammatic representation of (a) axial flow and (b) radial flow in a mixing vessel. In (a) axial flow has been obtained using a three-blade propeller whereas in (b) radial flow has been obtained using a turbine.

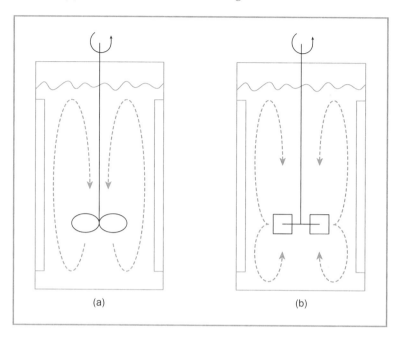

(a)　　　　　　　　　　(b)

**Figure 11.9** Flow patterns associated with a top-entry, off-centred propeller mixer.

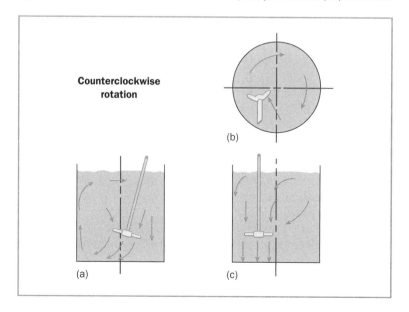

**Counterclockwise rotation**

(b)

(a)　　　　　　　　　　(c)

**Figure 11.10** Diagrammatic representation of a side-entry propeller mixer housed within a mixing vessel.

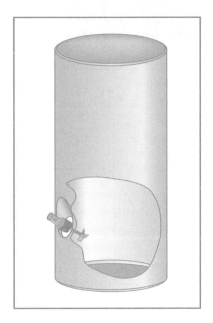

**Figure 11.11** A flat-bladed turbine.

## Design of mixing vessels

Mixing of pharmaceutical products (from liquid to solid) occurs in a mixing vessel (sometimes referred to as a reactor or a vat). The design of these vessels is performed according to the product characteristics and hence the type of mixing required. An example of a generic mixing vessel is shown in Figure 11.12.

From this diagram several key design features may be identified:

- the vessel itself
- the cooling jacket (used to heat and cool the contents of the vessel)

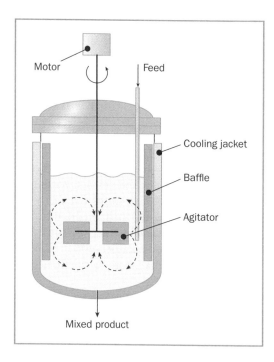

**Figure 11.12**   An illustration of a mixing vessel.

- the agitator (stirrer) (used to ensure flow of the contents)
- baffles (to facilitate homogeneous mixing).

### The vessel

The vessel is characterised in terms of the height to width ratio. In systems of low viscosity this ratio is normally circa 1.5:1 whereas in high viscosity systems (where the mixing patterns are primarily laminar) this ratio may reduce to 1:1. The vessels are commonly manufactured from stainless steel, the grade of which is selected to accommodate the material being mixed (e.g. acid resistant, alkali resistant, abrasion resistant). A common pharmaceutical grade is 316L stainless steel. Finally vessels are commonly manufactured with a curved bottom, although flat-bottomed variants are available.

In addition to the ratio of the height to the width of the vessel, there are several additional important parameters that directly affect mixing as follows:

- the height of the agitator from the base of the vessel
- the width of the agitator (and therefore the clearance between the agitator and the vessel wall)
- the speed of agitation
- the volume of material being mixed
- the design of the baffles (including the shape and number of baffles).

### The cooling jacket

In pharmaceutical processing it is common for temperature control of the material being mixed to be required. For example, the dissolution of certain drugs within liquid vehicles may require the temperature of the liquid vehicle to be increased, whereas increasing the temperature of the finished (mixed) product is frequently performed to enhance the subsequent rate of filtration of the product. Having heated the material in the vessel, the pharmaceutical engineer will ultimately wish to reduce the temperature of the product to room temperature in a rapid fashion. One of the methods by which this is achieved is by the use of jacketed vessels. Within the jacket water is circulated (at a chosen temperature) and acts a heat exchanger, whether reducing or increasing the temperature of the vessel contents. Steam is commonly used to increase the temperature of the vessel contents. The mode of operation of these systems is similar to that which operates in a laboratory condenser.

### The agitator

The agitator is normally composed of a shaft (which is connected to a motor) and a mixing head, the latter being classified according to their design. Examples of these designs include propellers, turbines, helical ribbons and Z blade mixers.

#### Propellers

Propellers are typically of the three-blade design (Figure 11.13) (frequently referred to as the marine design) and are used to mix low viscosity liquids at high rates of agitation. Typically the diameter of the propeller is within the range 0.13–0.67 of the tank diameter (this being dependent on the volume and properties of the material being mixed). The fluid flow patterns achieved by propellers are axial and may be directed towards the top or towards the bottom of the vessel (dependent on whether the drive is rotated clockwise or anticlockwise).

**Figure 11.13** A three bladed propeller.

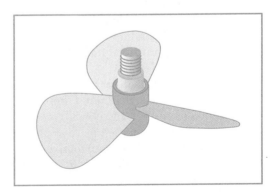

Whilst propellers are designed with three blades, other designs are available based on two and four blades. Propellers are designed to provide a defined pitch, which is the ratio of the distance moved by the fluid following one revolution of the propeller to the propeller diameter. The choice of propeller pitch affects the mixing properties of the mixer. When mixing large volumes of fluids (where typically the depth of the fluid is considerably greater than the width of the vessel), it is common to use an agitator system consisting of two propellers, one situated at the end of the shaft and the other situated further up the shaft. Under certain conditions the mixing of fluids using this type of agitator may be supplemented by side entry propellers. Alternatively the mixing of large volumes of liquids may be accomplished using more than one side-entry propeller, arranged at different heights within the mixing vessel.

Within the pharmaceutical industry portable mixers are also used to mix fluids. These mixers are clamped on to the vessel housing or are suspended into the vessel by a portable arm. Such mixers are commonly fitted with two propellers (arranged at different vertical distances along the shaft), which force the liquid to flow in different directions (thus enhancing mixing). Portable mixers are used at high speeds of operation (typically 15 Hz). The use of portable mixers allows mixing vessels to be used for the manufacture of a range of products and does not restrict the use of the vessel to the manufacture of a certain product type.

Propeller mixers are frequently used for mixing low viscosity fluids (<400 Pa s) at high rotation speeds (up to 3000 rpm). Examples of fluids that are mixed using propeller mixers include liquids and drug suspensions.

Ideally, propellers should offer the following properties:

- provide a wide range of operating speeds and hence shearing rates
- ensure complete (homogeneous) mixing
- be economical in use (although this will be compromised by the use of baffles and by the incorrect selection of propeller pitch)
- provide effective mixing (dispersion) of solids in the production of pharmaceutical suspensions (without damaging the particles being mixed)
- allow for different mounting types, e.g. off-centre, side-entry
- be easy to clear after use.

### Turbines

Turbines are used as an alternative to propellers for the mixing of low viscosity liquids; however, these systems may also be used for the effective mixing of medium viscosity liquids. The velocity of mixing of turbine systems is typically less than for propellers and is circa 3 m s⁻¹. Akin to propellers the diameter of turbines is approximately 0.13–0.67 that of the diameter of the vessel, with 0.33:1 being most typical. There is a wide range of turbine designs as shown in Figures 11.14 and 11.15.

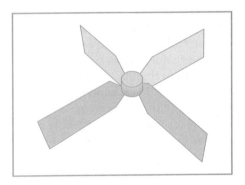

**Figure 11.14**  An axial flow four pitch bladed turbine.

**Figure 11.15**  Different types of propellers: (a) a three blade propeller; (b) a six flat-blade disc turbine (the Rushton turbine); (c) an angled (45°) six blade disc turbine; (d) a six curved-blade disc turbine.

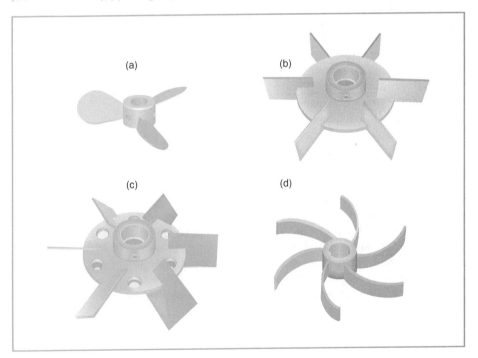

(a)

(b)

(c)

(d)

Flat blade radial turbines typically consist of a number of blades that are located equally around a horizontal disc. The circular movement of the flat horizontal disc requires comparatively little energy and thus the energy consumption of these turbines is associated with the impeller blades. The blades are designed to provide high shear conditions thereby optimising the use of energy for mixing and ensuring maximum turbulence. As illustrated in Figure 11.8 mixing of liquids using such systems involves the radial movement of the liquid both upwards and downwards. Curved blade turbines are also available and in many situations these offer better dispersion properties and, in the case of suspensions, less product build-up on the curved blades. As in the case of propellers, mixing of large volume of liquids may require the use of more than one propeller as shown in Figure 11.16.

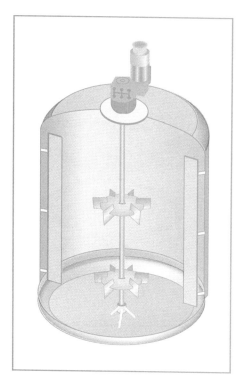

**Figure 11.16** Two six flat-blade (Rushton) disc turbines within a mixing vessel.

**Use of turbines**

Within the pharmaceutical industry turbines are employed for:
- mixing of liquids
- dissolving solids in liquids
- dispersion of solids in liquids (particularly angled blade systems)
- dispersion of gases in liquids
- heat transfer.

The systems described above exhibit low to medium viscosity.

### Anchor and helical ribbon impellers

Anchor and helical ribbon impellers are used for the mixing
of high viscosity systems (Figure 11.17). Key to the operation
of these systems is the ratio of the impeller diameter to the
diameter of the vessel, which is typically 0.95:1 for both systems.
Accordingly there is little clearance between the agitator and
the wall of the mixing vessel. This limited clearance allows the
energy supplied by the mixer to be effectively transferred to the
product that is adjacent to the impeller and vessel. Due to issues
regarding transference of energy, mixing of high viscosity systems
using anchor or helical ribbon impellers is performed at low shear
rates and thus the mixing of these systems occurs by laminar flow.

**Figure 11.17**  Two impellers that are commonly used for mixing high viscosity
systems: (a) the anchor impeller; (b) the double helical ribbon impeller.

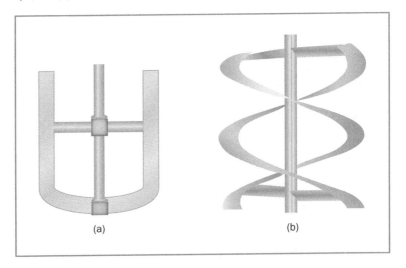

(a)  (b)

Other impeller geometries are available for the mixing of high
viscosity systems (e.g. liquids, gels, pastes) and include the sigma
blade, the Z blade mixer and the Banbury mixer. The mixing
components of the Z-blade mixer are shown in Figure 11.18. In
these systems the clearance between the impeller and the walls
of the mixing vessel is small, thereby optimising mixing of high
viscosity systems at low operative shear rates.

### In-line mixers

The mixing systems described thus far in this chapter are systems
in which the mixing of the fluid is facilitated by the movement

**Figure 11.18**   The mixing component of the Z-blade mixer.

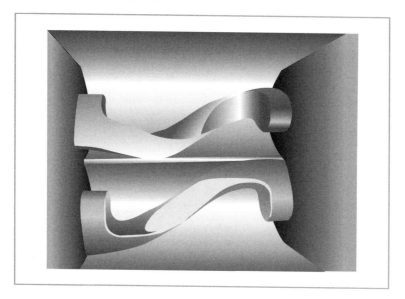

of an impeller, the resultant fluid flow ensuring homogeneity of mixing. Within the pharmaceutical industry mixing of products (low to medium viscosity) may be performed using static mixers. In these systems the mixing component shown in Figure 11.19 is housed within a metal casing and the product is passed through the mixer.

**Figure 11.19**   Two designs for static mixers.

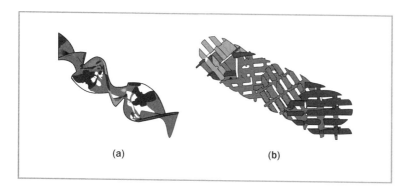

(a)                                          (b)

   Static mixers are composed of a number of helical elements, which give rise to two patterns, illustrated in Figure 11.20,

**Figure 11.20**   Mixing within a static mixer.

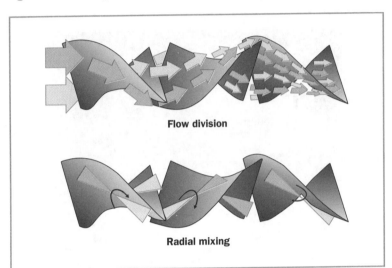

**Flow division**

**Radial mixing**

that result in mixing. In *flow division*, under laminar flow the fluid encounters the leading edge of a mixing element and is divided into two; contact with subsequent elements results in similar divisions. Accordingly, the divisions (termed striations) increase exponentially according to the number of mixing elements in the static mixer (i.e. the number of divisions is equal to $2^n$, where $n$ is the number of elements in the mixer). *Radial mixing* occurs under both laminar flow and turbulent flow and results from the circulation of the fluid around the mixing elements.

## Filtration of pharmaceutical products

### Introduction

Filtration is a unit operation that is employed in the pharmaceutical industry to separate particles from a product or to separate particles that are subsequently recovered and used in the manufacture of the product. In basic terms that process involves the passage of a fluid across the surface of a permeable membrane of defined pore size (termed the filter). The particulate materials are retained on the surface of the filter and are termed a cake. The liquid that is collected following passage across the filter is referred to as the filtrate.

There are two primary classifications of filtration dependent on whether the aim of the process is to collect and subsequently use the filtrate or particulate material (cake): clarification filtration and cake filtration.

## Clarification filtration

In this type of filtration the aim is to remove particulate materials from the liquid prior to the use of this liquid. For example in the manufacture of a solution product it is essential to ensure that all undissolved materials have been removed from the finished product to fulfil the product description in the finished product specification. The filtrate in this example will be subsequently filled into a bottle and packaged. A filter is chosen to have a pore size that is smaller than the particles within the liquid (typically 0.45 µm). In certain pharmaceutical products there is a requirement for the product to be sterile and this may require sterile filtration. Thus following filtration to remove the larger (undissolved drug and or excipient) particles, the solution is terminally filtered under aseptic conditions using a 0.22 µm filter.

## Cake filtration

Cake filtration is the process in which the solids are collected on the surface of the filter (termed a cake), removed and then subsequently used downstream in product manufacture. Within the pharmaceutical industry this is frequently used to separate precipitated drug from a solution (following a recrystallisation step) that is subsequently used in the production of a finished drug product.

## Theoretical aspects of filtration

The aims of a filtration process are as follows:
- to effectively remove all particles, either for disposal or for use downstream (e.g. drug substances)
- to ensure that the time required for filtration is minimised to facilitate batch production.

To achieve these aims it is initially important to consider the theoretical aspects of filtration. The rate of filtration may be mathematically defined as follows:

$$\left(\frac{\mathrm{d}V}{\mathrm{d}t}\right) = \frac{\Delta P \times A}{\eta \times R_{\text{filter}}^{\text{cake}}} \tag{11.5}$$

where $\left(\dfrac{\mathrm{d}V}{\mathrm{d}t}\right)$ refers to the rate of filtration (the volume of filtrate produced per unit time), $\Delta P$ refers to the pressure drop across the cake and filter, $A$ refers to the surface area of the filter, and $R_{\text{filter}}^{\text{cake}}$

refers to the resistance to filtration offered by the cake and the filter itself. The term $R_{\text{filter}}^{\text{cake}}$ may be further discerned as follows:

$$R_{\text{filter}}^{\text{cake}} = \tau \left( \frac{W}{A} \right) + R \qquad (11.6)$$

where $\tau$ refers to the average resistance of the cake, $W$ refers to the dry weight of the cake, $A$ refers to the surface area of the filter, and $R$ refers to the resistance of the filter (and filter medium).

Reference to these equations provides an insight into the filtration process and how this may be optimised. The following points are therefore of note:

1. *Increasing the pressure across the filter will increase the rate of filtration.* In certain situations, e.g. if the cake on the surface of the filter effectively blocks fluid flow across the underlying filter, increasing the pressure across the filter will not result in an increased filtration rate. Examples where this occurs include highly compressible cakes (in which there is limited cake porosity) and cakes that are slimy in nature.
2. *Increasing the surface area of the filter will increase the rate of filtration.* This is due to the greater surface area for filtration and the associated cake thickness. An additional effect of an increased surface area is an increase in the lifetime of the filter.
3. *The rate of filtration is inversely proportional to the viscosity of the product.* In situations where the filtration of more viscous liquids is required, it is commonplace to increase the temperature of the product as this will in turn reduce product viscosity. The relationship between viscosity and temperature is defined as follows:

$$\eta = Ae^{\frac{-E}{RT}} \qquad (11.7)$$

where $\eta$ refers to the viscosity of the liquid, $E$ refers to the activation energy (the energy to initiate molecular flow), $R$ refers to the gas constant, and $T$ refers to the temperature.
4. *Decreasing the resistance to flow across the filter increases the rate of filtration.* As previously stated, the resistance to flow across the filter is a composite of two resistance terms, namely the resistance to flow across the cake and the resistance to flow across the filter itself. Resistance to filtration resultant from specific cake resistance ($\alpha$) is affected by several factors, including:
   - *The thickness of the cake*: as the cake thickness increases, the filtration rate decreases. Therefore, it is important that during cake filtration, the cake is routinely removed to

minimise the resistance to filtration offered by the cake that has been deposited on to the filter surface.

- *The porosity of the cake*: as the porosity of the cake decreases so does the rate of filtration. Ultimately if the cake is highly compressible, the rate of filtration will be severely compromised. Particle size of the solids in the cake affect porosity and hence the rate of filtration.
- *As the resistance offered by the filter (R) to filtration increases, the rate of filtration decreases.* The resistance of the filter to filtration is a minimal contributor to the overall resistance term in cake filtration, contributing ≤10% of the overall resistance. Conversely the resistance to filtration offered by the filter itself is the sole contributor to the resistance to filtration. In clarification filtration, equation (11.5) may be rewritten as:

$$\frac{dV}{dt} = \frac{\varnothing P \times A}{\eta R_{filter}}$$
(11.8)

## The filter medium

The filter medium (filter) is the component within the filtration process that is used to separate solids from the system and, accordingly, the properties of this component are key to the overall success of the unit operation. A key design criterion is the pore size of the filter medium as this defines the nature of the filtration process. Filter media possessing larger pore sizes (>1 μm) are used for separation of large particles from a liquid, whereas smaller pore sizes are available and are used for a more complete clarification (0.45 μm) and sterilisation (0.22 μm). The pore sizes of the filter media used for clarification and for cake filtration differ. In clarification, the aim is to efficiently remove the (limited number of) particles/microorganisms from the solution. Thus filtration of these systems involves the use of a series of filters, each with decreasing pore sizes, until clarification is achieved. The sequential nature of the filtration process using filter media of different pore sizes is performed to ensure the correct balance between particle removal and resistance to blockage of the filter by large particles. In cake filtration the pore size of the filter is specifically chosen to ensure the retention of all particles on the surface of the filter and to prevent bleeding of the filter (forced movement of particles through the pores of the filter) whilst ensuring optimal removal of the eluted liquid.

There are a number of different materials of construction of filter media including woven cloth (principally cotton), woven synthetic materials (e.g. nylon, polytetrafluoroethylene, Teflon™) or metals.

Each type of filtration medium has its advantages and disadvantages. For example coarse filtration is performed using woven filters whereas removal of sub-micron particles from liquids is usually performed using porous plastics. Teflon is often used whenever the pressure applied to the filter medium is large due to its excellent mechanical properties. The importance of the mechanical properties of the filter medium must be stressed. Rupture of the filter medium will lead to failure of the filtration process.

## Filter aids

Filter aids are particulate systems that are used to coat the surface of the filter medium. These systems are applied to the surface of the filter medium as a suspension, the particle size of the filter aids being larger than that of the pores in the filter medium, thereby ensuring surface retention. Filter aids are designed to form a highly porous and non-compressible cake on the surface of the filter that retains the particles in cake filtration whilst optimising flow of liquid across the filter medium. In so doing the filter aids act to reduce the resistance of the system to fluid flow and hence (according to equations 11.5 and 11.6) the filtration rate is optimised. The physical properties of the filter aid are key to the success of cake filtration. Filter aids are not required for clarification filtration.

Ideally, filter aids should exhibit the following properties:

■ After successful deposition on to the filter surface, the filter aid should form a cake that facilitates liquid flow.
■ The particle size distribution of the filter aid should be specifically chosen to ensure that the material to be filtered and collected is retained.
■ The filter aid should not interact with the material being filtered.
■ The filter aid that is to be deposited on to the surface of the filter should be easily dispersed in an appropriate solvent prior to the uniform deposition of the filter aid on to the filter surface.

The mass of filter aid applied to the surface of the filter is an important determinant in the overall performance of filtration, with an optimum mass of filter aid being required for each filtration process (Figure 11.21). At quantities less than the optimum value, the filter aid may adversely affect filtration by adding extra thickness to the filtration surface without enhancing filtration performance. Conversely, if an excessive mass of filter aid is used, the benefits associated with the use of the filter (namely increased porosity) are obviated by the extra thickness of the filtration cake.

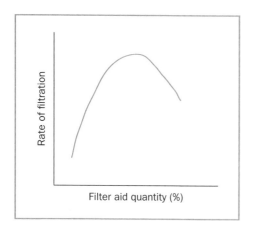

**Figure 11.21**    The effect of quantity of filtration aid on the rate of filtration in cake filtration.

Commonly employed filter aids include diatomaceous earth (silica), perlite (silica and aluminosilicates) and carbon. Coating of the filter with filter aid is commonly performed by circulating a suspension of the filter aid in a solvent across the filter until the required thickness of coating of the filter aid on the surface of the filter medium has been attained.

## Selection of the type of filtration process

Selection of the correct process for the filtration of a given pharmaceutical product involves identification of the key design criteria. Examples of these include:

- Is the filtration process clarification or cake filtration?
- Is this a batch or a continuous process?
- If the filtration method is clarification filtration, does the product have to be terminally sterilised by filtration?
- If the filtration method is clarification filtration, what is the viscosity of the product?
- Does the product require to be heated to aid processing (reduce viscosity)?
- What is the allocated time for production (and hence what is the required filtration rate)?
- What volume is required to be filtered?
- What pressure is required for filtration?

There are several methods that may be employed to filter pharmaceutical products, dependent on whether clarification filtration or cake filtration is required.

### Clarification filtration

In clarification filtration, suspended solids/microorganisms are collected as the product is passed across the filter medium. There are two main types of clarification filtration that are dependent on the design of the filter medium.

### Cartridge filtration

In this the filter medium is highly pleated and is typically cylindrical in design (Figure 11.22). The composition of the filter medium is dependent on the material being filtered but examples of materials that are used for the filtration of pharmaceutical products include:

- cellulose nitrate
- polyamide
- polyvinylidene chloride
- polytetrafluoroethylene
- nylon
- polyproplyene
- polyethersulphone.

**Figure 11.22**   Examples of filter cartridges.

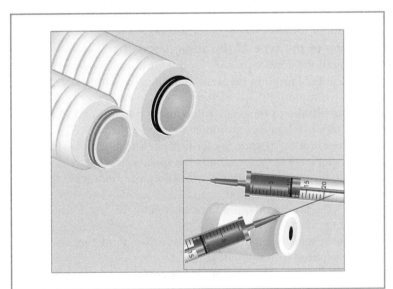

The highly pleated nature of the filter medium in the cartridge is a key design criterion and offers two key advantages. Firstly, the filtration area may be maximised whilst minimising the overall size of the filtration platform. This maximisation of surface area will in turn optimally enhance the rate of filtration, as defined in equation (11.6). Secondly, in practice, the cartridge arrangement ensures that the filtration medium may be quickly changed when needed thereby saving production time. The cartridges are held within metallic housings that may be designed to hold either one or more than one filter cartridges (Figure 11.23).

**Figure 11.23**   Examples of housings for filter cartridges, illustrating housings for single cartridges and for multiple cartridges.

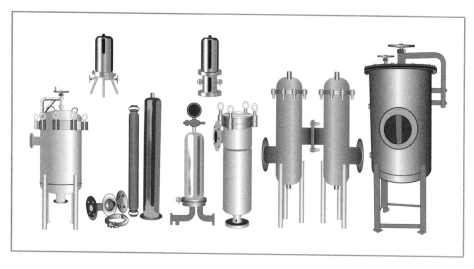

In these housings there is an inlet by which the liquid to be filtered enters the housing and an outlet that allows the liquid to leave the housing following passage across the filtration medium. To optimise the life of the filter medium, it is important that the pore size of the filter medium is correctly chosen. Therefore, clarification filtration of a liquid is performed using a series of filter media, each offering a different pore size. Filtration commences using filter media with larger pores sizes and finishes with the desired level of filtration, e.g. 0.22 µm for sterile products and 0.45–1 µm for non-sterile products. This will ensure that the life of the filter medium will be prolonged and not unduly shortened due to blockage of fluid flow.

### Filter press
The filter press is a large filtration system that is employed either for clarification or cake filtration. The three main components of a filter press are as follows:
- the outer casing/frame of the press
- the filter medium
- filter plates.

These features are shown in Figures 11.24, 11.25 and 11.26, respectively.

Figure 11.24   Illustration of the housing for the filter press showing the frame and the hydraulic system that are used to hold the filter plates and filter media in the correct location for optimal filtration.

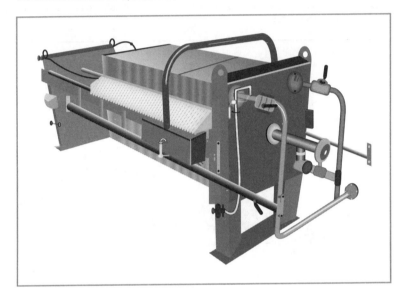

Figure 11.25   Illustration showing the filter plates that are arranged vertically in the filter press.

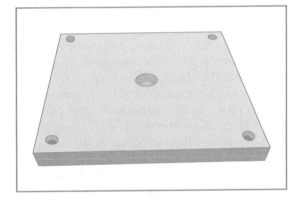

The filter medium is aligned with the filter plates (which are typically concave to allow fluid flow) and then vertically aligned into the filter press frame. The frame is held together both by a pressure applied horizontally and by assistance of a hydraulic system. The operation of the filter press is depicted in Figure 11.27.

As may be observed, the liquid to be filtered enters into the filter press through an inlet and passes through the filter medium

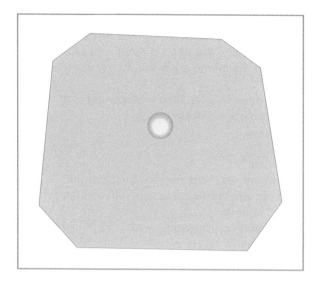

**Figure 11.26** Illustration showing the filter medium that is placed and held adjacent to the filter plates and housed within the filter press.

**Figure 11.27** Depiction of the operation of the filter press.

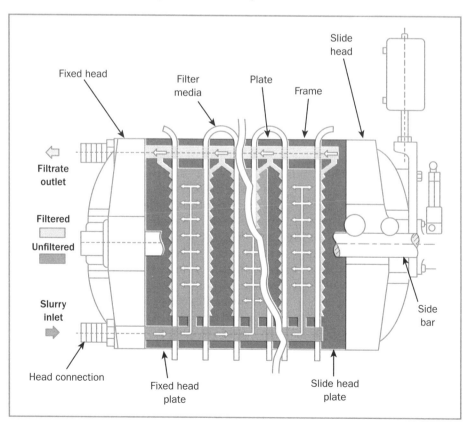

located between filter plates, prior to leaving the press through the outlet. The suspended solids are captured on the filter medium between the plates. When the filter press is employed as a cake filter, the spaces between the plates are generally greater than for clarification filtration.

The use of the filter press is associated with a number of advantages and disadvantages, as follows:

### Advantages
- It is simple in design.
- It can be used for both cake and clarification filtration.
- There is a relatively large filtration area per area of floor coverage.
- A wide range of operation pressures may be employed (to facilitate rapid filtration).

### Disadvantages
- It is mostly used for batch processes.
- Long dismantling and cleaning time.
- There is the possibility of leakage.
- Not used for sterile filtration but is primarily used in clarification filtration to remove particles of larger size.

## Cake filtration

As the reader will recall, the purpose of cake filtration is the removal of the insoluble fraction of a slurry, which will subsequently be employed in the manufacture of the dosage form. Typically this type of filtration is used to recover an active pharmaceutical ingredient following recrystallisation. One of the key features of filtration systems that are used for cake filters is that they operate as vacuum filters and accordingly the pressure difference across the filter is markedly greater than that achieved by gravity alone. As a result the rate of filtration is dramatically enhanced. Students will already have operated this type of filtration system using a Buchner funnel for the collection of purified synthetic chemicals. The Buchner filter is an example of batch equipment. This system, whilst suitable for laboratory use, is impractical at the industrial scale and, accordingly, the large-scale cake filtration of pharmaceutical systems involves the use of continuous vacuum filtration systems. For simplicity the design and operation of two continuous vacuum filters will be described in this chapter.

### Rotary drum filters

This type of filter is widely used in the pharmaceutical industry. There are basically two designs that differ in the nature of how the vacuum is applied to the filter. In the first (and most commonly

used) type, the vacuum is applied within sections on the
periphery of the rotating drum whereas in the second design the
vacuum is applied to the entire interior of the rotating drum.

**Figure 11.28**   Diagrammatic representation of the operation of the rotary drum
filter.

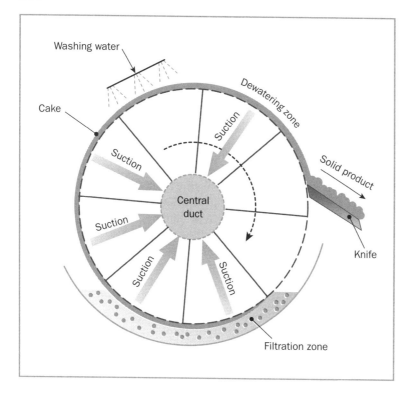

A diagrammatic representation of the operation of the rotary
drum filter is shown in Figure 11.28 in which a number of key
features may be observed:

- As in all cake filtration the filter medium, which rests on the
  rotating drum, has been coated with the appropriate mass of
  filter aid.
- The filtration zone contains the slurry of drug that has been
  dispersed in a liquid solvent, most typically water.
- The vacuum is drawn through sections (compartments) on the
  drum towards the central zone.
- A washing facility.
- A knife (or string) that is used to collect the dried solids.

The operation of the rotary drum filter is relatively
straightforward. Initially the slurry, a concentrated suspension

of solid in an appropriate solvent (usually water) is placed into a trough at the base of the filtration apparatus. The rotating drum (with the filter medium and filter aid) is partially immersed in this slurry. The drum is then rotated (usually from 0.1 rpm to 3 rpm) during which time the wet slurry is deposited on to the surface of the filter. Simultaneously, the vacuum is drawn through discrete sections on the outside of the drum, which acts to draw the liquid from the surface of the drum. As the rotation proceeds the now dried cake is washed by the application of water from a spray device, this water being removed (with the water soluble impurities) by the further application of the vacuum as the drum rotates. The final stage of the process involves the physical/ mechanical removal of the cake. This cycle continues until the solid material has been completely removed from the slurry.

The choice of the mechanism by which the solid cake is removed from the filter medium depends both on the types of systems available to the manufacturing company and also on the nature of the cake being removed. As shown in Figure 11.29, there are a range of different cake discharge systems including scraper, roller, and string and belt systems. The scraper system (sometimes referred to as a knife discharge) (Figure 11.29a) is located prior to the re-entry of the drum into the slurry and acts to mechanically remove the cake (with the assistance of compressed air). In the roller discharge system (Figure 11.29b) the dried cake is compressed against a roller on to which the cake is deposited. The cake is removed from the roller using a knife/scraper. In so doing the lifetime of the filter medium is increased because direct contact of the knife with the filter medium is avoided. The string discharge system (Figure 11.29c) involves a series of strings (metallic wires) that are located around the filter drum but are tangentially separated from the drum at the point of cake removal. The tangential separation of the strings from the drum causes the cake to be removed from the filter medium, this subsequently breaking away and being collected.

The nature of the cake material is an important consideration in the selection of the discharge mechanism. For example, string filters are particularly effective at removing cakes that are thin and sticky in nature. In addition the use of the string system enhances the lifetime of the filter medium, as it is not as mechanically demanding on the medium. The knife discharge system is particularly effective for the removal of friable cakes that would not have the required mechanical properties to be applicable for string discharge. The roller discharge system is particularly useful for the removal of cakes in which the particle size is small or for cakes that are sticky in nature.

**Figure 11.29**   Examples of discharge mechanisms used in rotary drum filtration: (a) knife/scraper; (b) roller; (c) string.

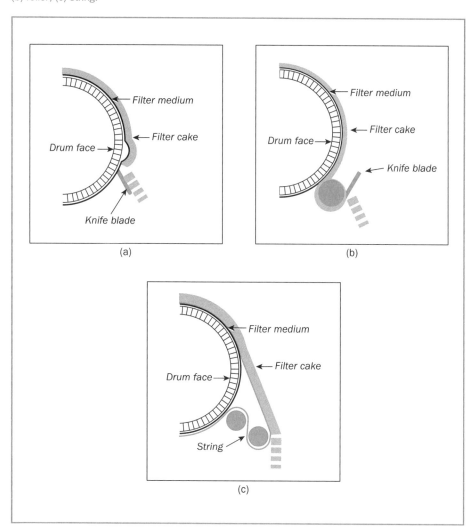

(a)

(b)

(c)

## Milling of pharmaceutical systems

### Introduction

The particle size of solids and dispersed liquids (within emulsions, creams and ointments) is recognised to be an important consideration in the formulation of certain types of dosage forms. As a result, control of the particle size range of therapeutic agents is frequently performed. Particle size range has implications in the non-clinical and clinical performance of dosage forms for a range of reasons, including:

- the rate of dissolution and hence bioavailability of poorly soluble therapeutic agents
- the physical stability of disperse systems (suspensions, emulsions, creams)
- the flow properties of pharmaceutical solids
- compression properties of pharmaceutical solids
- uniformity of content of therapeutic agents within dosage forms
- acceptability/palatability of dosage forms, e.g. suspensions.

## Overview of the milling process

Milling is a unit operation in which the size of solid particles is mechanically reduced via the attrition of particles following collision with other particles or with the hard surfaces of the mill chamber. It is an energy inefficient process, with an estimated circa 1% of energy used in the process being used directly to reduce particle size. Energy is lost from the milling process by a number of means, including:

- generation of heat associated with frictional forces
- noise and vibration
- movement of the powders into the mill chamber
- elastic deformation of the powders
- inefficiencies associated with the operation of the motor.

To understand the process of milling it is initially important to consider the effect of stress (force per unit area) on the deformation of particles. Figure 11.30 graphically displays the relationship between stress and strain in a solid material. Application of a stress within the elastic region (denoted by the black line in Figure 11.30) results in linear deformation of the solid. Subsequent removal of this stress allows the solid to return to its original dimensions.

If the applied stress exceeds the yield tensile stress (denoted by (a) in Figure 11.30), the solid undergoes plastic irreversible deformation (denoted by the green line in Figure 11.30). Following removal of the stress in this scenario, the solid does not return to its original dimensions. Finally, the application of a defined minimum stress, termed the ultimate tensile strength, leads to permanent deformation (fracture). Accordingly, fracture of particles (and hence particle size reduction) occurs whenever the stress applied to the particles during the milling process is sufficient to exceed the ultimate tensile strength of the particle (denoted by point (b) in Figure 11.30).

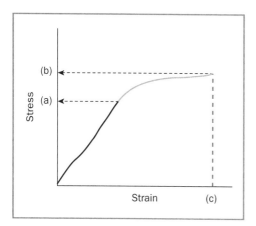

**Figure 11.30** Graphical representation of a stress curve for a solid material: (a) refers to the yield tensile strength, (b) refers to the ultimate tensile strength (c) refers to the elongation at break. The black section denotes the elastic region whereas the green section denotes the plastic region.

Milling is a random process and there is an uneven distribution of energy and hence stress to each particle undergoing size reduction. Therefore, some particles may be exposed to substantially greater stresses than is required for fracture whilst other particles may not receive sufficient energy to undergo fracture. As described above there is a minimum energy for each material that is required to ensure fracture. Coarse fracture, i.e. size reduction of large particles, usually occurs across mechanical defects in the material (cracks); particles containing a greater number of cracks will result in the production of a greater number of smaller particles. Increasing the applied stress within the mill results in greater particle size reduction, i.e. the size of the larger particles is further decreased and the number of fine particles (low particle size) is increased but their size is not. The reduction in the size of fine particles is controlled by the internal (molecular) structure of the material whereas the reduction in size of large particles is affected by both the magnitude of the applied stress and the method by which the stress is applied. As may be expected particle size reduction of fine particles (typically <50 μm) requires more energy than the size reduction of large particles due to both the greater surface area of fine particles and the use of energy to initiate and propagate mechanical defects in the particle structure.

## Modelling the relationship between energy provided and particle size reduction

Milling is a technique that uses considerable energy and therefore there is an interest in understanding the relationship between the energy applied during milling and the associated reduction in particle size. In light of the high and variable (non-useful) dissipation of energy in the milling process, the derived mathematical models are half-empirical. A generic relationship between energy applied and the resultant size of particles has been defined as follows:

$$\frac{dE}{dD} = -\frac{C}{D^n} \tag{11.9}$$

where $dE$ is the amount of energy required to reduce the particle size ($dD$) by a defined amount, $C$ is a milling coefficient and $n$ is an exponent.

Using this equation, three half-empirical equations may be derived that describe the relationship between the energy for particle size reduction within three defined particle size ranges, namely:

**(i)** *Kick's equation*

This equation is relevant for large particles ($d > 50$ mm) and is derived from the generic equation whenever the exponent $n$ is equal to 1, as follows:

$$E = C_k \ln \frac{d_1}{d_2} \tag{11.10}$$

where $E$ is the energy, $C_k$ is a milling coefficient and $d_1$ and $d_2$ are the initial (pre-milling) and final (post-milling) particle sizes.

**(ii)** *Bond's equation*

This equation is relevant for large particles whose diameter is less than 50 mm but is greater than 50 μm and is derived from the generic equation whenever the exponent $n$ is equal to 1.5, as follows:

$$E = C_b \left( \frac{1}{\sqrt{d_2}} - \frac{1}{\sqrt{d_1}} \right) \tag{11.11}$$

where $E$ is the energy, $C_b$ is a milling coefficient and $d_1$ and $d_2$ are the initial (pre-milling) and final (post-milling) particle sizes.

**(iii)** *Von Rittinger's equation*

This equation is relevant for large particles whose diameter is less than 50 μm and is derived from the generic equation whenever the exponent $n$ is equal to 2, as follows:

$$E = C_r \left( \frac{1}{d_2} - \frac{1}{d_1} \right) \tag{11.12}$$

where $E$ is the energy, $C_r$ is a milling coefficient and $d_1$ and $d_2$ are the initial (pre-milling) and final (post-milling) particle sizes.

## Operation of pharmaceutical mills

There is a wide range of pharmaceutical mills that are used to reduce the particle size of pharmaceutical systems; however, there are general similarities in their operation and design. All pharmaceutical mills may be considered to be composed of three separate parts, namely:

- *feed chute*: to facilitate the delivery of the material to the mill
- *grinding mechanism*: to mechanically reduce the size of the particles
- *discharge mechanism*: to facilitate the removal of the particles after size reduction.

There are four general mechanisms by which particle size reduction is achieved, the relative contributions of each being dependent on the nature of the individual machine being used. These are:

- *attrition*: particle size reduction following particles rubbing against either one another and/or the surface of the chamber of the mill
- *shear*: particle size reduction following the interaction of particle–particle interactions
- *impact*: particle size reduction following exposure to a single impact force
- *compression*: particle size reduction following exposure to two impact forces.

## Types of mills

There are wide range of pharmaceutical mills that are commonly employed to reduce the size of solid materials, either when presented as a solid mass or as a suspension of solid particles dispersed within a fluid, e.g. a pharmaceutical suspension. Examples of these and their operation are described in this section.

### The hammer mill

The hammer mill (Figure 11.31) is used to reduce the size of particles from approximately 5–50 mm to as low as 100 µm and, as such; it is referred to as an intermediate crusher. The mill consists of a metallic chamber that contains a rotating shaft on to which hammers are attached. The shaft is rotated at a chosen angular velocity and particle size reduction is achieved by the crushing of particles between the hammers and the internal surface of the grinding chamber. Powders are vertically fed into the grinding chamber through the feed chute and, following size reduction, the particles pass through a metallic screen and are removed (usually by gravity) from an exit chute.

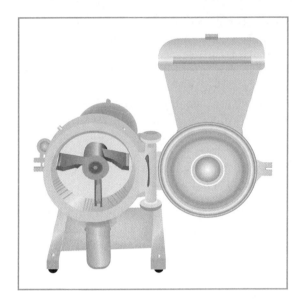

**Figure 11.31** A laboratory hammer mill showing the rotating shaft, hammers, metallic screen and the grinding chamber. In operation the mill is closed.

The size of the particles after milling is dependent on both the angular velocity of the drive shaft and the mesh size of the screen.

## The conical mill

The conical mill (Figure 11.32) is a grinding mill that is a common variant of the hammer mill. The name is derived from the conical shape of the grinding chamber. The powder is fed into the grinding chamber through the feed chute, where it contacts a rotating impeller. The particle size of the powder is reduced following crushing of the powder between the impeller and the conical chamber. Whenever the particle size is smaller than the screen mesh size, the powder passes through the screen and is discharged using an exit chute. In addition, the conical mill may be used for other operations, e.g. powder mixing, granulation, powder deagglomeration.

Pharmaceutically the conical mill generates less noise, less heat and less dust than the hammer mill. Furthermore, the conical mill has been stated to generate particles of a more uniform particle size.

## The ball mill

Ball mills (Figure 11.33) consist of a central grinding chamber, usually cylindrical or conical in shape, that contains the grinding medium (typically steel balls) and which is rotated on a horizontal or near horizontal axis. The inner surface of the grinding chamber is manufactured from an abrasion resistant material to enhance durability. The material to undergo size

**Figure 11.32** Images of the Quadro conical mill: (a) the laboratory mill unit with drive motor; (b) the impeller; (c) the screen through which the milled particles pass.

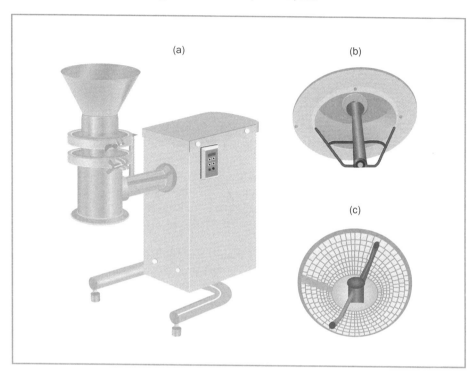

(a)

(b)

(c)

**Figure 11.33** Images of cylindrical and conical ball mills.

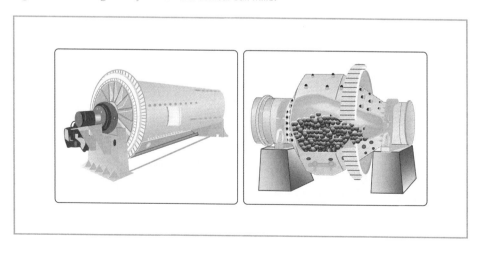

reduction is entered through one end of the mill and is exited at the other end of the mill. A metallic screen is placed over the exit chute from the mill to ensure that only particles of the required size are exited from the mill and that the grinding medium is retained within the grinding chamber.

The ball mill is used within the pharmaceutical industry to reduce the particle size of solids either when dispersed within a solvent (i.e. suspensions) or in their native state. The choice of the size of the grinding medium is dependent on the initial size of the particles and, in practice, balls of a range of sizes are used during size reduction. The larger balls reduce the size of the larger particles with the smaller balls being responsible for the reduction of particles within a lower particle size range. Particle size reduction occurs due to the crushing action (attrition) of the grinding medium on the particles, either between a ball(s) and two particles and/or between a ball, the particle and the wall of the grinding chamber.

In certain designs the chamber of the ball mill may be divided into different sections (Figure 11.34). The sample enters into the first chamber where the size of the larger particles is reduced. In this chamber the diameter of the grinding medium balls is relatively large. A screen separates this section from the second section of the grinding chamber through which particles pass whenever the size of the particles in the first chamber has been

**Figure 11.34** Cylindrical ball mill showing a two compartment grinding chamber. Size reduction of larger particles is performed in the first chamber (left) whereas the subsequent reduction of these size reduced particles to a smaller size fraction is achieved in the second (right) chamber.

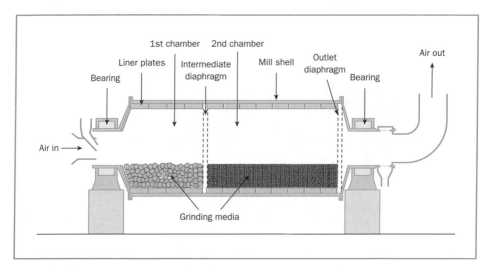

sufficiently reduced. In the second chamber the diameter of the grinding medium balls is smaller than that present in the first chamber and, as a result, the particle size of the material in this chamber is further reduced.

The operating conditions of the ball mill directly affect the resultant particle size. Factors that have been reported to affect particle size reduction include the diameter of the grinding medium, the rate of feed of product into the ball mill, mass of the grinding medium, the rotation speed of the grinding chamber and the physical properties of the solid particles.

### The diameter of the grinding medium

As described above selection of the diameter of the grinding medium is an essential milling parameter. The size of large particles is reduced using balls of a greater diameter. If the diameter of the grinding medium is too small compared to the size of the particles then the optimum particle size reduction is not achieved. Similarly if the diameter of the grinding medium is large then the particle size of the finished product will not be small. In practice to optimise the milling process, it is important to understand: (a) the initial particle size of the material to be size reduced; and (b) the required terminal particle size. This knowledge will allow the pharmaceutical scientist to choose the most appropriate diameter of grinding medium.

### The rate of feed of product into the ball mill

The extent of particle size reduction is directly affected by the residence time of the material within the grinding chamber. Therefore, as the rate of feed of the material into the grinding chamber is increased the resultant size reduction will be comparatively lower than whenever the feed rate is lower. The rate of feed is also affected by the angle of rotation of the grinding chamber. Increasing the angle at which the grinding chamber is rotated (i.e. the slope) will reduce the retention time of the product in the grinding chamber and result in a product of a greater particle size in comparison to horizontal operation of the milling process.

### Mass of the grinding medium

As the mass of the grinding medium is increased, e.g. by changing the density and/or increasing the number of balls, the extent of particle size reduction is increased. In practice, the volume of the grinding chamber occupied by the grinding medium in most ball milling procedures is circa 50% and therefore changing the density of the balls alters the mass of grinding medium.

### The rotation speed of the grinding chamber

As may be expected there is an optimum rotation speed for particle size reduction. If the rotation speed is low, then the balls have limited momentum, the stresses imparted to the particles are lower and therefore poor particle size reduction results. Conversely, if the rotation speed is too high then the balls are forced against the walls of the grinding chamber due to the resulting centrifugal force, again with limited particle size reduction. In this scenario the balls remain in contact with the walls of the grinding chamber during rotation. In addition, large rotation speeds will increase the wear on the walls of the grinding chamber due to the increased momentum of the balls and its effect upon impact with the walls.

The rotation speed of the ball mill that is associated with the location of the grinding medium with the walls of the grinding chamber is referred to as the *critical speed*. Rotation speeds that exceed the critical speed will result in no/minimal reduction in particle size of the solid material. The critical speed may be calculated using the following equation:

$$\omega_c = \frac{76.6}{\sqrt{D_{mill}}} \tag{11.13}$$

where $\omega_c$ refers to the critical rotation speed and $D_{mill}$ refers to the diameter of the mill.

From this equation it may be noted that the critical rotation speed of the mill decreases as the diameter of the mill increases. Typically ball mills operate at a rotation speed that lies between 60% and 85% of the critical speed.

### Physical properties of the solid particles

The physical properties of the solid particles directly affect the resultant particle size reduction. Hard materials or materials that exhibit significant plastic deformation will require greater stresses to induce fracture. Therefore, modifications to the milling method, e.g. rate of feed, mass of grinding medium, will be required to ensure that the particle size of such materials is suitably reduced.

**Note:** A primary advantage concerning the use of the ball mill is the ability to reduce the particle size of native powders and also particles within a suspension

## Colloid mill

The colloid mill (Figure 11.35) is used within the pharmaceutical industry to reduce the size of solid particles within a liquid phase (suspension) and, additionally, to reduce the size of liquid

**Figure 11.35**   Images of a colloid mill.

droplets within an emulsion to enhance the stability and clinical properties of these systems. Using this technique, the disperse phase can be easily reduced to particle sizes less than 10 µm.

In use, a thin film of the solid dispersion or emulsion is passed into the gap between the rotor and stator of the colloid mill (Figure 11.36) where it is exposed to a high shear rate due to the angular movement of the rotor. The rotation speed employed in the colloid mill usually ranges between 2000 and 18000 rpm. At this rotation speed the solid dispersion/emulsion is exposed to high levels of hydraulic shear which results in reduction of the particle size of the dispersed solids/dispersed droplets to low sizes.

Due to the application of a high shear rate to a small volume of material, there may be generation of heat and therefore colloid mills are frequently designed to accommodate a cooling jacket. This may also be used to heat the sample, if required. It should be noted that the power consumption of the colloid mill is high and therefore it is recommended that the starting particle size of the solid dispersion/emulsion should be low (i.e. reduced by other methods) to ensure that the power consumption is minimised.

**Figure 11.36**   Image of the (upper) rotor and (lower) stator of the colloid mill.

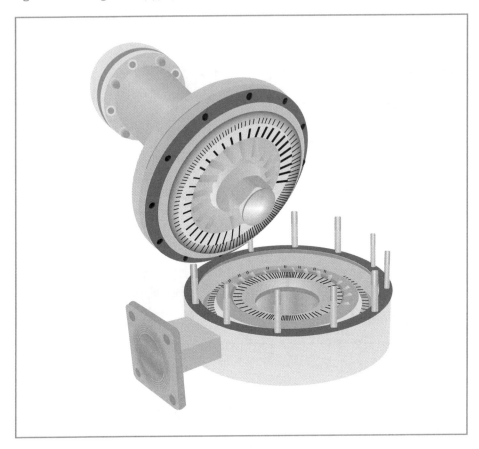

### Fluid energy mills

Fluid energy mills are high-energy mills that are used to reduce the particle size of solids within the range 1–20 µm. Whilst there are different designs, the basic principles of these systems involve particle impaction and attrition within the grinding chamber of the mill due to the movement of particles at high velocities. Typically, the particles are entered into the mill via a solids inlet and are drawn into the grinding chamber, usually by the vacuum generated by the Venturi system. The particles are then exposed to a high-pressure stream of air (at a pressure of approximately 8 bar) or superheated steam (at a pressure of approximately 8–16 bar). Within the chamber the movement of the fluid (e.g. air containing dispersed solids) is turbulent and this leads to particle size reduction, primarily through particle–particle collisions. The design of fluid energy mills ensures that the large particles are retained in the fluid energy mill due to the action of centrifugal forces pushing the particles to the periphery of the grinding chamber. Whenever the size of these particles has been sufficiently reduced, they are then carried out of the mill as a solid dispersion within the fluid vortex.

Two designs of fluid energy mills are shown in Figures 11.37 and 11.38.

**Figure 11.37**  Representation of a fluid energy mill.

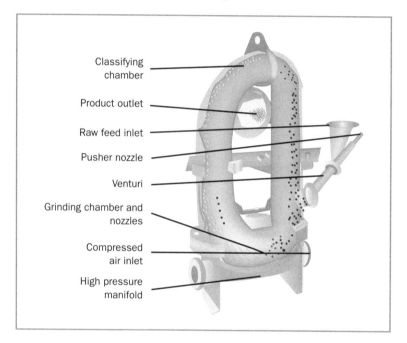

- Classifying chamber
- Product outlet
- Raw feed inlet
- Pusher nozzle
- Venturi
- Grinding chamber and nozzles
- Compressed air inlet
- High pressure manifold

## Drying of pharmaceutical systems

### Introduction

Drying is a unit operation that is commonly employed within the pharmaceutical industry and which involves the removal of a liquid phase from pharmaceutical systems by the application of heat. The reader should note that there are other, isothermal, methods by which liquids are removed from solid systems, including liquid extraction, adsorption of liquids from gases and the use of desiccants. This chapter will focus only on the application of heat to pharmaceutical systems for the purpose of liquid removal. Within the pharmaceutical industry liquid removal primarily refers to the removal of water; however, there are situations where the removal of organic, volatile solvents is required.

**Figure 11.38** Representation of a microniser fluid energy mill.

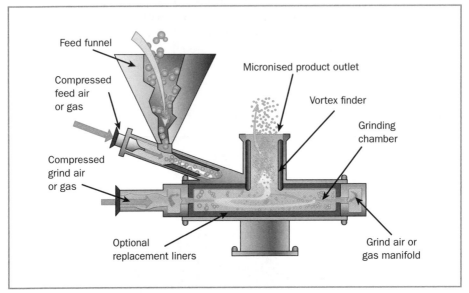

There is a wide range of manufacturing situations where drying is required, including:

- the removal of water from raw materials prior to inclusion into a manufacturing process, e.g. spray-dried lactose, hot-melt extrusion of pharmaceutical systems
- the removal of water from granules prior to further processing as tablets
- water removal from raw materials and/or pharmaceutical products to enhance their chemical stability, e.g. penicillin products and vaccines are provided in the dry state and are reconstituted prior to clinical use
- water removal from products to reduce the cost of transportation
- solvent (organic or aqueous) removal in tablet coating processes.

## Drying terminology

Drying involves the transfer of the liquid within the pharmaceutical system to the gaseous phase and subsequent removal by the adjacent air stream. In the case of water removal, water is removed as water vapour. The capacity of the adjacent gas (e.g. air) stream to remove this generated water vapour is dependent on the humidity of the gas, a measure of the water content in the gas. The water content in air is commonly calculated as follows:

Relative humidity =

$$\frac{(\text{Vapour pressure of water in air})}{(\text{Saturated vapour pressure of water in air})} \times 100 \quad (11.14)$$

An alternative expression is the relative saturation, which is defined as the ratio of the mass of water vapour present per known mass of air to the mass of vapour required to saturate a known mass of air at the same temperature.

Measurement of the humidity of air is performed by the determination of the wet bulb and dry bulb temperatures and the subsequent use of psychrometric charts. The *dry bulb* temperature is the temperature of air whereas the *wet bulb* temperature refers to the temperature whenever the bulb of the thermometer is covered by a cotton wick that is immersed in water. To discern the differences in these two temperatures, it is necessary to understand the process of drying. Whenever a sample is dried, evaporation of the liquid occurs due to the difference in the vapour pressure of the surface water and the vapour pressure of air. Molecules escape from the surface of the liquid and the latent heat of vaporisation results in a reduction of the temperature of the surface of liquid. At this stage there is a difference in the temperatures of air and the liquid surface and, consequently, there is a transfer in heat from the air to the liquid surface until there is a stabilisation of the temperature across the two phases. The temperature at this point is referred to as the wet bulb temperature.

An example of a psychrometric chart for the determination of the relative humidity is shown in Figure 11.39.

The use of this chart to determine the relative humidity is straightforward and requires knowledge of the wet bulb and dry bulb temperatures. Consider a situation in which the dry bulb temperature is 25°C and the wet bulb temperature is 20°C, the relative humidity is calculated as follows:

1. The intersection of the vertical lines associated with the dry bulb temperature (25°C) and the wet bulb temperature (20°C) is determined and a horizontal line that intersects the *y* axis (humidity ratio) drawn. The humidity ratio under these conditions is 0.0126 g water/g dry air.
2. The point of intersection of the vertical line associated with the dry bulb temperature (25°C) and the 100% relative humidity (outer) curve is defined. The humidity ratio is determined by drawing a horizontal line from this point

**Figure 11.39** A psychrometric chart used to calculate the relative humidity of air.

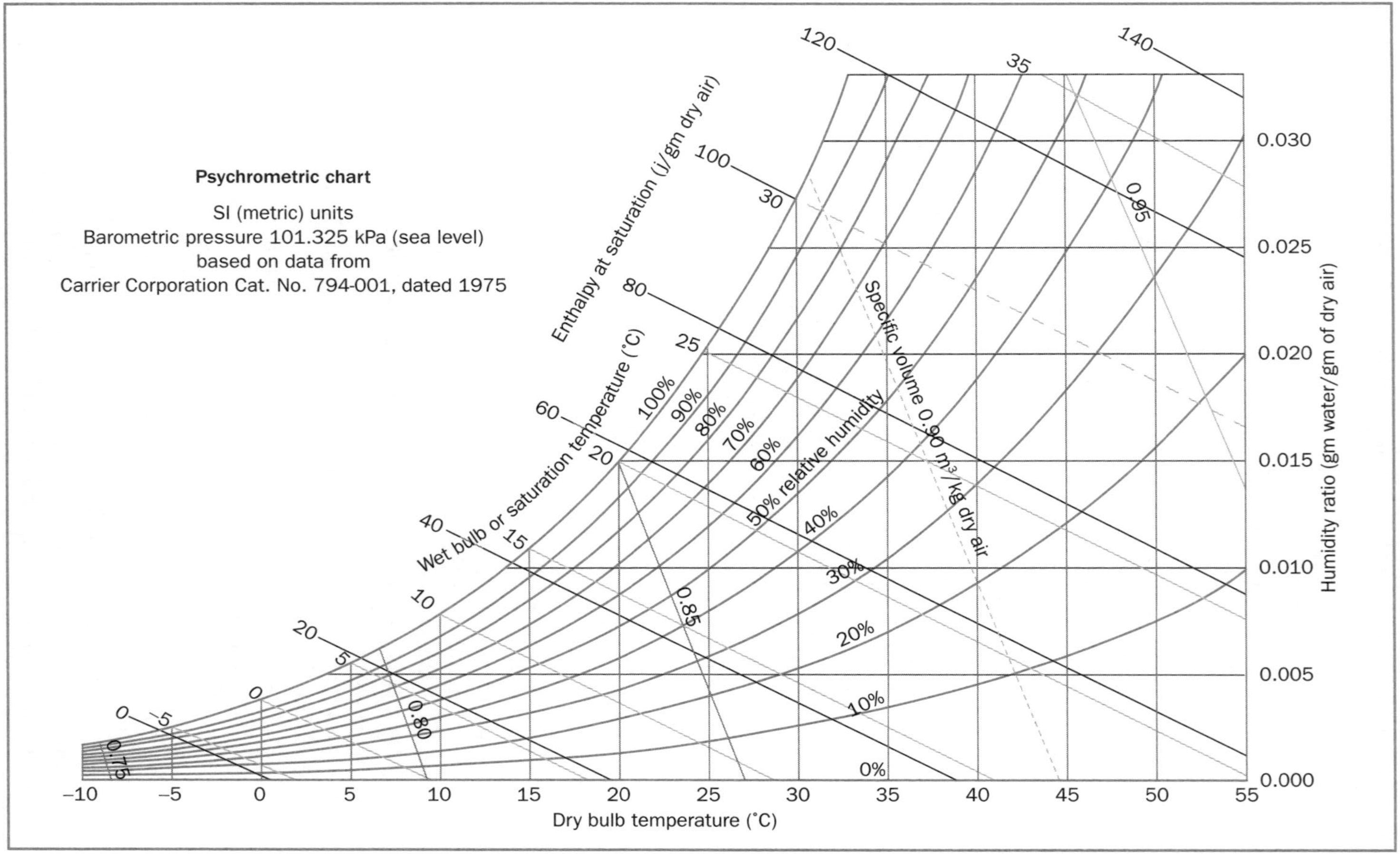

Psychrometric chart
SI (metric) units
Barometric pressure 101.325 kPa (sea level)
based on data from
Carrier Corporation Cat. No. 794-001, dated 1975
Enthalpy at saturation (j/gm dry air)
Wet bulb or saturation temperature (°C)
Specific volume 0.90 m³/kg dry air
50% relative humidity
100%
90%
80%
70%
60%
40%
30%
20%
10%
0%
0.75
0.80
0.85
0.95
0
20
40
60
80
100
120
140
–5
0
5
10
15
20
25
30
35
Dry bulb temperature (°C)
–10
–5
0
5
10
15
20
25
30
35
40
45
50
55
Humidity ratio (gm water/gm of dry air)
0.000
0.005
0.010
0.015
0.020
0.025
0.030

until it intersects with the $y$ axis (humidity ratio). The humidity ratio under these conditions is 0.020 g water/g dry air.

3. The percentage relative humidity is determined as follows:

$$\text{Relative humidity (\%)} = \frac{(0.0126)}{(0.020)} \times 100 = 63\% \qquad (11.15)$$

A further term that is commonly used in drying is the equilibrium water content of a material. This is the ratio of the mass of water associated with a known mass of material that is devoid of moisture. The equilibrium water content is affected by several parameters including the chemical properties and the particle size of the material. For example, most drugs will have an equilibrium moisture content between 0.20% and 2.5% w/w whereas the comparator values for kaolin and starch are approximately 1% and 30%, respectively.

## Types of water in solid materials

There are three types of water that are associated with solid materials, as follows:

- water of crystallisation (e.g. hydrates, as shown in Figure 11.40)

**Figure 11.40**  Chemical structures of (a) oxytetracycline dihydrate and (b) ampicillin trihydrate.

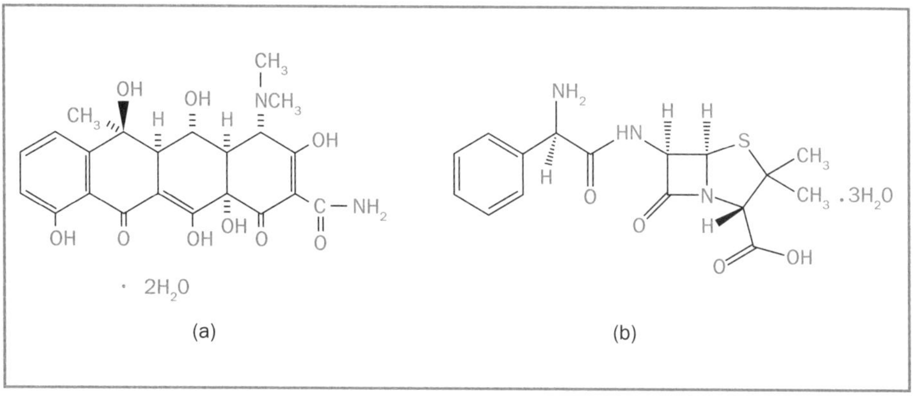

- water that is bound to the solid material, usually via hydrogen bonding
- water that exists as a free liquid (and hence exerts a vapour pressure).

During drying the free water is readily removed; however further energy is required to remove the other, chemically associated water molecules.

## Theory of drying

Drying of a liquid from a material involves both heat and mass transfer. Heat is transferred to the material from the drier and this is used to generate the latent heat of vaporisation that is required to remove (vaporise) the liquid. Mass transfer during drying occurs during the diffusion of water through the material being dried to the surface, the evaporation of the liquid from the surface and the subsequent removal of the molecules by the flow of air across the surface of the material. As a result of these two transfer phenomena, it is possible to generate mathematical expressions that describe these phenomena. It is important to note that in heat transfer during drying there may be different contributions due to conduction, convection and radiation, the relative importance of these being determined by the drying method used. As a result the specific equations related to each method of drying may differ.

### Heat transfer

In air drying the rate of heat transfer across the layer of liquid at the surface of the material may be described as follows:

$$q = h_{surface} A(T_{air} - T_{material\ surface}) \tag{11.16}$$

where $q$ refers to the heat transfer rate (J s$^{-1}$), $h_{surface}$ refers to the surface heat transfer coefficient (J m$^{-1}$ s$^{-1}$ °C$^{-1}$), $A$ refers to the area for heat transfer (m$^2$), $T_{air}$ refers to the air temperature (°C), and $T_{material\ surface}$ refers to the temperature of the surface of the material (°C).

As may be discerned from equation (11.17), the overall driving force for heat transfer is the temperature gradient that exists across the material surface and air.

An alternative expression that is commonly used is:

$$\frac{dM}{dt} = \frac{q}{\lambda} \tag{11.17}$$

where $\dfrac{dM}{dt}$ refers to the rate of evaporation (kg s$^{-1}$), $\lambda$ refers to the latent heat of vaporisation of water (J kg$^{-1}$) and $q$ refers to the heat transfer rate (J s$^{-1}$). The heat transfer rate $q$ is composed of the sum of heat transfer rates from conduction, convection and radiation, i.e.

$$q = (q_{conduction} + q_{convection} + q_{radiation}) \tag{11.18}$$

All three types of heat transfer may occur at the same time, or in various combinations. For example, in convection drying, hot gas flows past the wet surface of a granule. However, at the immediate surface of the granule, there is a relatively quiet layer of gas known as the film or stagnant layer. Heat is transferred from the bulk gas through the film to the granule via molecular conduction. The resistance of this stagnant layer or film to heat flow depends primarily on its thickness. This is one of the reasons why increases in the velocity of the drying air increase the heat-transfer coefficient. As the velocity of the drying air increases, the stagnant layer becomes thinner. However, under the conditions used in the convective drying of granulations, there will always be a thin film of stagnant air surrounding each granule.

## Mass transfer

The driving force for mass transfer is the difference in the humidity between the liquid surface and the air stream passing above, as follows:

$$\frac{dM}{dt} = kA(H_{material\ surface} - H_{air}) \tag{11.19}$$

where $\dfrac{dM}{dt}$ refers to the rate of mass transfer (kg h$^{-1}$), $k$ refers to the coefficient of mass transfer (kg h$^{-1}$ m$^{-2}$), the value of which is dependent on the velocity of air flow above the liquid surface. $H_{air}$ refers to the absolute humidity of air and $H_{material\ surface}$ refers to the absolute humidity at the surface of the material.

At steady state, the rate of evaporation is equal to the rate of mass transfer and therefore a composite equation may be defined, namely:

$$\frac{dM}{dt} = \frac{q}{\lambda} = kA\left(H_{material\ surface} - H_{air}\right) =$$
$$\left(\frac{q_{conduction} + q_{convection} + q_{radiation}}{\lambda}\right) \tag{11.20}$$

Equation (11.21) may then be used to manipulate the drying process. In particular the rate of drying $\dfrac{dM}{dt}$ may be enhanced by:

- *Increasing the heat transfer rate (q).* This is performed by consideration of the individual heat transfer rates for conduction, convection and radiation. The convective heat transfer rate is increased by increasing the air flow rate and temperature. The radiation heat transfer rate is increased by the use of a radiation drier whereas the conductive heat transfer rate is increased by reducing the thickness of the material that is being dried.

- *Increasing the coefficient of mass transfer.* This is performed by increasing the air flow rate in the drying chamber.
- *Increasing the humidity difference between the absolute humidity of air and the surface of the material to be dried,* i.e. $(H_{material\ surface} - H_{air})$. This is performed in practice by dehumidifying the inlet air in the drying chamber.

### The drying curve

The drying curve is a graphical expression of the rate of change in the moisture/solvent content of a material as a function of moisture/solvent concentration, as shown in Figure 11.41.

Alternatively, a drying curve may be presented in terms of the reduction in mass of the material as a function of moisture/solvent concentration or time, as represented in Figure 11.42.

**Figure 11.41** Diagrammatic representation of a drying curve.

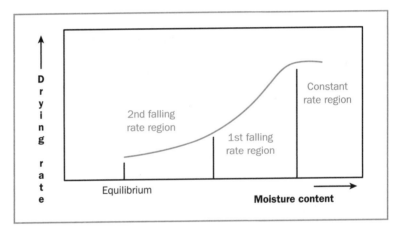

In Figures 11.41 and 11.42 there are several key regions that may be identified, as follows:

### The constant rate period

This period occurs in the initial stages of drying and corresponds to the evaporation of solvent and the subsequent diffusion of the solvent through the stationary film of air adjacent to the material, prior to subsequent removal by the flow of air across the face of the material. Mathematically this stage may be expressed as follows:

$$\frac{dM}{dt} = kA\left(P_{solvent} - P_{solvent/air}\right) \tag{11.21}$$

**Figure 11.42**   Diagrammatic representation of a drying curve, expressing the reduction in moisture/solvent content as a function of moisture content or time.

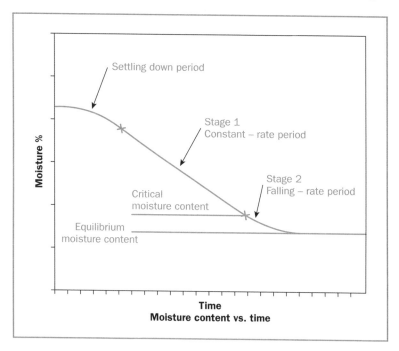

where $\dfrac{\mathrm{d}M}{\mathrm{d}t}$ refers to the rate of drying, $k$ is a mass transfer coefficient from the material through the air interface, $P_{\text{solvent}}$ and $P_{\text{solvent/air}}$ refer to the vapour pressure of the solvent and the partial pressure of the solvent in the air stream, respectively.

In the case where the solvent is water, the nomenclature within this equation is modified to:

$$\frac{\mathrm{d}M}{\mathrm{d}t} = kA\left(P_{\text{water}} - P_{\text{water/air}}\right) \tag{11.22}$$

where $P_{\text{water}}$ and $P_{\text{water/air}}$ refer to the vapour pressure of water and the partial pressure of water in the air stream, respectively.

In this stage of drying, the rate of drying may also be expressed by:

$$\frac{\mathrm{d}M}{\mathrm{d}t} = \frac{hA\emptyset T}{\lambda} kA\left(P_{\text{solvent}} - P_{\text{solvent/air}}\right) = kA\left(H_s - H\right) \tag{11.23}$$

where (in addition to the notation described above):

- $h$ refers to the heat transfer coefficient from the air stream to the surface of the material being dried. This coefficient is dependent on the velocity and direction of the air stream. Typical values for this coefficient range between 10 and 20 W m$^{-2}$ K$^{-1}$.
- $\Delta T$ refers to the temperature difference between the air stream and the surface of the material
- $\lambda$ refers to the latent heat of vaporisation
- $A$ is the surface area available for drying
- $(H_s - H)$ refers to the humidity gradient.

Removal of solvent from the material being dried within the constant rate period is dependent on the evaporation rate of the solvent and is similar in magnitude to the rate of evaporation of the pure solvent. As the observed rate of drying is constant within this period, this infers that as the solvent evaporates from the material being dried, this volume of liquid is replaced by the movement of solvent from the material into the bulk phase. The critical moisture content/critical solvent content is commonly used to define the moisture/solvent content at the end of the constant rate period.

### The first falling period

The equilibrium that existed in the constant rate period regarding the evaporation of solvent and the movement of solvent from the material to be dried no longer operates. During this phase the rate of removal of solvent is not sufficient to ensure that the air stream that flows above the surface of the material is saturated with the solvent. The rate of drying within this phase is controlled by the rate of movement of the solvent to the surface of the material being dried. This process becomes increasingly difficult as drying progresses due to the increasing tortuosity of travel, thereby accounting for the observed reduction in the rate of drying. This phase terminates whenever the solvent is unable to move to the surface of the material, and therefore drying from the surface of the material no longer occurs. Thus at the end of this phase the surface of the material being dried may be assumed to be free of solvent (dry if water is the solvent).

### The second falling period

In this period solvent that is located within the material is removed. Typically, the liquid in the solid is converted to vapour within the bulk of the material being dried, the vapour diffuses to the material surface and finally the flowing air stream that operates above the material–air interface removes the vapour.

The diffusion of the vapour through the material is controlled, at least in part, by the physicochemical properties of the material being dried, e.g. porosity, but there is a lower dependence on the external conditions. In practice, as the solid material becomes drier, the thermal conductivity of the solid material decreases, and therefore the temperature of the heat source will have to be increased to compensate. Pharmaceutically, this is a concern whenever the material being dried is thermolabile.

## Types of pharmaceutical driers

There are a range of driers that are employed to dry pharmaceutical systems and these may be classified as follows:

- *Static bed convective driers, e.g. the tray drier.* In these systems, there is no/limited movement of the material being dried during the drying process. Due to this there is a limit to the mass of material that is directly exposed to the heated air stream.
- *Dynamic convective driers, e.g. the fluid bed drier.* In these systems, there is movement of the individual particles of the material being dried, resulting in the individual exposure of each particle to the heat source.
- *Pneumatic driers, e.g. spray driers.* In these systems the individual particles of material are carried in a stream of gas (e.g. following nebulisation) and are therefore completely exposed to the heated air.
- *Freeze driers, e.g. microwave driers.* These systems use freezing and then removal of the frozen water through sublimation.

Examples of the operation of these are provided in the following sections. Given the nature of this text, it is not possible to cover the design and operation of all driers. Accordingly, in this regard radiation drying is not covered in this chapter. The interested reader should consult a specialist text to gain information on this topic.

### Tray/shelf driers (static driers)

A schematic of the design and operation of tray driers is shown in Figure 11.43. As may be observed air enters the drier and is warmed by a heat source, prior to being passed over the surface of the material to be dried which is located on a series of trays. Drying of the material occurs by convection, exhaust air (containing solvent/water) being removed from the drier via an outlet. The efficacy of drying within these driers is dependent on the surface area of the material exposed to the drying air. To enhance the surface area of the material exposed to the drying air, these driers normally contain a number of trays on to which the material is placed, as shown in Figure 11.44. Whilst this does increase the efficacy of drying, in general tray driers are not

**Figure 11.43**  Diagrammatic representation of the tray drier.

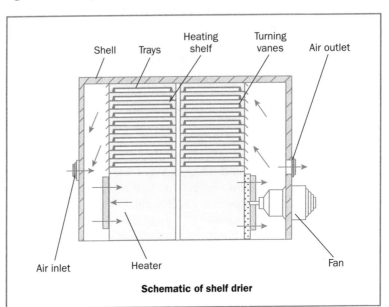

Schematic of shelf drier

**Figure 11.44**  A tray drier.

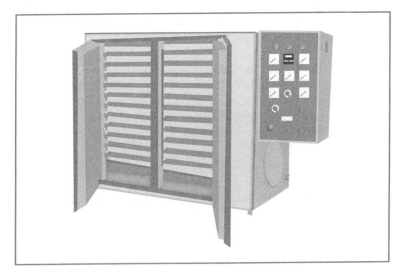

particularly efficient due to the long drying times required. Other problems associated with the use of tray driers include:

■   Inappropriate choice of drying conditions may lead to caking of the material being dried. In this there is rapid loss of the solvent/water from the surface of the material which may lead to the formation of a hard crust (cake), which subsequently provides a resistance barrier to the removal of the solvent underneath the cake.

- If there are components of the material being dried that are soluble in the solvent, these may diffuse through the solvent to the surface of the material being dried. As a result these components will be present on the surface of the dried solid, e.g. coatings, granules. This may lead to problems regarding the pharmaceutical and/or clinical performance of these systems, e.g. inappropriate drug release kinetics, poor compression characteristics, mottled effect due to inconsistent distribution of a colour.

## Fluid-bed driers (dynamic driers)

In fluid-bed driers, the individual solid particles are fluidised (suspended) within a heated air stream, allowing individual intimate contact with the drying fluid. The movement of the suspended particles in the air stream is turbulent and, using this technique, the drying of granular solids is more rapid than is observed using static driers. A schematic of the operation of the fluid-bed drier is shown in Figure 11.45.

In this schematic diagram it may be observed that air is drawn into the heating chamber where it is warmed to the required temperature, filtered and passed into the drying chamber. Within the drying chamber the particles are suspended within the air

**Figure 11.45**   Diagrammatic representation of the fluid-bed drier.

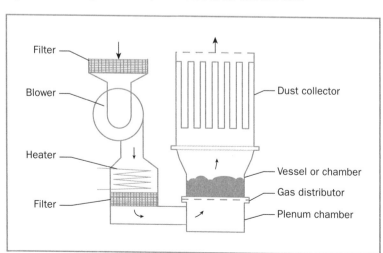

stream and dried. Air (containing the solvent) exits the fluidised bed drier through a filter bag. The particles are retained within the chamber via the presence of a filter bag at the exit of the drier and a perforated metal plate, which allows air to pass into the chamber but prevents the particles from leaving the chamber through the action of gravity.

To ensure optimal drying it is important to optimise the fluid (air) kinetics within the drying chamber as these affect the movement of particles within the drying chamber. Accordingly, in fluid bed drying, the relationship between the pressure drop across the bed of particles to be dried and air velocity is measured. This relationship is shown in Figure 11.46.

**Figure 11.46**   The relationship between log pressure drop across a fluidised bed (y axis) and log air velocity (x axis). $v_{mf}$ refers to the *minimum fluidisation velocity*. At this stage the relationship between the log pressure drop and log air velocity arches/curves. At point d and beyond, there is no change in the pressure drop despite an increased fluid velocity.

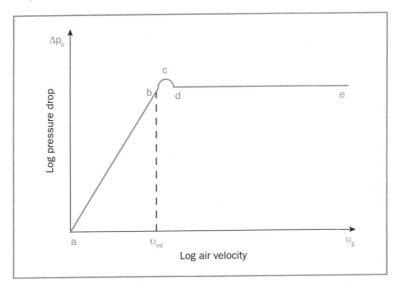

Increasing the air velocity at the start of the operation (a) leads to an increase in the pressure drop across the particle bed. Initially, at low velocities, flow occurs within the interparticle volume and there is little particle rearrangement. As the air velocity is further increased a point is reached ($v_{mf}$, b) in Figure 11.46 where a pressure drop has been reached at which point the gravitational forces and the frictional drag on the particles are equal. At this stage further increases in air velocity lead to a reduction in the pressure drop due to a spatial rearrangement of the particle to increase porosity, signified by the region commencing at point (d). As the air velocity is increased the particles are individually suspended (fluidised) in the air flow. As stated previously the air flow within the region d–e is turbulent and this leads to high mass and heat transfer and hence more rapid drying. The use of excessive air velocities (i.e. >e) may lead to problems regarding the retention of the suspended, dried particles within the drying chamber.

### Advantages

There are several advantages associated with the use of fluidised bed drying, including:

- high heat and mass transfer rates, resulting in more rapid drying in comparison with shelf/tray driers
- individual drying of the particles, resulting in a constant rate of drying from the surface of each particle
- precise control of the temperature of the fluidised bed
- the turbulent movement of the particles within the drying chamber will lead to particle–particle interactions/collisions which may in turn reduce the irregular shape of particles
- reduced migration of solvent-soluble components during the drying process.

### Disadvantatges

The disadvantages of fluidised bed drying include:

- the generation of fine particles may result from particle–particle collisions within the drying chamber, which may not be pharmaceutically useful
- if there are fine particles generated during fluidisation, these may subsequently lodge within the filter bags, resulting in loss of material and reduction in the efficiency of the drying process due to a reduction in the rate of removal of exhaust air from the drying chamber.

## Spray driers (pneumatic driers)

Spray driers are specifically used to dry liquid materials, e.g. protein solutions, polymeric solutions, suspensions of low viscosity. In these the liquid solution/suspension is atomised and dispersed within the heated air stream. In light of the small droplet size (and hence large surface area), evaporation of the solvent occurs rapidly to generate a fine, spherical solid powder, which is subsequently collected.

The key design features in spray drying are shown in Figure 11.47. As may be observed the spray drying process occurs within the drying chamber. The other components are designed to separate and collect the particles and to remove the moist air.

**Figure 11.47**    Schematic of the operation of a spray drier.

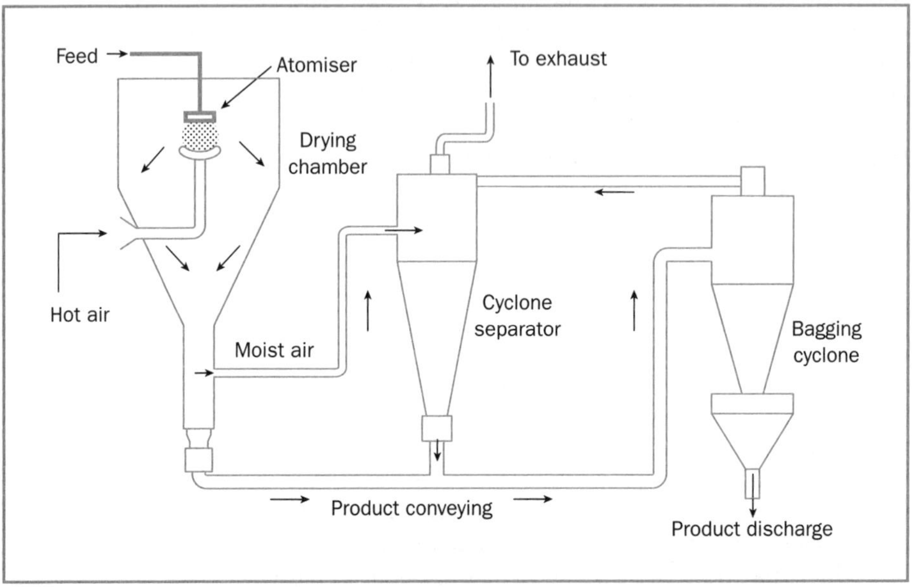

As introduced above, there are several key steps within the drying process, as follows:

1.  The solution/suspension is atomised and introduced into the warm air stream. This is normally produced by blowing air over a heat exchanger.
2.  The temperature of the droplet rapidly and temporarily reaches and exceeds the wet bulb temperature of the air stream.
3.  The liquid at the surface of the droplet evaporates (to maintain the thermal equilibrium) and a solid layer forms around the droplet.
4.  The solid layer acts as a barrier through which diffusion of the solvent must occur.
5.  The diffusion of the liquid through the barrier occurs at a slower rate than the transfer of heat across the barrier to the liquid underneath.
6.  The subsequent rate of evaporation of the solvent is greater than the diffusion rate across the barrier and therefore there is an increase in the internal pressure within the droplet. As a result the volume of the droplet increases, the thickness of the barrier layer subsequently decreases and there is a resultant increase in the rate of diffusion across the barrier.
7.  The mechanical properties of the barrier directly affect the morphology of the solid particles produced by spray drying. If the internal pressure exceeds the ultimate tensile strength of the barrier, then the particles will rupture. Conversely, a mechanically robust barrier will be able to withstand the deformation resulting from the increased internal pressure and upon drying will adopt a spherical shape.

Spray driers have several notable design features that affect their operation. The feed of the product into the drying chamber occurs either through gravity or by the use of a mechanical pump; the rate of delivery of the fluid is controlled so that drying of the droplets occurs prior to contact with the walls of the drying chamber. The generation of droplets occurs using an atomiser. This is a critical step as the atomiser controls the size and surface area of the droplets, which, in turn, affects the drying kinetics and the size (and the distribution of size) of the dried particles. There are three main atomiser types:

### Pressure nozzle atomisers
In this method the liquid feed is pumped through an orifice, generally within a pressure range of 200–2000 kPa using a positive displacement, high pressure pump (Figure 11.48). Using this method leads to the narrowest particle size distribution. The particle size of the droplets produced is dependent on the fluid flow and the pressure used to move the fluid through the orifice.

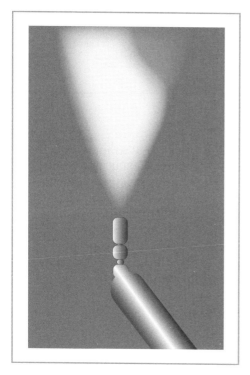

**Figure 11.48** Droplet production using a pressure atomiser.

### Pneumatic (two-fluid or gas-atomising) atomisers
In this method there are two fluid feeds, namely the liquid fluid and a high velocity gas, which acts to disrupt the fluid flow into droplets. Atomisation can occur within the atomiser or external to the atomiser depending on the fluid feed arrangements (Figure 11.49). The particle size distribution associated with

this method is broader than that associated with pressure nozzle atomisers; however, this method may be efficiently used both to produce small particles and, additionally, to atomise more viscous solutions. The power requirement for this method is higher than for other methods (and may be up to 48 MPa). The particle size (and distribution) is dependent on the pressure of the compressed gas feed and the flow rate of fluid through the nozzle.

**Figure 11.49**   Droplet production using a pneumatic atomiser. In the diagram on the left atomisation occurs within the atomiser (termed internal atomisation), whereas in the diagram on the right atomisation occurs outside the atomiser (termed external atomisation).

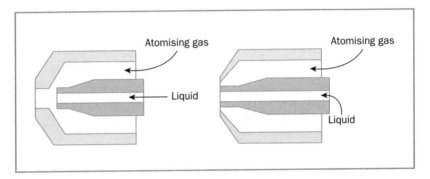

### Spinning disc atomisers

In this method the liquid is delivered across a rapidly rotating disc or wheel (typically rotating at between 3000 and 50 000 rpm), the centrifugal force within the chamber being responsible for the generation of droplets. The size distribution of particles produced using this method is usually broad; however, one advantage of this method is the flexibility in the types of liquid feeds that may be atomised, e.g. high viscosity liquids and suspensions that would normally be difficult to process using other techniques due to nozzle blockage. The average particle size and size distribution may be engineered in this technique by control of the rotation speed and diameter of the rotating disc/wheel.

Separation of the dried particles from the gas flow is achieved using a *cyclone separator*. This is a method that does not require the use of filters/meshes but instead uses rotational and gravitational effects to separate the solid particles from the gas flow. The separator is normally cylindrical or conical in design and within this chamber the gas flow follows a helical pattern before exiting at the base of the separator chamber.

The main pharmaceutical applications of spray drying are as follows:

- Due to the rapid drying time associated with spray drying, this method may be successfully used to dry therapeutic agents that are thermolabile or may undergo degradation by oxidation.
- Modification of the morphology of therapeutic agents to enhance powder flow and/or compression in the manufacture of solid dosage forms or to simply improve powder flow in the manufacture of other dosage forms, e.g. injections for reconstitution. The enhanced flow properties are due to the spherical shape and controlled size distribution of the particles resultant from spray drying.
- Production of solid particles containing entrapped drug that may be used for controlled release or targeted release of therapeutic agents. Furthermore, spray drying may be used to generate the amorphous form of a therapeutic agent within a polymeric particle. This approach may then be used to enhance the solubility and hence absorption of therapeutic agents when administered orally.

## Freeze driers

Freeze drying (lyophilisation) is a drying (dehydration) technique that results in the production of solid materials and which has a wide range of pharmaceutical applications. These include:

- Drying of therapeutic agents that undergo degradation (hydrolysis) in the liquid state, even when dried under normal drying conditions.
- Drying of products of pharmaceutical importance to the solid state. These can then be reconstituted with the addition of water prior to use. Examples of these include:
  - blood plasma
  - blood serum
  - vaccines
  - antibiotics
  - hormones
  - proteins (and enzymes).
- The enhancement of the shelf-life of products that may undergo degradation within conventional platforms.
- To enable accurate filling of small masses of therapeutic agents designed for use as parenteral formulations. In this approach the therapeutic agent is dissolved in solution and filled into the final packaging. Following freeze drying the required mass of therapeutic agent will have been accurately dispensed into the final packaging.

### Theoretical aspects of freeze drying

There are number of steps in the freeze drying process, which are described below (with particular reference to the simplified phase diagram for water, Figure 11.50).

**Figure 11.50** Phase diagram for water and the relevance to the freeze drying process.

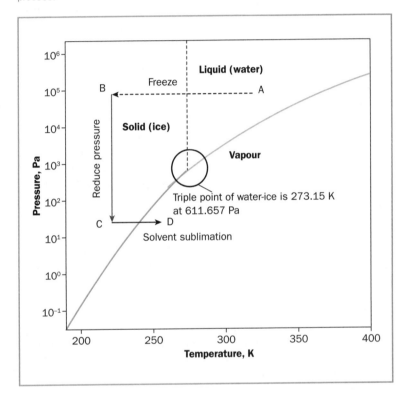

1. *Pretreatment of the solution.* This stage, which may not be needed for all systems, involves any process that is necessary to optimise the freeze drying process. Examples include the incorporation of agents that may enhance the stability of the material being freeze dried, the concentration of the product by removal of liquid.

2. *Freezing of the solution.* This stage involves freezing of the solution by reduction of the temperature of the system. In Figure 11.50 the start of this process is denoted by point A whereas the end of the process is denoted by point B. This is an important stage in the freeze drying process as the frozen structure produced will not be changed over the rest of the drying process and, indeed, the properties of the finished

product are dependent on this frozen structure. The freezing process may be differentiated into two key steps. Initially as the product is cooled to 273.15 K (0°C), freezing is facilitated by ice nucleation, i.e. initial production of ice particles serves to act as the focus for further ice production. The removal of the liquid water phase will then serve to increase the concentration of solutes within the remaining liquid phase. As the temperature is further reduced to between 233.15 K and 263.25 K two major events may occur, namely:

- a eutectic mixture between the ice phase and therapeutic agent (and other components) may form
- an amorphous (glassy) matrix, possessing a distinctive glass transition temperature may form.

3. *Primary drying.* Once the system has frozen the pressure within the freeze drier is reduced to pressures close to a vacuum formation (point C in Figure 11.50). Sublimation of ice is then facilitated by increasing the temperature of the system. In practice the primary drying phase is conducted at a temperature that is close to the glass transition temperature of the amorphous (glassy) matrix and results in a porous matrix. Drying at temperatures above the glass transition temperature is disadvantageous. Under such conditions the system exists in a rubbery/liquid state and collapse of the matrix occurs.

4. *Secondary drying.* In this stage the drying temperature is increased (298.15–338.15 K) to remove any residual water within the matrix. Following removal of residual water the matrix retains an amorphous, porous structure, which, when required, may be rapidly reconstituted by the addition of water. Storage of the freeze dried product must be performed below the glass transition temperature of the glassy matrix in order to retain the structure of the porous solid matrix.

Despite the apparent differences between freeze drying and other drying methods that have been described in this chapter, all drying methods, including freeze drying, fulfil several common requirements. Firstly, the vapour pressure at the surface of the material being dried must be greater than the partial vapour pressure of the drying environment. This positive vapour pressure difference is the thermodynamic driving force that facilitates drying. Secondly, control of the rate of introduction of the latent heat of vaporisation to the drying surface is important. In so doing, temperatures at the material surface and within the matrix may be accurately controlled which, in turn, optimises the properties of the dried solid matrix. Thirdly, there must be removal of the evaporated moisture from the drying chamber.

### Design and operation of freeze driers

Whilst their individual designs may differ, all freeze driers have the same general operational components, as follows:

1. a compartment that allows vacuum drying
2. pumps (or alternative mechanisms) that are responsible for generating a vacuum
3. a source of heat
4. a mechanism to enable vapour removal. These are typically condensers that condense water vapour as ice; the ice being removed by subsequent heating or mechanical scraping, desiccants or manual removal using pumps.

## Multiple choice questions

1. **Which of the following statements are true concerning freeze drying?**
   a. At the end of the freeze drying process the product is a free-flowing powder.
   b. Cooling below zero causes freeze concentration.
   c. During primary drying the water is removed by sublimation of ice.
   d. Removal of ice results in the formation of pores throughout the product.

2. **Which of the following apply during spray drying?**
   a. Wet particles are sprayed into a stream of hot air.
   b. The product is a porous cake with good dissolution properties.
   c. Spray dried products are often used for tablet and capsule formulations.
   d. Heat input goes into evaporation rather than temperature increase.

3. **Which of the following apply during fixed bed drying?**
   a. The inappropriate choice of drying conditions can lead to formation of a hard dry crust at the surface of material.
   b. Intergranular migration may lead to deposition of water soluble excipients on the bed surface.
   c. Intragranular migration may lead to deposition of water soluble excipients on the bed surface.
   d. Material is spread directly on to shelves or on trays resting on the shelves.

4. **Which of the following statements are true concerning the milling of pharmaceutical solids?**
   a. It is an energy inefficient process.

b. The resultant particle size of the milled solids is dependent on the initial (starting) particle size of the solids.

c. Conventional milling methods may be used to generate nanoparticles.

d. The elastic properties of the solids will influence the milling efficiency.

5. **Which of the following statements are true concerning the filtration of pharmaceutical solids?**

a. The rate of filtration is proportional to the viscosity of the liquid to be filtered.

b. The rate of filtration is proportional to the pressure to be applied to the liquid to be filtered.

c. Filtration efficiency may be improved by reducing the pore size of the filtration medium.

d. The pore size of clarification filters is greater than that of 'cake' filters.

6. **The advantages of filter cakes include which of the following?**

a. The efficiency of sterilisation of solutions is greater whenever a filter cake is used.

b. Filter cakes may increase the lifetime of the filter medium.

c. Improved flow rates during the filtration process.

d. Improved separation.

7. **Which of the following statements are true concerning the mixing of pharmaceutical liquids?**

a. High speed mixing, resulting in a laminar flow mixing profile, is frequently used.

b. As the viscosity of the liquid increases the degree of turbulent mixing increases.

c. Increasing the temperature of pharmaceutical liquids decreases the required mixing speed.

d. The nature of the mixing profile may be determined with reference to Reynolds number.

8. **Which of the following statements are true with reference to the water loss profile of a material during drying?**

a. In the first falling rate period, water is still being lost from the surface.

b. As the solid becomes drier the thermal conductivity decreases.

c. Solvent evaporation cools the material.

d. During the second falling rate period water is removed from within the bed.

9. **Which of the following statements apply during the constant rate period of drying?**
   a. The rate is constant, as it is only limited by the rate of migration within the sample.
   b. Solvent is removed from the surface.
   c. The moisture content will decrease, at the end of this period the moisture content will be equal to the critical moisture content of the material.
   d. Both the fastest drying rate and the highest temperatures are exhibited.

10. **As a pharmaceutical scientist you have been asked to reduce the particle size of a drug to enable its use within an aerosol designed for the treatment of asthma. Which of the following particle size reduction methods may be used?**
    a. The hammer mill.
    b. The conical mill.
    c. Fluid-energy mill.
    d. None of the above.

11. **As a pharmaceutical scientist you have been asked to separate drug particles from a concentrated drug suspension that has been produced as part of the synthesis stage of drug manufacture. The volume of the batch is 2000 litres and the concentration of drug is 55% w/w. Which of the following separation methods are appropriate?**
    a. Cartridge filtration.
    b. Filter press.
    c. Rotary drum (vacuum) filtration.
    d. Centrifugation.

12. **Which is/are suitable methods for clarification filtration of a liquid product of batch size 100 litres?**
    a. Cartridge filtration using a 0.22 µm filter cartridge.
    b. Cartridge filtration using a single 10 µm filter.
    c. Rotary drum (vacuum) filtration through a 1 µm filter.
    d. Cartridge filtration using a sequence of filters, i.e. 10 µm filter followed by second filtration using a 1 µm cartridge filter and a third filtration using a 0.45 µm cartridge filter.

13. **A sterile productions company manufactures injections, which are terminally sterilised by filtration. The batch size is 2000 litres. Which is the most appropriate method for sterile filtration of the batch?**
    a. Cartridge filtration using a 0.22 µm filter cartridge.

b. Cartridge filtration using a sequence of two filters, i.e. 1 μm filter followed by filtration using a 0.45 μm cartridge filter.
c. Rotary drum (vacuum) filtration through a 0.22 μm filter.
d. Cartridge filtration using a sequence of filters, i.e. 1 μm filter followed by second filtration using a 0.45 μm cartridge filter and a third filtration using a 0.22 μm cartridge filter.

14. **Which of the following factors are important regarding the selection of the most appropriate filtration method for the sterile filtration of eye drop formulations?**
a. pH of the solution.
b. Temperature at which filtration is performed.
c. The evaporation rate of the formulation.
d. The (aqueous) contact angle of the filter.

15. **Which of the following statements are true regarding freeze drying?**
a. Drying occurs by sublimation.
b. The rate of freezing affects the properties of the dried product.
c. Sublimation requires an increased pressure.
d. The process is isothermal.

16. **Which of the following statements are true regarding mixing of low viscosity liquids?**
a. The mechanism of mixing is predominantly laminar.
b. A suitable mixing vessel should have a length to diameter ratio of at least 2:1.
c. Mixing is optimal at low shear rates.
d. Ideally anchor impellers will provide suitable mixing kinetics.

17. **As a pharmaceutical scientist you have been asked to mix a concentrated aqueous drug suspension (50% w/w) prior to downstream processing. Optimal mixing performance will be achieved by which of the following?**
a. Mixing the suspension at high shear rates.
b. Mixing the suspension at elevated temperature.
c. Mixing the suspension using a three-blade propeller.
d. Using an in line mixing system at low shear rates.

18. **Concerning the rate of filtration of pharmaceutical liquids, which of the following statements is/are true?**
a. It is increased if the surface area of the filter is increased.
b. It is increased if the pressure applied to the filter is decreased.

c. It is increased if the viscosity of the liquid is increased.
d. It is dependent on the type of filter used.

19. **Concerning pharmaceutical mixing, which of the following statements is/are true?**
a. At high shear rates laminar flow is the predominant mechanism of mixing of solutions.
b. Molecular diffusion is responsible for complete mixing of suspensions.
c. Impeller design directly affects the mixing pattern of solutions.
d. Low viscosity liquids are mixed using a turbulent mechanism.

20. **Which of the following statements is/are true concerning mixing?**
a. Reynolds number is a measure of the viscosity of the systems.
b. Reynolds number provides basic information regarding the possible type of mixing equipment.
c. Propeller mixing systems are used to mix high viscosity liquids.
d. The power requirement of the propeller mixer is greater than that of the Z blade mixer.

21. **Size reduction of the internal phase of emulsions may be performed using which of the following?**
a. Ball mill.
b. Hammer mill.
c. Fluid energy mill.
d. Colloid mill.

# Multiple choice answers

## Chapter 1

| 1. | **a.** False. | **b.** True. | **c.** True. | **d.** False. |
| 2. | **a.** True. | **b.** False. | **c.** True. | **d.** False. |
| 3. | **a.** True. | **b.** False. | **c.** True. | **d.** False. |
| 4. | **a.** True. | **b.** False. | **c.** True. | **d.** True. |
| 5. | **a.** False. | **b.** True. | **c.** False. | **d.** True. |
| 6. | **a.** False. | **b.** True. | **c.** False. | **d.** True. |
| 7. | **a.** True. | **b.** True. | **c.** True. | **d.** False. |
| 8. | **a.** False. | **b.** True. | **c.** False. | **d.** False. |
| 9. | **a.** False. | **b.** False. | **c.** True. | **d.** False. |
| 10. | **a.** False. | **b.** True. | **c.** True. | **d.** False. |
| 11. | **a.** True. | **b.** False. | **c.** False. | **d.** True. |
| 12. | **a.** False. | **b.** True. | **c.** False. | **d.** False. |
| 13. | **a.** True. | **b.** False. | **c.** True. | **d.** False. |
| 14. | **a.** False. | **b.** True. | **c.** True. | **d.** False. |
| 15. | **a.** True. | **b.** False. | **c.** True. | **d.** False. |

## Chapter 2

| 1. | **a.** True. | **b.** True. | **c.** True. | **d.** True. |
| 2. | **a.** True. | **b.** False. | **c.** True. | **d.** True. |
| 3. | **a.** False. | **b.** False. | **c.** True. | **d.** False. |
| 4. | **a.** True. | **b.** False. | **c.** True. | **d.** False. |
| 5. | **a.** True. | **b.** True. | **c.** False. | **d.** False. |
| 6. | **a.** False. | **b.** True. | **c.** False. | **d.** True. |
| 7. | **a.** False. | **b.** True. | **c.** True. | **d.** True. |
| 8. | **a.** False. | **b.** False. | **c.** True. | **d.** True. |
| 9. | **a.** True. | **b.** True. | **c.** False. | **d.** False. |
| 10. | **a.** True. | **b.** True. | **c.** True. | **d.** False. |
| 11. | **a.** True. | **b.** True. | **c.** False. | **d.** False. |
| 12. | **a.** False. | **b.** False. | **c.** True. | **d.** False. |
| 13. | **a.** False. | **b.** False. | **c.** True. | **d.** True. |
| 14. | **a.** False. | **b.** False. | **c.** False. | **d.** True. |
| 15. | **a.** True. | **b.** False. | **c.** False. | **d.** True. |

## Chapter 3

| | | | | |
|---|---|---|---|---|
| 1. | **a.** True. | **b.** True. | **c.** True. | **d.** True. |
| 2. | **a.** True. | **b.** False. | **c.** True. | **d.** False. |
| 3. | **a.** False. | **b.** False. | **c.** True. | **d.** False. |
| 4. | **a.** True. | **b.** True. | **c.** False. | **d.** False. |
| 5. | **a.** False. | **b.** True. | **c.** True. | **d.** True. |
| 6. | **a.** True. | **b.** False. | **c.** False. | **d.** True. |
| 7. | **a.** True. | **b.** False. | **c.** True. | **d.** True. |
| 8. | **a.** False. | **b.** False. | **c.** False. | **d.** True. |
| 9. | **a.** True. | **b.** True. | **c.** False. | **d.** False. |
| 10. | **a.** True. | **b.** True. | **c.** True. | **d.** True. |
| 11. | **a.** True. | **b.** False. | **c.** False. | **d.** False. |
| 12. | **a.** True. | **b.** True. | **c.** False. | **d.** False. |
| 13. | **a.** True. | **b.** False. | **c.** False. | **d.** False. |
| 14. | **a.** True. | **b.** True. | **c.** False. | **d.** False. |
| 15. | **a.** False. | **b.** False. | **c.** False. | **d.** False. |

## Chapter 4

| | | | | |
|---|---|---|---|---|
| 1. | **a.** False. | **b.** False. | **c.** False. | **d.** True. |
| 2. | **a.** True. | **b.** True. | **c.** True. | **d.** True. |
| 3. | **a.** True. | **b.** True. | **c.** True. | **d.** False. |
| 4. | **a.** True. | **b.** False. | **c.** True. | **d.** True. |
| 5. | **a.** False. | **b.** False. | **c.** True. | **d.** True. |
| 6. | **a.** True. | **b.** False. | **c.** True. | **d.** False. |
| 7. | **a.** False. | **b.** True. | **c.** True. | **d.** False. |
| 8. | **a.** False. | **b.** True. | **c.** False. | **d.** True. |
| 9. | **a.** True. | **b.** True. | **c.** False. | **d.** True. |
| 10. | **a.** True. | **b.** True. | **c.** True. | **d.** False. |
| 11. | **a.** True. | **b.** False. | **c.** False. | **d.** False. |
| 12. | **a.** False. | **b.** False. | **c.** False. | **d.** True. |
| 13. | **a.** True. | **b.** True. | **c.** False. | **d.** False. |
| 14. | **a.** False. | **b.** True. | **c.** False. | **d.** False. |
| 15. | **a.** True. | **b.** True. | **c.** False. | **d.** True. |
| 16. | **a.** True. | **b.** True. | **c.** False. | **d.** True. |

## Chapter 5

| | | | | |
|---|---|---|---|---|
| 1. | **a.** True. | **b.** True. | **c.** False. | **d.** True. |
| 2. | **a.** True. | **b.** True. | **c.** False. | **d.** False. |
| 3. | **a.** True. | **b.** True. | **c.** True. | **d.** False. |
| 4. | **a.** True. | **b.** True. | **c.** False. | **d.** False. |
| 5. | **a.** False. | **b.** True. | **c.** True. | **d.** True. |
| 6. | **a.** False. | **b.** False. | **c.** True. | **d.** True. |

| 7. | **a.** False. | **b.** False. | **c.** False. | **d.** True. |
| 8. | **a.** False. | **b.** True. | **c.** True. | **d.** False. |
| 9. | **a.** True. | **b.** False. | **c.** True. | **d.** False. |
| 10. | **a.** False. | **b.** False. | **c.** False. | **d.** False. |
| 11. | **a.** True. | **b.** False. | **c.** True. | **d.** False. |
| 12. | **a.** False. | **b.** True. | **c.** False. | **d.** True. |
| 13. | **a.** True. | **b.** False. | **c.** False. | **d.** True. |
| 14. | **a.** False. | **b.** False. | **c.** True. | **d.** False. |
| 15. | **a.** False. | **b.** True. | **c.** False. | **d.** True. |

## Chapter 6

| 1. | **a.** True. | **b.** True. | **c.** True. | **d.** False. |
| 2. | **a.** False. | **b.** True. | **c.** True. | **d.** True. |
| 3. | **a.** True. | **b.** True. | **c.** False. | **d.** False. |
| 4. | **a.** False. | **b.** True. | **c.** True. | **d.** True. |
| 5. | **a.** False. | **b.** True. | **c.** False. | **d.** False. |
| 6. | **a.** True. | **b.** True. | **c.** True. | **d.** False. |
| 7. | **a.** True. | **b.** False. | **c.** True. | **d.** False. |
| 8. | **a.** True. | **b.** False. | **c.** True. | **d.** True. |
| 9. | **a.** True. | **b.** False. | **c.** False. | **d.** True. |
| 10. | **a.** True. | **b.** True. | **c.** True. | **d.** False. |
| 11. | **a.** True. | **b.** True. | **c.** False. | **d.** True. |
| 12. | **a.** False. | **b.** True. | **c.** False. | **d.** True. |
| 13. | **a.** True. | **b.** False. | **c.** False. | **d.** False. |
| 14. | **a.** True. | **b.** False. | **c.** False. | **d.** False. |
| 15. | **a.** False. | **b.** True. | **c.** True. | **d.** True. |

## Chapter 7

| 1. | **a.** True. | **b.** True. | **c.** False. | **d.** False. |
| 2. | **a.** True. | **b.** True. | **c.** False. | **d.** True. |
| 3. | **a.** True. | **b.** False. | **c.** False. | **d.** True. |
| 4. | **a.** False. | **b.** True. | **c.** False. | **d.** True. |
| 5. | **a.** True. | **b.** True. | **c.** True. | **d.** False. |
| 6. | **a.** True. | **b.** False. | **c.** True. | **d.** True. |
| 7. | **a.** False. | **b.** True. | **c.** True. | **d.** False. |
| 8. | **a.** True. | **b.** False. | **c.** True. | **d.** True. |
| 9. | **a.** True. | **b.** False. | **c.** True. | **d.** True. |
| 10. | **a.** True. | **b.** False. | **c.** True. | **d.** False. |
| 11. | **a.** True. | **b.** True. | **c.** False. | **d.** True. |
| 12. | **a.** False. | **b.** True. | **c.** False. | **d.** True. |
| 13. | **a.** True. | **b.** True. | **c.** False. | **d.** False. |
| 14. | **a.** True. | **b.** False. | **c.** False. | **d.** True. |
| 15. | **a.** False. | **b.** True. | **c.** False. | **d.** False. |

## Chapter 8

| | | | | |
|---|---|---|---|---|
| 1. | a. True. | b. False. | c. True. | d. False. |
| 2. | a. True. | b. True. | c. False. | d. False. |
| 3. | a. False. | b. True. | c. True. | d. True. |
| 4. | a. True. | b. False. | c. False. | d. False. |
| 5. | a. True. | b. False. | c. True. | d. False. |
| 6. | a. False. | b. False. | c. True. | d. False. |
| 7. | a. True. | b. True. | c. True. | d. False. |
| 8. | a. False. | b. True. | c. True. | d. True. |
| 9. | a. True. | b. False. | c. False. | d. False. |
| 10. | a. True. | b. True. | c. False. | d. True. |
| 11. | a. True. | b. False. | c. False. | d. True. |
| 12. | a. True. | b. True. | c. False. | d. False. |
| 13. | a. False. | b. True. | c. False. | d. True. |
| 14. | a. False. | b. True. | c. True. | d. False. |
| 15. | a. True. | b. False. | c. False. | d. False. |

## Chapter 9

| | | | | |
|---|---|---|---|---|
| 1. | a. False. | b. False. | c. True. | d. True. |
| 2. | a. True. | b. False. | c. True. | d. False. |
| 3. | a. True. | b. True. | c. True. | d. False. |
| 4. | a. True. | b. False. | c. True. | d. False. |
| 5. | a. False. | b. True. | c. False. | d. True. |
| 6. | a. False. | b. False. | c. True. | d. False. |
| 7. | a. False. | b. False. | c. False. | d. True. |
| 8. | a. False. | b. False. | c. False. | d. True. |
| 9. | a. True. | b. True. | c. False. | d. True. |
| 10. | a. True. | b. True. | c. False. | d. True. |
| 11. | a. True. | b. False. | c. False. | d. False. |
| 12. | a. False. | b. True. | c. False. | d. False. |
| 13. | a. False. | b. False. | c. True. | d. False. |
| 14. | a. False. | b. True. | c. True. | d. True. |
| 15. | a. True. | b. True. | c. False. | d. True. |

## Chapter 10

| | | | | |
|---|---|---|---|---|
| 1. | a. True. | b. True. | c. True. | d. False. |
| 2. | a. False. | b. True. | c. True. | d. False. |
| 3. | a. True. | b. False. | c. True. | d. False. |
| 4. | a. False. | b. False. | c. False. | d. True. |
| 5. | a. True. | b. True. | c. False. | d. False. |
| 6. | a. True. | b. True. | c. True. | d. True. |
| 7. | a. True. | b. True. | c. False. | d. True. |

| | | | | |
|---|---|---|---|---|
| 8. | a. True. | b. True. | c. True. | d. False. |
| 9. | a. True. | b. False. | c. False. | d. True. |
| 10. | a. False. | b. True. | c. True. | d. False. |
| 11. | a. False. | b. False. | c. False. | d. True. |
| 12. | a. False. | b. True. | c. False. | d. False. |
| 13. | a. False. | b. True. | c. True. | d. False. |
| 14. | a. False. | b. True. | c. True. | d. True. |
| 15. | a. False. | b. True. | c. False. | d. True. |

## Chapter 11

| | | | | |
|---|---|---|---|---|
| 1. | a. False. | b. True. | c. True. | d. True. |
| 2. | a. False. | b. False. | c. True. | d. False. |
| 3. | a. True. | b. True. | c. False. | d. True. |
| 4. | a. True. | b. True. | c. False. | d. True. |
| 5. | a. False. | b. True. | c. False. | d. False. |
| 6. | a. False. | b. True. | c. True. | d. True. |
| 7. | a. False. | b. False. | c. False. | d. True. |
| 8. | a. True. | b. True. | c. True. | d. True. |
| 9. | a. False. | b. True. | c. True. | d. False. |
| 10. | a. False. | b. False. | c. True. | d. False. |
| 11. | a. False. | b. True. | c. True. | d. False. |
| 12. | a. False. | b. False. | c. True. | d. False. |
| 13. | a. False. | b. False. | c. False. | d. True. |
| 14. | a. True. | b. True. | c. False. | d. True. |
| 15. | a. True. | b. True. | c. False. | d. False. |
| 16. | a. False. | b. True. | c. False. | d. False. |
| 17. | a. False. | b. False. | c. False. | d. True. |
| 18. | a. True. | b. True. | c. False. | d. True. |
| 19. | a. False. | b. False. | c. True. | d. True. |
| 20. | a. False. | b. True. | c. True. | d. False. |
| 21. | a. False. | b. False. | c. False. | d. True. |

# Index